NATO ASI Series

Advanced Science Institutes Series

A Series presenting the results of activities sponsored by the NATO Science Committee, which aims at the dissemination of advanced scientific and technological knowledge, with a view to strengthening links between scientific communities.

The Series is published by an international board of publishers in conjunction with the NATO Scientific Affairs Division

A	Life Sciences	Plenum Publishing Corporation
B	Physics	London and New York
C	Mathematical and Physical Sciences	D. Reidel Publishing Company Dordrecht and Boston
D	Behavioural and Social Sciences	Martinus Nijhoff Publishers Dordrecht/Boston/Lancaster
E	Applied Sciences	
F	Computer and Systems Sciences	Springer-Verlag Berlin/Heidelberg/New York
G	Ecological Sciences	

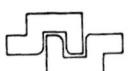

Series E: Applied Sciences – No. 111

Advanced Agricultural Instrumentation

Advanced Agricultural Instrumentation

Design and Use

edited by

William G. Gensler

The University of Arizona
Department of Electrical
and Computer Engineering
Tuscon, Arizona 85721
USA

1986 **Martinus Nijhoff Publishers**
Dordrecht / Boston / Lancaster
Published in cooperation with NATO Scientific Affairs Division

Proceedings of the NATO Advanced Study Institute on 'Advanced Agricultural Instrumentation', Il Ciocco (Pisa), Italy, May 27–June 9, 1984

Library of Congress Cataloging in Publication Data

NATO Advanced Study Institute on Advanced Agriculutral
   Instrumentation (1984 : Pisa, Italy)
   Advanced agricultural instrumentation.

   (NATO ASI series. Series E, Applied sciences ; no.
111)
   "Proceedings of the NATO Advanced Study Institute on
Advanced Agricultural Instrumentation, Il Ciocco (Pisa),
Italy, May 27–June 9, 1984"--T.p. verso.
   "Published in cooperation with NATO Scientific
Affairs Division."
   Includes index.
   1. Agricultural instruments--Congresses. I. Gensler,
William G. II. North Atlantic Treaty Organization.
Scientific Affairs Division. III. Title. IV. Series.
S676.5.N37 1984   631.3   86-5384
ISBN-13:978-94-010-8459-8

ISBN-13:978-94-010-8459-8   e-ISBN-13:978-94-009-4404-6
DOI:10.1007/978-94-009-4404-6

Distributors for the United States and Canada: Kluwer Academic Publishers, 190 Old Derby Street, Hingham, MA 02043, USA

Distributors for the UK and Ireland: Kluwer Academic Publishers, MTP Press Ltd, Falcon House, Queen Square, Lancaster LA1 1RN, UK

Distributors for all other countries: Kluwer Academic Publishers Group, Distribution Center, P.O. Box 322, 3300 AH Dordrecht, The Netherlands

ADVANCED AGRICULTURAL INSTRUMENTATION

DESIGN AND USE

PREFACE

The basic approach of the NATO Advanced Studies Institute on Advanced Agricultural Instrumentation was to bring together a heterogeneous combination of Instrument Designers and Instrument Users from Industry, Government and Academia. By mixing together these three groups with diverse backgrounds in a lecture and laboratory environment, the individual instruments were analyzed from virtually their conception to their fruition in the field. The lecture hall, greenhouse and grapevines at Il Ciocco were an excellent implementation of this approach. The members of the Institute not only discussed the instrument principles, but assessed the advantages and disadvantages of each during the laboratory sessions. This volume bears the imprint of this approach. It emphasizes not only the instrument design, but also the use of the instrument under field conditions.

The volume is divided into several natural groupings. In the first section, the instruments for photosynthetic measurements are examined in the papers of Briggs and Welles, and Long. Vredenberg and Robinson address the specific topic of fluorometry. These papers offer contrasting viewpoints and instrument approaches.

The second section focuses on environmental measurements. Soil moisture is considered from the dual viewpoint of neutron probes and soil psychrometers. Non contact radiation thermometry and humidity are examined by Everest and Salasmaa.

At this point in the volume the theoretical analysis of Norman brings together the measurements of the plant-environment continuum.

Nutrient analysis is the topic of the next grouping. Jones and Albery consider the detailed measurement of plant nutrient status in both open field and greenhouse. Lucas presents specific

instrumentation to measure ion transfer at the cellular level.
McDonald addresses the question of the growth and nutrient measurements from a more general viewpoint.

In the final grouping the specific topic of plant water status is considered. Schurer examines water status measurements in general and specific heat pulse instrumentation to measure the kinetics of water transfer. Gensler presents direct biophysical and electrochemical methods of assessing water status of open field crops.

There is at present a strong contrast between the degree of quantification present in the agricultural research community and the quantification practiced by the production agriculture community. Hopefully, this volume will help to bring these two groups together.

The editor wishes to express his appreciation to the NATO Scientific Affairs Division - in particular to Drs. Mario Di Lullo and Craig Sinclair for their support to this Advanced Study Institute.

The editor would also like to thank Monica Badilla for her editing and Sally Anderson for her excellent typing and appraisal of the manuscript.

W. Gensler, Editor
Tucson, Arizona, U.S.A.

TABLE OF CONTENTS

TABLE OF CONTENTS -- Continued

PHOTOSYNTHESIS RELATED INSTRUMENTATION

RADIATION MEASUREMENT

William Biggs
LI-COR, Inc.
Lincoln, Nebraska

Much confusion has existed regarding the measurement of radiation. This paper presents a comprehensive summary of the terminology and units used in radiometry, photometry, and the measurement of photosynthetically active radiation (PAR). Measurement errors can arise from a number of sources, and these are explained in detail. Finally, the conversion of radiometric and photometric units to photon units is discussed. In this report, the International System of Units (SI) is used unless noted otherwise (1).

1. RADIOMETRY

Radiometry (2) is the measurement of the properties of radiant energy (SI unit: joule, J), which is one of the many interchangeable forms of energy. The rate of flow of radiant energy, in the form of an electromagnetic wave, is called the radiant flux (unit: watt, W; $1W = 1Js^{-1}$). Radiant flux can be measured as it flows from the source (the sun, in natural conditions) through one or more reflecting, absorbing, scattering and transmitting media (the Earth's atmosphere, a plant canopy) to the receiving surface of interest (a photosynthesizing leaf) (3).

1.1 Terminology and Units

Radiant flux is the amount of radiation coming from a source per unit time. Unit: watt, W.

Radiant intensity is the radiant flux leaving a point on the source, per unit solid angle of space surrounding the point. Unit: watts per steridian, $W\ sr^{-1}$

Radiance is the radiant flux emitted by a unit area of a source or scattered by a unit area of a surface Unit: $W\ m^{-2}\ sr^{-1}$.

Irradiance is the radiant flux incident on a receiving surface from all directions, per unit area of surface. Unit: W m^{-2}.

Absorptance is the fraction of the incident flux that is absorbed by a medium.

Reflectance and transmittance are equivalent terms for the fractions that are reflected or transmitted.

Spectroradiometry. All the properties of the radiant flux depend on the wavelength of the radiation. The prefix spectral is added when the wavelength dependency is being described. Thus, the spectral irradiance is the irradiance at a given wavelength, per unit wavelength interval. The irradiance within a given waveband is the integral of the spectral irradiance with respect to wavelength (3). Unit: W m^{-2} nm^{-1}. Spectral measurements can be made using the LI-1800 Portable Spectroradiometer.

Global solar radiation is the solar irradiance received on a horizontal surface (also referred to as the direct component of sunlight plus the diffuse component of skylight received together on a horizontal surface). This physical quantity is measured by a pyranometer such as the LI-200SA. Unit: W m^{-2}.

Direct solar radiation is the radiation emitted from the solid angle of the sun's disc, received on a surface perpendicular to the axis of this cone, comprising mainly unscattered and unreflected solar radiation. This physical quantity is measured by a pyrheliometer. Unit: W m^{-2}.

Diffuse solar radiation (sky radiation) is the downward scattered and reflected radiation coming from the whole hemisphere, with the exception of the solid angle subtended by the sun's disc. Diffuse radiation can be measured by a pyranometer mounted on a shadow band, or calculated using global solar radiation and direct solar radiation. Unit: W m^{-2}.

2. PHOTOSYNTHETICALLY ACTIVE RADIATION

In the past there has been disagreement concerning units and terminology used in radiation measurements in conjunction with the plant sciences. It is LI-COR's policy to adopt the recommendations of the international committees, such as the Commission Internationale de l'Eclairage(CIE), the International Bureau of Weights and Measures, and the International Committee on Radiation Units. The International System of Units(SI) should be used whenever a suitable unit exists (1).

2.1 Units

The SI unit of radiant energy flux is the watt (W). There is no official SI unit of photon flux. The mole of photons and einstein are commonly used to designate Avogadro's number of photons (6.022 10^{23} photons). The einstein has been used in the past in plant science, however, most societies now recommend the use of the mole since the mole is an SI unit. When either of these definitions are used, the quantity of photons in a mole is equal to the quantity of photons in an einstein (1 mole = 1 einstein = 6.022 10^{23} photons). Note: The einstein has also been used in books on photochemistry, photobiology and radiation physics as the quantity of radiant energy in Avogadro's number of photons (4). This definition is not used in photosynthesis studies.

2.2 Terminology

LI-COR continues to follow the lead of the Crop Science Society of America, Committee on Terminology (5) and other societies, until international committees put forth recommendations. LI-COR has introduced the term Photosynthetic Photon Flux Fluence Rate (PPFFR). Oceanographers and limnologists have sometimes called this Quantum Scaler Irradiance or Photon Spherical Irradiance.

Photosynthetically Active Radiation (PAR) is defined as radiation in the 400 to 700 nm waveband. PAR is the general radiation term which covers both photon terms (6) and energy terms.

Photosynthetic Photon Flux Density (PPFD) is defined as the photon flux density of PAR, also referred to as Quantum Flux Density. This is the number of photons in the 400-700 nm waveband incident per unit time on a unit surface. The ideal PPFD sensor responds equally to all photons in the 400-700 nm waveband and has a cosine response. This physical quantity is measured by a cosine (180°) quantum sensor such as the LI-190SA or LI-192SA. The LI-191SA Line Quantum Sensor also measures PPFD. Figure 1 shows an ideal quantum response curve and the typical spectral response curve of LI-COR quantum sensors. Units: 1 μmol s^{-1} m^{-2} = 1 μE s^{-1} m^{-2} = 6.022 10^{17} photons s^{-1} m^{-2}.

Photosynthetic Photon Flux Fluence Rate (PPFFR) is defined as the photon flux fluence rate of PAR, also referred to as Quantum Scaler Irradiance or Photon Spherical Irradiance. This is the integral of photon flux radiance at a point over all directions about the point. The ideal PPFFR sensor has a spherical collecting surface which exhibits the properties of a cosine receiver at every point on its surface (Figure 2) and responds equally to all photons in the 400-700 nm waveband (Figure 1). This physical quantity is measured by a spherical (4 π collector) quantum sensor such as the LI-193SA. Units: 1 μmol s^{-1} m^{-2} = 1 μE s^{-1} m^{-2} = 6.022 10^{17} photons s^{-1} m^{-2} = 6.022 10^{17} quanta s^{-1} m^{-2}. Note: There is no unique relationship between the PPFD and the PPFFR. For a collimated beam at normal incidence, they are equal; while for perfectly diffuse radiation, the PPFFR is 4 times the PPFD. In practical situations the ratio will be somewhere between 1 and 4.

Photosynthetic Irradiance (PI) is defined as the radiant energy flux density of PAR. This is the radiant energy incident per unit time on a unit surface. The ideal PI Sensor responds equally to energy in the 400-700 nm waveband (Figure 1) and has a cosine response. Unit: W m^{-2}. Note: There is no unique relationship between PI and PPFD. This depends on the spectral properties of the light source (3).

6

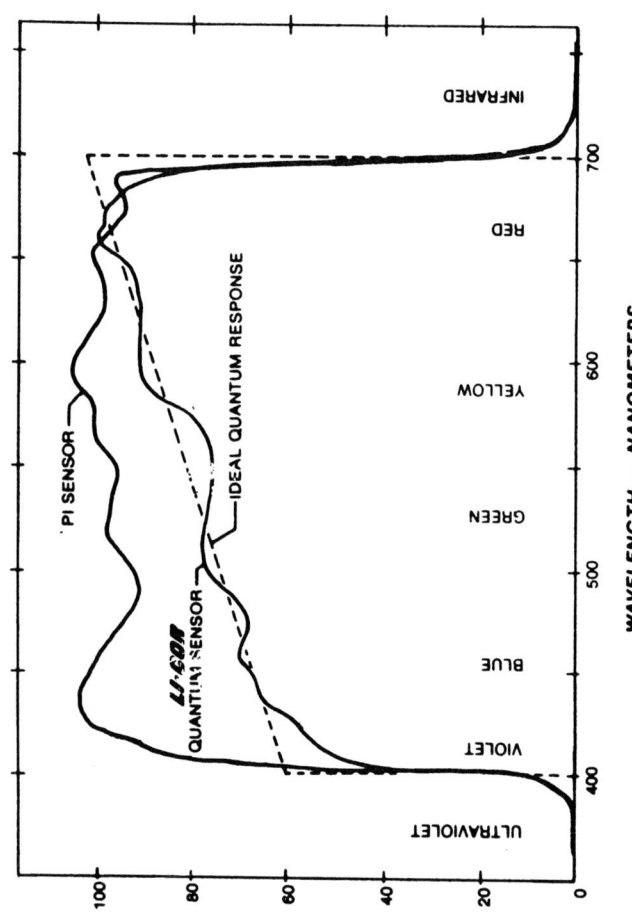

Figure 1. Typical spectral response of LI-COR Quantum Sensors and Photosynthetic Irradiance Sensors vs. wavelength and ideal quantum response.

3. PHOTOMETRY

Photometry refers to the measurement of visible radiation (light) with a sensor having a spectral responsivity curve equal to the average human eye. Photometry is used to describe lighting conditions where the eye is the primary sensor such as illumination of work areas, interior lighting, television screens, etc. Although photometric measurements have been used in the past in plant science, PPFD and irradiance are the preferred measurements. The use of the word "light" is inappropriate in plant research. The terms "ultraviolet light" and "infrared light" clearly are contradictory (3). The spectral responsivity curve of the standard human eye at typical light levels is called the CIE Standard Observer Curve (photopic curve) and covers the waveband of 380-770 nm. The human eye responds differently to light of different colors and has maximum sensitivity to yellow and green (Figure 3). In order to make accurate photometric measurements of various colors of light or from differing types of light sources, a photometric sensor's spectral responsivity curve must match the CIE photopic curve very closely.

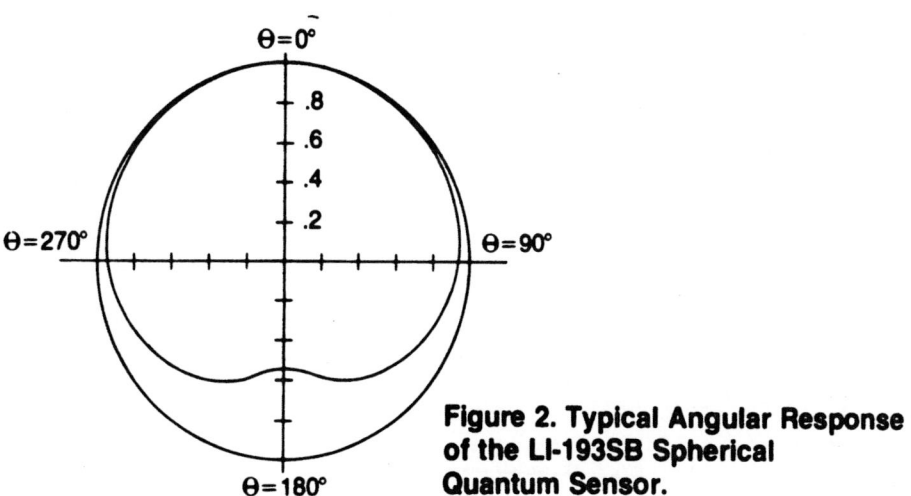

Figure 2. Typical Angular Response of the LI-193SB Spherical Quantum Sensor.

3.1 Terminology and Units (7)

Luminous flux is the amount of radiation coming from a source per unit time, evaluated in terms of a standardized visual response. Unit: lumen, lm.

Luminous intensity is the luminous flux per unit solid angle in the direction in question. Unit: candela, cd. One candela is one lm sr^{-1}.

Luminance is the quotient of the luminous flux at an element of the surface surrounding the point and propagated in directions defined by an elementary cone containing the given direction, by the product of the solid angle of the cone and the area of the orthogonal projection of the element of the surface on a plane perpendicular to the given direction. Unit: cd m^{-2}; also, lm sr^{-1} m^{-2}. This unit is also called the nit.

Illuminance is defined as the density of the luminous flux incident at a point on a surface. Average illuminance is the quotient of the luminous flux incident on a surface by the area of the surface. This physical quantity is measurd by a cosine photometric sensor such as the LI-210SA. Unit lux, lx. One lux is one lm m^{-2}.

4. MEASUREMENT ERRORS

At a Controlled Environments Working Conference in Madison, Wisconsin USA (1979), an official from the U.S. National Bureau of Standards (NBS) stated that one could not expect less than 10 to 25% error in radiation measurements made under non-ideal conditions. In order to clarify this area, the sources of errors which the researcher must be aware of when making radiation measurements have been tabulated. Refer also to the specifications given with each sensor for further details.

4.1 Absolute Calibration Error

Absolute calibration error depends on the source of the lamp standard and its estimated uncertainty at the time of calibration, accuracy of filament to sensor distance, alignment accuracy, stray light, and the lamp current measurement accuracy. Where it is necessary to use a transfer sensor (such as for solar calibrations) additional error will be introduced. LI-COR quantum and photometric sensors are calibrated against a working quartz halogen lamp. These working quartz halogen lamps have been calibrated against laboratory standards traceable to the NBS. Standard lamp current is metered to 0.035% accuracy. Microscope and laser alignment in the calibration setup reduce alignment errors to less than 0.1%. Stray light is reduced by black velvet to less than 0.1%. The absolute calibration accuracy is limited to the uncertainty of the NBS traceable standard lamp. The absolute calibration specification for LI-COR sensors is ± 5% traceable to the NBS. This accuracy is conservatively stated and the error is typically ± 3%. Absolute calibrations

and spectral responses of LI-COR sensors have been checked by the
National Research Council of Canada (NRC) to insure the accuracy
and quality of LI-COR calibrations.

4.2 Relative Error (Spectral Response Error)

This error is also called actinity error or spectral correction
error. This error is due to the spectral response of the sensor
not conforming to the ideal spectral response. This error occurs
when measuring radiation from any source which is spectrally different than the calibration source.

The quantum and photometric sensor spectral response conformity
is checked by LI-COR using a monochromator with a 2.5 nm bandwidth
and a blackened thermopile-calibrated silicon photodiode. The
spectral response of the sensor is achieved by the use of computer-
tailored filter glass. Relative errors for various sources due to
a non-ideal spectral response are checked both by actual measurement
and by a computer program which utilizes the source spectral irradiance data and the sensor spectral response data.

All LI-COR quantum and photometric sensors have relative errors
<7% when used in growth chambers, daylight, greenhouses, plant
canopies and aquatic conditions. When used with sources that have
strong spectral lines such as gas lamps or lasers, this error could
be larger depending on the location of the lines.

The LI-COR pyranometer measures irradiance from the sun plus
sky. The LI-200SA is not spectrally ideal (equal spectral response
from 280-2800 nm). See Figure 4. However, it is calibrated for
this wavelength range using a thermopile pyranometer (Eppley PSP).
NOAA states in a test report that for clear, unobstructed daylight
conditions, the LI-COR pyranometer compares very well with class
one thermopile pyranometers (8). The LI-200SA should not be used
under spectrally different radiation (than the sun), such as in
growth chambers, greenhouses, or plant canopies. Under such artificial or shaded conditions, a thermopile pyranometer should be used.

4.3 Spatial Error

This error is caused by a sensor not responding to radiation
at various incident angles. Spatial error consists of the cosine
error (or the angular error) and the azimuth error.

Cosine Error. A sensor with a cosine response (follows
Lambert's cosine law) allows measurement of flux densities through
a given plane, i.e. flux densities per unit area. When a parallel
beam of radiation of a given cross-sectional area spreads over a
flat surface, the area that it covers is inversely proportional to
the cosine of the angle between the beam and a plane normal to the
surface. Therefore, the irradiance due to the beam is proportional
to the cosine of the angle. A radiometer, whose response to beams
coming from different directions follows the same relationship, is
said to be "cosine-corrected" (9). A sensor without an accurate

cosine correction can give a severe error under diffuse radiation conditions within a plant canopy, at low solar elevation angels, under fluorescent lighting, etc.

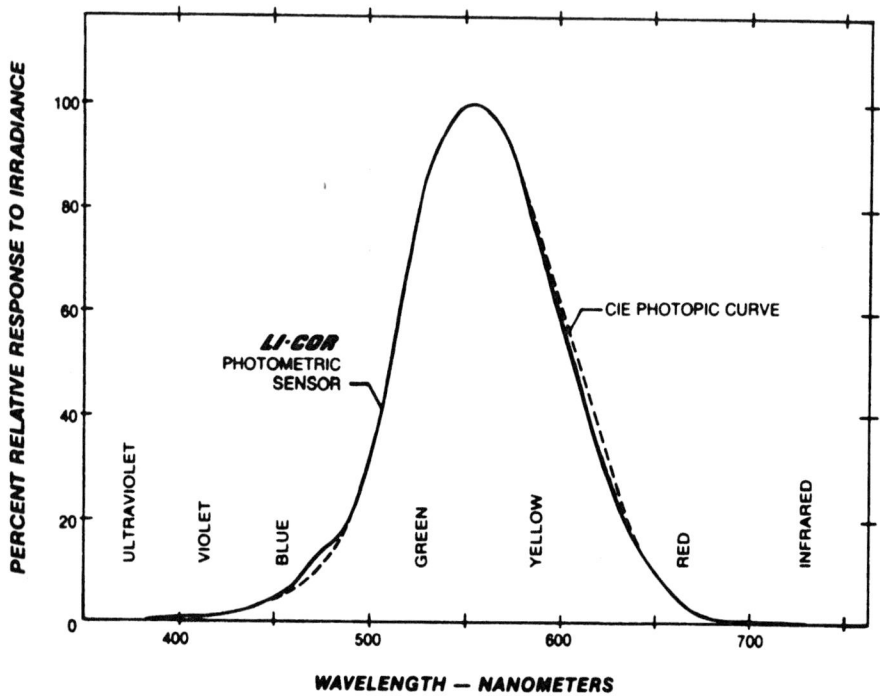

Figure 3. Typical spectral response of LI-COR Photometric Sensors vs. the CIE photopic response curve.

Cosine response is measured by placing the sensor on a platform which can be adjusted to rotate the sensor about an axis placed across the center of the measuring surface. A collimated source is directed at normal incidence, and the output of the sensor is measured as the angle of incidence is varied. The cosine error at angle θ is the percent difference of the ratio of the measured output at angle θ and normal incidence (angle 0°) as compared to the cosine of angle θ. This is repeated for various azimuth angles as necessary.

The LI-190SA, LI-200SA and LI-210SA are fully corrected cosine sensors. These sensors have a typical cosine error of less than 5% up to an 80° angle of incidence. Totally diffuse radiation introduces a cosine error of approximately 2.5%. For sun plus sky at a sun elevation of 30° (60° angle of incidence), the error is 2%.

The LI-192SA sensor has a slightly greater cosine error since this sensor has a cosine response optimized for both air and water.

The LI-191SA uses uncorrected acrylic diffusers and has a greater error at high angles of incidence. For totally diffuse

radiation, the error is 8%. For conditions within canopies, the
error is less because the radiation is not totally diffuse.

Angular Error. A spherical PPFFR sensor measures the total
flux incidence on its spherical surface divided by the cross-sectional area of the sphere. Angular error is measured by directing
a collimated source at normal incidence and rotating the sensor
360° about an axis directly through the center of the sphere at 90°
from normal incidence. This is repeated for various azimuth angles
as necessary to characterize the sensor.

The LI-193SA angular error is due to variations in density in
the diffusion sphere and the sphere area lost because of the sensor
base (Figure 2). This error is less than -10% for totally diffuse
radiation, but is usually much less than -10% because the upwelling
radiation is much smaller than the downwelling radiation.

Azimuth Error. This is a subcategory of both cosine and
angular error. It is specified separately at a particular angle
of incidence. This error is the percent change of the sensor output as the sensor is rotated about the normal axis at a particular
angle of incident radiation. This error is less than ± 1% at 45°
for the LI-190SA, LI-192SA, LI-200SA and LI-210SA sensors. The
error is less than ± 3% for the LI-191SA and LI-192SA sensors.

4.4 Displacement Error

In highly turbid waters, the LI-192SA Spherical Sensor will
indicate high PPFFR values due to the displacement of water by the
sensor sphere volume. This is because the point of measurement is
taken to be at the center of the sphere, but the attenuation which
would have been provided by the water within the sphere is absent.
This error is typically + 6% for water with an attenuation coefficient of 3 m^{-1} (10).

4.5 Tilt Error

Tilt error exists when a sensor is sensitive to orientation
due to the effects of gravity. This exists primarily in thermopile
type detectors. Silicon type detectors do not have this error.
All LI-COR sensors are of the latter type and have no tilt error.
This error in the LI-200SA Pyranometer is nonexistent and has an
advantage over thermopile type detectors for solar radiation
measurements (8).

4.6 Linearity Error

Linearity error exists when a sensor is not able to follow
proportionate changes in radiation. The type of silicon detectors
used in LI-COR sensors have a linearity error of less than ± 1%
over seven decades of dynamic range.

4.7 Fatigue Error

Fatigue error exists when a sensor exhibits hysteresis. This is common in selenium-based illumination meters and can add a considerable error. For this reason, LI-COR sensors incorporate only silicon detectors which exhibit no fatigue error.

4.8 Temperature Coefficient Error

Temperature coefficient error exists when the output of a sensor changes with a constant input. This error is typically less than ± 0.1% per °C for the LI-190SA, LI-191SA, LI-192SA, LI-193SA and LI-210SA sensors. This error is slightly higher for the LI-200SA.

4.9 Response Time Error

This error exists when the source being measured changes rapidly during the period of measurement. <u>Averaging</u>: Large errors can exist when measuring radiation under rapidly changing conditions such as changing cloud cover and wind if measuring within a crop canopy, and waves if measuring underwater. The use of an integrating meter to average the reading will eliminate this error.
<u>Instantaneous</u>: When radiation measurements are desired over a period of time (much less than the response time of the system), large errors can exist. For example, if one were to measure the radiation from a pulsed source (such as a gas discharge flash lamp) with a typical system designed for environmental measurements, the reading would be meaningless. Such a measurement should not be made with LI-COR instruments without consultation with LI-COR.

4.10 Long-Term Stability Error

This error exists when the calibration of a sensor changes with time. This error is usually low for sensors using high quality silicon photovoltaic/photodiodes and glass filers. LI-COR uses only these high quality components. The use of Wratten filters and/or inexpensive silicon or selenium cells add significantly to long-term stability error. The stability error of LI-COR sensors is typically ± 2% per year.

4.11 Immersion Effect Error

A sensor with a diffuser for cosine correction will have an immersion effect when immersed in water. Radiation entering the diffuser scatters in all directions within the diffuser, with more radiation lost through the water-diffuser interface than in the case when the sensor is in the air. This is because the air-diffuser interface offers a greater ratio of indices of refraction than the water-diffuser interface. LI-COR provides a typical immersion effect correction factor for the underwater sensors. Immersion effect error is the difference between this typical figure and the actual figure for a given sensor in a particular environment. Since LI-COR test measurements are done in clear water, the error is also dependent on other variables such as turbidity, salinity, etc. Immersion effect error is typically ± 2% or less. A complete report on the immersion effect properties of LI-COR underwater sensors is available from LI-COR.

4.12 Surface Variation Error

In general, the absolute responsivity and the relative spectral responsivity are not constant over the radiation-sensitive surface of sensors. This error has little effect in environmental measurements except for spatial averaging sensors such as the LI-191SA. This error is < ± 7% for the LI-191SA.

4.13 User Errors

Spatial User Error. This is different than sensor-caused spatial error. Spatial user error can be introduced by using a single small sensor to characterize the radiation profile within a crop canopy or growth chamber. The flux density measured on a given plane can vary considerably due to shadows and sunflecks. To neglect this in measurements can introduce errors up to 1000%. Multiple sensors or sensors on track scanners can be used to minimize this error. If track scanners are used, the output of the sensors must be integrated.

The LI-191SA Line Quantum Sensor, which spatially averages radiation over its one meter length, minimizes this error and allows one person to easily make many measurements in a short period of time. Another method, although not as accurate, is to use an integrating meter and the LI-190SA Quantum Sensor and physically scan the sensor by hand within the canopy while integrating the output with the meter.

Another type of spatial user error can be caused by misapplication of a cosine-corrected sensor where a spherical sensor would give a more accurate measurement. An example is in underwater photosynthetic radiation measurements when studying phytoplankton.

User Setup/Application Errors. These errors include such causes as:
- Reflections or obstructions from clothing, buildings, boats, etc.
- Dust, flyspecks, seaweeds, bird droppings, etc.
- Shock, causing permanent damage of optics within the sensor.
- Submersion of terrestrial sensors in water for an extended period (partial or total). Rain does not affect the sensors since they are completely weatherproof.
- Use of the incorrect calibration constant.

- Incorrect interpolation of analog meters.
- Using the wrong meter function.
- Failure to have sensors recalibrated periodically.

4.14 Readout Error

This error is due to the readout instrument as distinguished from the sensor. Zero drift, temperature, battery voltage, electronic stability, line voltage, humidity and shock are all factors which can contribute to readout error. The use of electronic circuitry such as chopper-stabilized amplifiers and voltage regulators in LI-COR meters largely eliminates many of these problems: zero drift, temperature, battery voltage, electronic stability, line voltage.

4.15 Total Error

The errors given are largely independent of each other and are random in polarity and magnitude. Therefore, they can be summed in quadrature (the square root of the sum of the squares). The total error is shown below for an LI-190SB Quantum Sensor and LI-COR meters when used for measuring lighting in a typical growth chamber or natural daylight over a temperature range of 15° to 35°C.

	Typical Error
Absolute Error	5% max., 3% typical
Relative (spectral response) Error	5%
Spatial (cosine) Error	2%
Displacement Error	0%
Tilt Error	0%
Linearity Error	0%
Fatigue Error	0%
Sensor Temperature Coefficient Error	1% (0.1% per °C)
Response Time Error	0%
Long-term Stability Error	2% (2% per year)
Immersion Effect Error	0%
Surface Variation Error	0%
Readout Error	1%
User Error	?

The total error = Square Root $(5 \times 5 + 5 \times 5 + 2 \times 2 + 1 \times 1) = 7.6\%$. All of the above errors are minimized by LI-COR through design and calibration. While this error seems reasonably low, it must be remembered that no user error has been added, and that statistically it is possible that all the errors could be of the same polarity. The sum of the errors (less the user error) could equal 16% in the worst case. The absolute error is conservatively stated and is more typically ± 3%. User error in vegetation canopies where shadows and sunflects exist can be very large (1000%) and one of the methods described under Spatial User Error should be employed.

When purchasing a radiation measuring system, it is necessary to insure that the spatial (cosine, etc.) and relative spectral response errors are as low as possible. These two errors depend upon the skill and expertise of the designer and manufacturer. Some manufacturers deliberately do not give specifications for these errors, and the user can expect large errors. The absolute error is largely dependent on the NBS lamp standard. Minimization of this error can be achieved by the more experienced companies through the use of precise techniques and expensive capital equipment. In order to insure long-term stability, it is necessary that the manufacturer use the highest quality silicon photovoltaic/photodiodes and only the best glass filters. Modern electronic readout instruments virtually eliminate readout error.

The user should be aware of all the types of errors that can occur, particularly the relative and spatial errors, since these can add considerably to the total error. LI-COR has and continues to put forth a considerable effort to insure that the spectral and cosine response of all quantum and photometric sensors are as nearly ideal as optically possible. This assures LI-COR customers of the best possible accuracy.

5. CONVERSION OF UNITS

5.1 Conversion of Photon Units to Radiometric Units

Conversion of quantum sensor output in μmol s^{-1} m^{-2} (400-700 nm) to radiometric units in W m^{-2} (400-700nm) is complicated. The conversion factor will be different for each light source, and the spectral distribution curve of the radiant output of the source (Wλ; W m^{-2} nm^{-1}) must be known in order to make the conversion. The accurate measurement of Wλ is a difficult task, which should not be attempted without adequate equipment and calibration facilities. The radiometric quantity desired is the integral of Wλ over the 400-700 nm range, or:

$$W_T = \int_{400}^{700} W\lambda \, d\lambda \qquad (1)$$

At a given wavelength λ, the number of photons per second is

$$\text{photons s}^{-1} = \frac{W\lambda}{hc/\lambda} \qquad (2)$$

where h = 6.33 x 10^{-34} joules - s (Planck's constant),
c = 3.00 x 10^{8} m s^{-1} (velocity of light) and λ is in nm.
hc/λ is the energy of one photon.

Then, the total number of photons per second in the 400-700 nm range is

$$\int_{400}^{700} \frac{W\lambda}{hc/\lambda} \, d\lambda \tag{3}$$

This is the integral which is measured by the sensor. If R is the reading of the quantum sensor in $\mu mol \ s^{-1} \ m^{-2}$ (1 $\mu mol \ s^{-1} \ m^{-2}$ = 6.022×10^{17} photons $s^{-1} \ m^{-2}$), then

$$6.022 \times 10^{17} \ (R) = \int_{400}^{700} \frac{W\lambda}{hc/\lambda} \, d\lambda \tag{4}$$

Combining Eq. (1) and Eq. (4) gives

$$W_T = 6.022 \times 10^{17} \ (Rhc) \frac{\int_{400}^{700} W\lambda \, d\lambda}{\int_{400}^{700} \lambda W\lambda \, d\lambda} \tag{5}$$

To achieve the two integrals, discrete summations are necessary. Also, since $W\lambda$ appears in both the numerator and the denominator, the normalized curve $N\lambda$ may be substituted for it. Then

$$W_T = 6.022 \times 17^{17} \ (Rhc) \ \frac{\sum_i N\lambda_i \Delta\lambda}{\sum_i \lambda_i N\lambda_i \Delta\lambda} \tag{6}$$

where $\Delta\lambda$ is any desired wavelength interval
         λ_i is the center wavelength of the interval and
         $N\lambda_i$ is the normalized radiant output of the source at the
             center wavelength.

In final form this becomes

$$W_T \approx 119.8(R) \ \frac{\sum_i N\lambda_i}{\sum_i \lambda_i N\lambda_i} \ W \ m^{-2} \tag{7}$$

where R is the reading in $\mu mol \ s^{-1} \ m^{-2}$.

The following procedure should be used in conjunction with Eq.(7)

1. Divide the 400-700 nm range in "i" intervals of equal wavelength spacing $\Delta\lambda$.
2. Determine the center wavelength (λ_i) of each interval.
3. Determine the normalized radiant output of the source ($N\lambda_i$) at each of the center wavelengths.
4. Sum the normalized radiant outputs as determined in Step 3 to find $\sum_i N\lambda_i$.
5. Multiply the center wavelength by the normalized radiant output at the wavelength for each interval.
6. Sum the products determined in Step 5 to find $\sum_i \lambda_i N\lambda_i$.
7. Use Eq. (7) to find W_T in W m^{-2}, where R is the quantum sensor output in µmol s^{-1} m^{-2}.

The following approximation assumes a flat spectral distribution curve of the source over the 400-700 nm range (equal spectral irradiance over the 400-700 nm range) and is shown as an example.

Given: i = 1
$\Delta\lambda$ = 300 nm
λ_i = 550 nm

$$W_T \approx 119.8(R) \left(\frac{N(550)}{550\ N(550)} \right) = \frac{119.8(R)}{550} = 0.22(R)\ W\ m^{-2}$$

or

1 W m^{-2} $\approx$ 4.6 µmol s^{-1} m^{-2}

This conversion factor is within ± 8.5% of the factors determined by McCree as listed in Table I (3).

5.2 Conversion of Photon Units to Photometric Units

To convert photon units (µmol s^{-1} m^{-2}, 400-700 nm) to photometric units (lux 400-700 nm), use the above procedure, except

a. Replace Eq. (1) with

$$Lux = 683 \int_{400}^{700} y\lambda\ W\lambda\ d\lambda$$

where $y\lambda$ is the luminosity coefficient of the standard CIE curve with $y\lambda$ = 1 at 550 nm
$W\lambda$ is the spectral irradiance (W m^{-2} nm^{-1}).

b. Replace Eq. (5) with

$$\text{Lux} = (683)(6.022 \times 10^{17})(\text{Rhc}) \frac{\int_{400}^{700} y\lambda \, W\lambda \, d\lambda}{\int_{400}^{700} \lambda W\lambda \, d\lambda}$$

c. Replace Eq. (6) with

$$\text{Lux} = (683)(6.022 \times 10^{17})(\text{Rhc}) \frac{\sum_i y\lambda_i N\lambda_i \Delta\lambda}{\sum_i \lambda_i N\lambda_i \Delta\lambda}$$

d. Replace Eq. (7) with

$$\text{Lux} = 8.17 \times 10^4 \, (R) \frac{\sum_i y\lambda_i \lambda_i}{\sum_i \lambda \, N\lambda_i}$$

e. Replace Step 4 with:

   i. Multiply the luminosity coefficient ($y\lambda$) of the center wavelength by the normalized radiant output ($N\lambda$) at that wavelength for each interval.

   ii. Sum the products determined in step i. above to find $\sum_i y\lambda_i \, N\lambda_i$.

The following approximation assumes a flat spectral distribution curve of the source over the 400-700 nm range (equal spectral irradiance over the 400-700 nm range) and is shown as an example.

Given: i = 1 to 31
        $\Delta\lambda$ = 10 nm
        λ_1 = 400, λ_2=410, λ_3=420 . . . λ_{31}=700
        $N\lambda$ = 1 for all wavelengths
        $y\lambda_1$ = 0.0004, $y\lambda_2$=0.0012, $y\lambda_3$=0.004 . . . $y\lambda_{31}$=0.0041

$$\text{Lux} = 8.17 \times 10^4(R) \frac{\sum_i y\lambda_i}{\sum_i \lambda_i} = 8.17 \times 10^4(R)\left(\frac{10.682}{17050}\right)$$

   Lux = 51.2R, where R is in μmol s^{-1} m^{-2}

Or,

   1000 lux = 1 klux = 19.5 μmol s^{-1} m^{-2}

Table 1. Approximate conversion factors for various light sources.
(3) (PAR waveband 400-700nm)

	Light Source					
	Daylight	Metal halide	Sodium (HP)	Mercury	White fluor.	Incand.
To Convert	Multiply by					
W m^{-2} (PAR) to μmol s^{-1} m^{-2} (PAR)	4.6	4.6	5.0	4.7	4.6	5.0
klux to μmol s^{-1} m^{-2} (PAR)	18	14	14	14	12	20
klux to W m^{-2} (PAR)	4.0	3.1	2.8	3.0	2.7	4.0

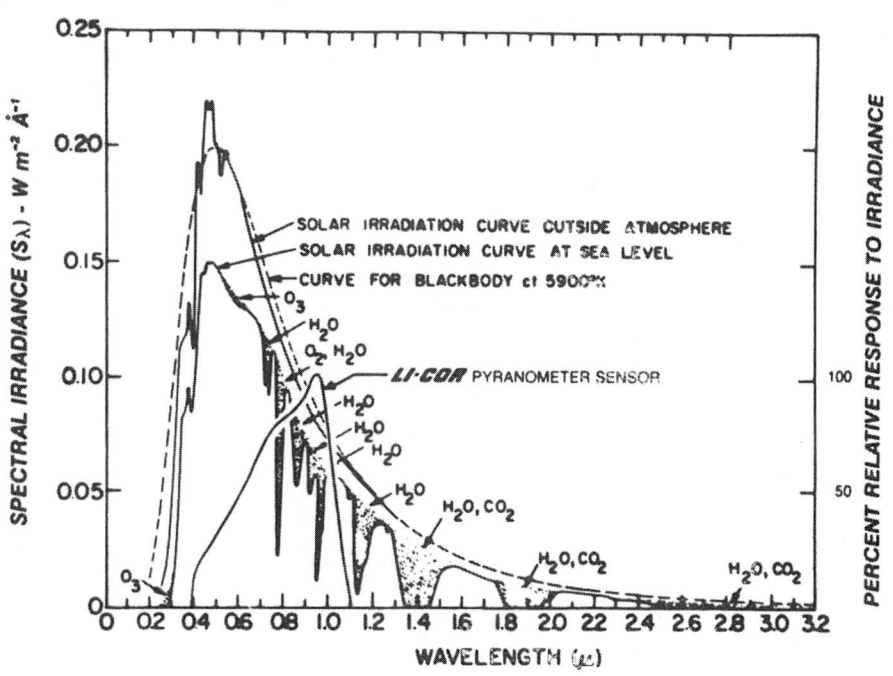

Figure 4. The LI-200SB Pyranometer spectral response is illustrated
along with the energy distribution in the solar spectrum (6).

CONVERSION FACTORS FOR RADIATION DATA

Type of Measurement	Instantaneous Measurements	Full Sun Plus Sky	Integrated Measurements	One Day's Integration
Quantum	$1\ \mu E\ s^{-1}\ m^{-2} =$ $1.0\ \mu mol\ s^{-1} m^{-2}$ $6.022\ 10^{17} photons\ s^{-1} m^{-2}$ $6.022\ 10^{17} quanta\ s^{-1} m^{-2}$ $6.022\ 10^{13} quanta\ s^{-1} cm^{-2}$	PPFD 2000 $\mu E\ s^{-1} m^{-2}$	$1\ E m^{-2} =$ $1\ mol\ m^{-2}$ $6.022\ 10^{23} photons\ m^{-2}$ $6.022\ 10^{23} quanta\ m^{-2}$	Photosyn Photon Expo 60 E m^{-2}
Radio-metric	$1\ W\ m^{-2} =$ $1.433\ 10^{-3} cal\ cm^{-2} min^{-1}$ $1.433\ 10^{-3} langley\ min^{-1}$ $0.100\ mW\ cm^{-2}$ $100\ \mu W\ cm^{-2}$ $1.0\ J\ s^{-1} m^{-2}$ $1000\ erg\ s^{-1} cm^{-2}$ $0.317\ BTU\ ft^{-2}\ h^{-1}$ $5.283\ 10^{-3} BTU\ ft^{-2}\ min^{-1}$	Total Irrad. 1000 Wm^{-2} Photosynthetic Irradiance 500 Wm^{-2} Near IR (745-815 nm) Irradiance 60 W m^{-2}	$1\ Wh\ m^{-2} =$ $0.0860\ cal\ cm^{-2}$ $0.0860\ langley$ $0.1\ mWh\ cm^{-2}$ $100\ \mu Wh\ cm^{-2}$ $3600\ J\ M^{-2}$ $0.3600\ J\ cm^{-2}$ $3.600\ 10^{6}\ erg\ cm^{-2}$ $0.317\ BTU\ ft^{-2}$	Total Rad. E 8000 Wh m^{-2} Photosyn. Rad. Exp. 4000 W Near (745-815) IR Rad.Exp. 500 Wh m^{-2}
Photo-metric	$1\ lux =$ $1\ lm\ m^{-2}$ $0.0929\ lm\ ft^{-2}$ $0.0929\ footcandle$ $0.001\ klux$	Illuminance 100 klux	$1\ lux\ h =$ $0.0929\ footcandle\ h$ $0.001\ klux\ h$	Lum.Exp. 800 klux h

The values shown for "full sun plus sky" and "one day's integration" are typical measurements taken in mid-summer in Lincoln, Nebraska, USA, using cosine type sensors. These values can vary greatly depending on atmospheric conditions, location (including latitude and elevation), time of year, and time of day.

A PORTABLE PHOTOSYNTHESIS SYSTEM

Jon Welles
LI-COR, Inc.

6. BACKGROUND

The ability to measure photosynthetic rates of leaves or whole plants in the field is crucial not only to plant physiology, but to many other disciplines, such as ecology and entomology. Historically, the limitation to field measurements of gas exchange has been the measurement of CO_2. The requirements involved either a mobile laboratory to get the analysis equipment into the field, or else techniques whereby samples (such as syringes containing gas samples or plant material labeled with $^{14}CO_2$) could be taken back to a laboratory for subsequent analysis (11).

Recent advances in infrared gas analysis and microcomputers now make it possible to make gas exchange measurements and analysis in the field with equipment small enough for one person to easily carry (13,15). This section details one such system, the LI-COR model LI-6000.

7. SYSTEM DESCRIPTION

The LI-6000 operates in a closed system mode. A leaf is enclosed in a chamber with no exchange of air with the outside. Transpiration will cause the humidity in the chamber to rise, and photosynthesis will cause the CO_2 concentration to fall. The conductance and photosynthetic rate is calculated from the rates of change of these quantities.

The system (Figures 5,6) consists of a control console, a CO_2 analyzer, and a sensor housing that can be attached to one of several different leaf chambers. Total system weight is 7 kg. Measurements are made of leaf temperature, chamber air temperature, relative humidity, CO_2 concentration, and photosynthetically active radiation (PAR). Calculations of stomatal conductance and apparent photosynthetic rate are made by the onboard microprocessor after the measurements on a leaf are completed. The measurement time is typically 30 seconds to 1 minute.

The system is powered by a 12 volt rechargeable battery that, when fully charged, provides 2 hours of operating time. The batteries are easily interchanged without loss of power to the instrument.

7.1 Carbon Dioxide Analyzer

The infrared gas analyzer is a non-dispersive infrared type capable of measuring 0 to 1100 vpm. A pump located in the analyzer housing withdraws air from the leaf chamber, passes it through the analyzer, and returns it to the chamber. A thermistor to measure air temperature is located in the path of the air that exists the analyzer's sample tube.

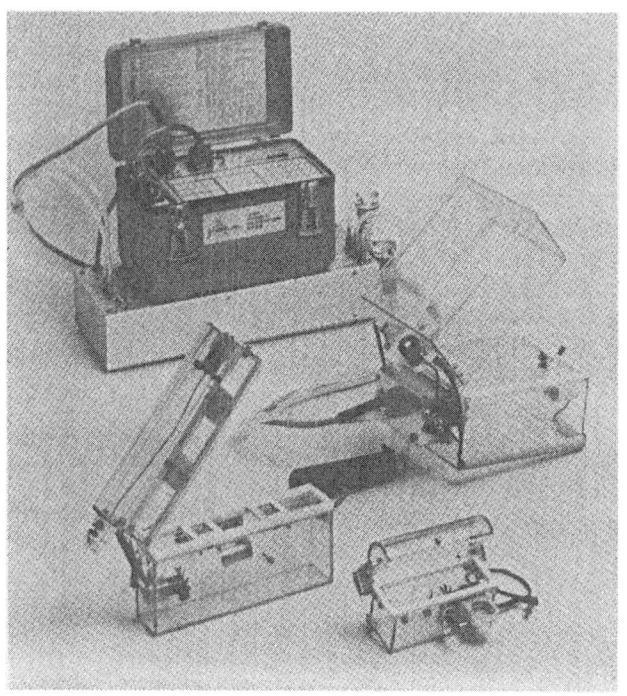

Figure 5.

The analyzer housing contains tubes for a soda lime trap for CO_2 and magnesium perchlorate $(Mg(ClO_4)_2)$. In normal operation, air from the leaf chamber is drawn through an air filter and the desiccant tube before passing through the gas analyzer and returning to the leaf chamber. The desiccant serves to eliminate uncertainties in the measurement due to the water sensitivity of the analyzer or dilution effects. In this mode of operation, the returning dry air flow to the chamber is accounted for when stomatal resistance is computed. If desiccant is not used, there is a correction to the photosynthesis calculation that can be enabled in the software to correct for water sensitivity or dilution.

7.2 Sensor Housing and Leaf Chamber

The sensor housing includes the handle, connectors, and electronic for the leaf chamber sensors and fans. The sensors include a thermocouple (leaf temperature), thermistor (air temperature), Vaisala HUMICAP (Humidity), and LI-COR Quantum sensor (PAR).

Figure 6.

The purpose of the leaf chamber is to provide a closed volume around the leaf within which the humidity and CO_2 changes can be measured. A closed pore foam gasket around the mating halves of the chamber helps provide a tight seal around the leaf. The interior of the chamber is coated with Teflon to minimize interaction with water vapor or CO_2.

Because it uses a closed system approach, the size and shape of the leaf chamber is very important. The enclosed leaf area, the chamber volume, and the photosynthetic rate all influence the rate at which the CO_2 will be drawn down in the system. Because of the analyzer's noise (less than 2 ppm peak to peak), a total drawdown of 20 to 30 ppm is recommended during a measurement.

The leaf chambers have one or two internal fans which serve to insure that the air extracted from the chamber is representative of the chamber, and to minimize the boundary layer resistance of the leaf.

7.3 System Console

The system console includes the microcomputer that handles the data logging, calculations, and data storage. The rechargeable battery that powers the entire system is also contained within the console. Battery voltage can be monitored on the console display. An internal backup battery preserves the instrument's internal memory for up to several months when power is off, or the main battery disconnected.

The internal microcomputer gives the system some powerful capabilities. There are 5 basic groups of tasks that the computer allows the user to perform:

1. Real time monitoring of any data channel.
2. Recording data. A measurement consists of two or more observations of each data channel.
3. Computing results. After data is recorded, conductance, photosynthesis, and several other results are computed and available for display before subsequent measurements are made.
4. Data storage and retrieval. 32K or 64K of random access memory is available for data storage. Stored data can be retrieved, viewed on the console, modified, re-computed, and restored at the user's discretion.
5. Data communications. Stored data can be transferred to RS-232 compatible devices. Data formats include ASCII, HEX, and BINARY. The binary format allows very rapid data transfer, compact storage on mass storage devices, and can be transferred back into the instrument.

8. THEORY OF OPERATION

The derivations below are of the equations used by the LI-6000 in its calculations, and consider a leaf of area "a" contained within a closed volume "v". It is assumed that the surface of the enclosing volume in no way interacts with the water vapor or CO_2 in the air within that volume. A complete list of symbols is given at the end of the chapter.

8.1 Stomatal Resistance

The transpiration rate E (mass H_2O area^{-1} time^{-1}) can be expressed in terms of the rate of change of vapor concentration with time:

$$E = \frac{v}{a} dq/dt \tag{8}$$

where q is the concentration (mass volume^{-1}) of the water vapor in the volume. Transpiration can also be expressed in terms of a leaf resistance r_L (time distance^{-1}) to the difference in vapor pressure between the leaf interior and ambient conditions. By approximating the vapor pressure difference with the vapor concentration difference, we can write

$$E = \frac{(q_s - q)}{r_L} \tag{9}$$

where q_s is the saturation vapor concentration for the leaf. Equating expressions (1) and (2), and solving for leaf resistance yields

$$r_L = \frac{a\,(q_s - q)}{v\,dq/dt} \tag{10}$$

The LI-6000 records data at discrete points in time, so the equation used to calculate leaf resistance for a pair of observations (denoted by subscripts 1 and 2) is in finite difference form:

$$r_L = \frac{a\,(t_2 - t_1)\,(q_s[T_L] - \overline{H}\,q_s[T_a])}{v\,(H_2\,q_s[T_{a2}] - H_1\,q_s[T_{a1}])} \tag{11}$$

T_L, T_a, and $\overline{H}$ refer to the mean of the two observations of leaf temperature (C), air temperature (C), and humidity (fraction between 0 and 1), respectively. The function $q_s[T]$ is the saturation vapor concentration (g m^{-3}) at temperature T. The LI-6000 contains this data in a table in its memory.

The stomatal resistance is usually arrived at by subtracting an assumed value of the boundary layer resistance from the measured leaf resistance. For systems in which only one side of the leaf is measured, this is quite correct. For systems such as the LI-6000 in which both sides of the leaf are measured simultaneously, the procedure is a bit more complicated. The calculated resistance r_L is the parallel combination of the top surface stomatal resistance r_{st} plus the boundary layer resistance r_h, and the bottom surface stomatal resistance r_{sb} plus the boundary layer resistance. Assuming that the boundary layer resistance is the same for the top and bottom of the leaf, we can write

$$r_L = \frac{(r_{st} + r_h)\,(r_{sb} + r_h)}{(r_{st} + r_{sb} + 2r_h)} \tag{12}$$

Rearranging equation (12) to solve for the parallel combination of the top and bottom surface stomatal resistance leads to

$$r_s = \frac{r_{st}\, r_{sb}}{r_{st} + r_{sb}}$$

$$= r_L - r_h \left(1 - \frac{2r_L - r_h}{r_{st} + r_{sb}}\right) \tag{13}$$

The term in brackets in equation (13) ranges in value from 0.5 to 1, depending on the ratio or r_{sb} and r_{st}. With leaves having no stomata on one surface ($r_{st} \gg r_{sb}$ or $r_{st} \ll r_{sb}$), the value of the term in brackets is 1, and $r_s = r_L - r_h$. At the other extreme, when $r_{sb} = r_{st}$, the value of the term in brackets is 0.5, and $r_s = r_L - (r_h/2)$. In general, simply subtracting the boundary layer resistance from the leaf resistance will not yield the correct stomatal resistance; the error can be as large as one half of the boundary layer resistance in magnitude.

Solving equation (13) requires some additional knowledge about r_{sb} or r_{st}. Perhaps the most practical is an estimate of their ratio. Defining

$$K = \frac{r_{st}}{r_{sb}} \tag{14}$$

equation (13) can be rewritten as

$$r_s = r_L - r_h \left(1 - \frac{2r_L - r_h}{(K + 1)r_{sb}}\right) \tag{15}$$

We now make an assumption based on the following observation:

When $K = 1$, $r_{sb} = 2r_L - r_h$.

When $K = $ infinity, $r_{sb} = r_L - r_h/2$.

In general, let $r_{sb} = (r_L - r_h/2)(1 + 1/K)$.

With this assumption, equation (15) becomes

$$r_s = r_L - r_h \left(\frac{K^2 + 1}{(K + 1)^2}\right) \tag{16}$$

There are two things to note at this point. One is that K is symmetric. That is, equation (16) yields the same result if $1/K$ is substituted for K. Thus, leaves with stomata on only one side can be characterized by K=0 or K=infinity. The second observation is of the error caused by the assumption used to get from equation

(15) to equation (16). As long as the boundary layer resistance is less than 1/5 of the leaf resistance, the error will be negligible (a fraction of a percent). If r_h approaches r_L, however, the error approaches 100% when K=0.

8.2 Photosynthesis

The apparent photosynthetic rate A (mass CO_2 area^{-1} time $^{-1}$) measured in a closed chamber is expressed by:

$$A = - \left(\frac{v \, dq_c}{a \, dt} \right) \tag{17}$$

where q_c is the concentration (mass volume^{-1}) of the CO_2 in the air. It is related to volumetric concentration C (μ liter liter^{-1} or ppm) by

$$q_c = \left(\frac{C}{10^6} \right) \left(\frac{44 \text{ (g/mole)} * 10^3 \text{ (liter m}^{-3})}{22.4 \text{ (liter/mole)} \frac{(T_a + 273)}{273} * \frac{1013}{P_a}} \right) \tag{18}$$

The finite difference form of equation (17) used by the LI-6000 is

$$A = \frac{v \, (C_1 - C_2) \, P_a \, 0.00529}{a \, (t_2 - t_1) \, (T_a + 273)} \tag{19}$$

where A is in (mg CO_2 m^{-2} s^{-1}),
C_1 and C_2 are initial and final concentrations of CO_2 in the chamber expressed in volumetric ppm,
T_a is the mean air temperature during the measurement interval.

8.3 Desiccant Correction

Normally, the LI-6000 is operated with magnesium perchlorate desiccant in the air stream ahead of the CO_2 analyzer. This is done to remove all water sensitivity and dilution effects from the CO_2 analyzer readings. Because the LI-6000 operates in a closed system, however, the presence of the desiccant causes dry air to be returned to the leaf chamber. This in turn tends to suppress to some degree the normal humidity increase. A correction is done when computing leaf resistance to compensate for this, when desiccant is used with the LI-6000.

Consider a container of volume v having a time dependent internal concentration q(t) of some non-interacting gas. When a steady flow f of concentration Q is introduced to the chamber, the following behavior is exhibited:

$$\frac{dq}{dt} = -\frac{f}{v}(Q - q) \tag{20}$$

The solution to equation (20) over time interval t_1 to t_2 with the boundary conditin $q(t_1) = q_1$ yields a final concentration q_2 of

$$q_2 = Q (1 - e^{-F}) + q_1 e^{-F}$$

where
$$F = (t_2 - t_1) \frac{f}{v} \tag{21}$$

When leaf resistance is calculated using equation (11), it is done on pairs of observations logged by the LI-6000. The denominator of equation (11) contains the vapor density difference between the two observations. When the desiccant correction is enabled, it is applied to this difference; the final humidity is corrected to be what it would have been had the desiccant not been in place. From equation (21), and assuming incoming vapor density Q is 0,

$$q_2 \text{ adjusted} = q_2 \text{ measured} + q_1(1 - e^{-F}) \tag{22}$$

Note that this correction does not actually change recorded data, but is done temporarily during computations of leaf resistance. Flow rate is measured automatically with an electronic flow meter.

8.4 Water Sensitivity and Dilution Correction

When desiccant is not used ahead of the CO_2 analyzer, a correction must be made in the photosynthesis calculation to account for any water sensitivity of the analyzer, and for dilution of the CO_2 due to the increase in water vapor in the air. Both of these possible effects are accounted for by adjusting the C_2 term (final CO_2 reading of a pair) in Equation (19) by an amount proportional to the vapor density change ($q_2 - q_1$) during that time interval:

$$C_2 \text{ adjusted} = C_2 - W (q_2 - q_1)$$

where
$$\tag{23}$$

$$W = W_0 + W_s C_2$$

W is the change in CO_2 per change in water vapor density, and is specified as a function of CO_2 concentration in terms of a slope W_s and offset W_0.

W_s and W_0 are obtained by a simple experimental procedure when using the LI-6000 without a desiccant. If it is necessary to correct for dilution alone, it can be done by setting W_0 to 0, and W_s to $-1/(\text{density of the air})$, or typically $-1.0 \text{ m}^3 \text{ kg}^{-1}$.

8.5 Internal Carbon Dioxide Concentration

A very useful physiological parameter is an estimation of the subtomatal CO_2 concentration. Photosynthesis can be expressed in terms of the concentration difference of CO_2 (ambient minus internal) and the leaf resistance to CO_2. This relationship (12,14) can be expressed (ignoring units for the moment):

$$A \frac{C_a - C_i}{r_{Lc}} - \overline{C}E \tag{24}$$

The $\overline{C}E$ term is a correction for the interaction of incoming CO_2 and outgoing water vapor. Equation (24) can be rearranged to be solved for the internal CO_2 concentration. In conventional units, the solution is:

$$C_i = \frac{C_a (Z - E/k) - A}{(Z + E/k)}$$

where $\tag{25}$

$$K = 1.8179 \times 10^5$$

$$Z = \frac{0.00529 \ P_a}{r_{Lc} (T_a + 273)}$$

$$r_{Lc} = 1.6 \ r_s + 1.37 \ r_h \frac{K^2 + 1}{(K + 1)^2}$$

C_i and C_a are in volumetric ppm. The relation between resistance to water vapor and resistance to CO_2 diffusion is 1.6 for stomatal, and 1.37 for boundary layer. These values represent ratios of the binary diffusion coefficients for CO_2 in air and water vapor in air.

8.6 Leaf Resistance Error Analysis

The fractional error in leaf resistance r_L due to humidity, leaf temperature, or air temperature measurement error can be calculated by differentiating r_L with respect to the parameter in question, dividing by r_L, and multiplying by the error in that parameter. Errors due to volume and leaf area are simply proportional. Details will not be presented here, only the final results.

The effect of a 1% (relative humidity units) error in the humidity measurement is indicated in Figure 7. The solution depends upon the ratio of air temperature change to humidity change. The upper curve represents no humidity change (ratio equals infinity), and would be caused by the returning dry air exactly balancing the humidity added by the leaf. The lower curve represents the other extreme in which there is no temperature change given some finite humidity change (ratio equals 0). Typically, the appropriate error curve for transient measurements is near the bottom curve.

ERROR IN LEAF RESISTANCE
caused by 1% humidity offset

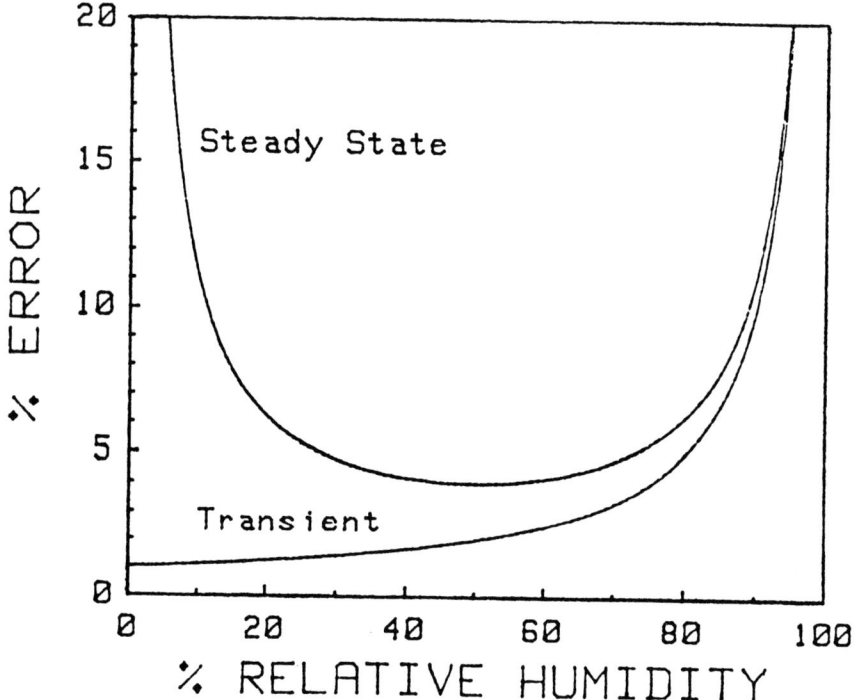

Figure 7.

The dominant feature of Figure 7 is that the error in leaf resistance calculations becomes infinite as humidity approaches 100%. Note that Figure 7 assumes a constant 1% humidity error over the full range of humidity. (For a 2% error, double each value on the curve, etc.) In reality, the error in the humidity measurement itself usually increases at higher humidities, making the leaf resistance error curve rise even faster. Thus, the accuracy of leaf resistances calculated under conditions of very high humidity is certainly questionable.

The error in leaf resistance due to leaf temperature error turns out to be very close to the same magnitude and opposite sign of the error due to air temperature. It is thus important that the leaf and air temperature sensors be matched. This is easily tested using an empty, closed, shaded leaf chamber with the fans running. The two temperature readings should be nearly identical. The error due to a difference of 0.1 C between the sensors results in a leaf resistance error of less than 2% at humidities less than 70%, independent of the temperature. This error also rapidly increases, though, as the humidity approaches 100%.

The sum of the errors leads one to expect about a 10% accuracy of leaf resistance measurements below 70% humidity.

8.7 Apparent Photosynthesis Error Analysis

The potentially dominant error in photosynthetic rate is due to CO_2 measurement. In a transient system such as the LI-6000, the photosynthetic rate is calculated from the difference in CO_2 over time. Errors in this difference can be due to noise and nonlinearity; errors in absolute accuracy are cancelled out.

The error due to noise is simply the noise divided by the total drawdown. Thus, with a 2 ppm peak-to-peak noise level, a 20 ppm change produces a 10% error, a 30 ppm change a 6% error, etc. The measurement time is ultimately determined by the desired drawdown, since the two are proportional.

Errors due to non-linearity are also proportional. That is, if the analyzer is accurate at 330 ppm, but in error by 1 ppm at 300 ppm, the error in photosynthesis on a measurement that starts at 330 ppm and ends at 300 ppm would be 1/30 or 3%. The linearity of the analyzer is typically 0.5%.

Overall accuracy for photosynthesis with a 30 ppm drawdown is typically 10%.

9. INSTRUMENT OPERATION

9.1 Calibration

The user can adjust the analyzer's zero and span through use of two potentiometers located on the analyzer housing. There are three software adjustments that are made in converting the analyzer's millivolt output into ppm.

1. Density correction. The analyzer's millivolt output is adjusted based upon the current temperature of the air passing through the analyzer, and the temperature of the air when the gain was last set. This correction helps reduce the need to adjust the gain (requiring a gas of known CO_2 concentration) as the temperature changes.

2. Linearization table. The temperature adjusted raw millivolt signal is linearized by use of a factory generated linearization lookup table. The table consists of 18 pairs of data

points that describe that particular analyzer's departure from being linear as a function of the raw signal. It also includes the observed linearity dependence on temperature for the analyzer.

3. Calibration table. The relationship between linearized milli-volts and ppm comes from the calibration table. In its simplest form, the table consists of two pairs of data points: 0 mV = 0 ppm, and 4096 mV = 1100 ppm. As with the linearization table, the LI-6000 interpolates between points. The table can consist of up to 5 pairs of points should the user feel that this is necessary due to, for example, a linearity change in the analyzer.

Once the calibration and the linearization tables are installed in the memory of the instrument, the calibration process consists of monitoring ppm on the console, setting the zero potentiometer while flowing zero gas through the analyzer, and setting the span potentiometer while flowing a span gas through the analyzer. The tables are retained as part of the memory.

9.2 Parameters

There are several parameters that the user needs to set before operating the instrument. Some of these are values needed for calculations, and others affect how the instrument will operate. These values are retained in permanent memory, and need to be accessed only when one or more need to be changed. Below is a partial list of the parameters:

- Atmospheric pressure (mb).
- Volume of the entire system (cm^3).
- Boundary layer resistance of one side of the leaf. This is primarily dependent on the chamber used.
- Estimate of the ratio of stomatal resistance of one side of the leaf to the other. This is a species dependent parameter and is used with the boundary layer parameter to arrive at the true boundary layer resistance for the leaf.
- The slope and intercept of the analyzer's water sensitivity curve if desiccant is not used with the analyzer.
- Time step. The time interval between observations. If zero, then data logging is manual (an observation is logged only in response to a user keystroke). If non-zero, the user initiates logging with a keystroke, but succeeding observations are logged automatically.

9.3 Tests

There are several tests that can be done on the system prior to taking measurements. Some need not be done more than once, and others should be done on a regular basis.

Leak Test. The degree to which the system has unwanted air
exchange with the environment can be tested by measuring the rate
of change of CO_2 within the closed system given a large differential
of CO_2 between the inside and outside. The differential is created
by flushing the system with CO_2 free air for a period of time. Given
a several hundred ppm gradient between the inside and outside of
the system, the rate of CO_2 increase should be on the order of 1
ppm per minute.

Volume Test. The volume of the system as supplied from the
manufacturer is known and need not be tested. A method of determining the volume of a system having user built or modified components is to inject a known amount of CO_2 into the chamber and
determining the volume based upon the resulting change in CO_2
concentration.

Boundary Layer Determination. This is readily done using a
moistened piece of blotter paper to simulate a leaf. The computed
resistance (assuming no boundary layer) is the boundary layer
resistance. The boundary layer resistance for one leaf side is
typically 0.5 s cm^{-1} for the 4 liter chamber and 0.3 s cm^{-1} for the
1 liter chamber. In the 0.25 liter chamber, the boundary layer
resistance can range from 0.2 to 1.0 s cm^{-1}, depending upon the
leaf size.

9.4 Measurement Process

The measurement process is outwardly very simple. The mechanical steps are:

1. Select the leaf and clamp it in place in the chamber.
2. Wait for the CO_2 to start dropping at a constant rate. There
   is a time lag of several seconds since the air needs to be
   pumped through the analyzer.
3. Initiate logging. After the key is pressed to do this, subsequent observations are recorded automatically at the predetermined time interval. After 10 observations are recorded,
   the leaf can be removed from the chamber while the instrument
   performs its computations.
4. View results. After about 10 seconds, the results are available for viewing on the display. By viewing the mean, range,
   intercept and standard error of the photosynthesis calculations,
   and the internal CO_2 calculations, the user can decide quite
   quickly if the measurement is valid or not. The user can choose
   whether or not to store the data in the instrument's memory.

One of the needed inputs is leaf area, since the results are
displayed on a leaf area basis. The 1 liter chamber has adjustable
inserts that can be used with many types of leaves to maintain a
constant leaf area exposed in the chamber. With this arrangement,
leaf area would only have to be entered once. With leaves whose
shape is such that the part exposed in the chamber forms a rectangle,

the instrument can be configured so that only the width need be
entered. Each width entry is automatically multiplied by the
chamber width (exposed leaf length) to arrive at the leaf area.
The leaf width can be obtained while waiting for the instrument to
log all of its observations. For irregularly shaped leaves, leaf
areas would have to be determined by some other method either before
or after the gas exchange measurement. Stored data is readily re-
called and recomputed, so proper leaf areas can be entered after
all other data is collected.

While the mechanics of collecting data have been simplified, it
should not be forgotten that the measurement is only as good as the
techniques and precautions exercised. This involves an understanding
of the plant and its behavior, so that care can be taken to avoid
doing those things that would unduly influence the final result.

9.5 Calculation and Analysis

Figure 8 is an example of the output of the LI-6000 for a
single measurement of a leaf. The major calculations listed below
are performed on the data set after the measurements are recorded.

9.6 Resistance (or Conductance) and Apparent Photosynthesis.

For each pair of observations collected, these two values are
calculated. This approach allows for nonlinear changes in humidity
or CO_2 concentration and thus linear changes in resistance or photosynthesis over the measurement period. Fitting a straight line
(this is not done) to the humidity or CO_2 values first would be to
assume that resistance or photosynthesis does not change over the
measurement period.

9.7 Mean and Range

The time weighted mean of each observation is computed. This
includes both measured (humidity, leaf temperature, air temperature,
CO_2, and PAR) and computed (resistance and photosynthesis) variables.
The range is simply the largest value minus the smallest value.

9.8 Initial Value and Standard Error of the Intercept

A straight line is fit to measured and computed variables, and
the intercept value is computed. This is done for two reasons. If
the presence of the chamber in fact caused the leaf to respond
either by closing its stomata or reducing photosynthesis, the computed intercept value will be different from the mean value. Closing
stomata, for example, will result in the intercept value of resistance to be lower than the mean value. Comparing the initial and
mean values is thus a quick diagnostic tool in the field. The
second reason is that the computed initial values are logical
choices for computing transpiration and internal CO_2 concentration.

LI-6000 Data Page

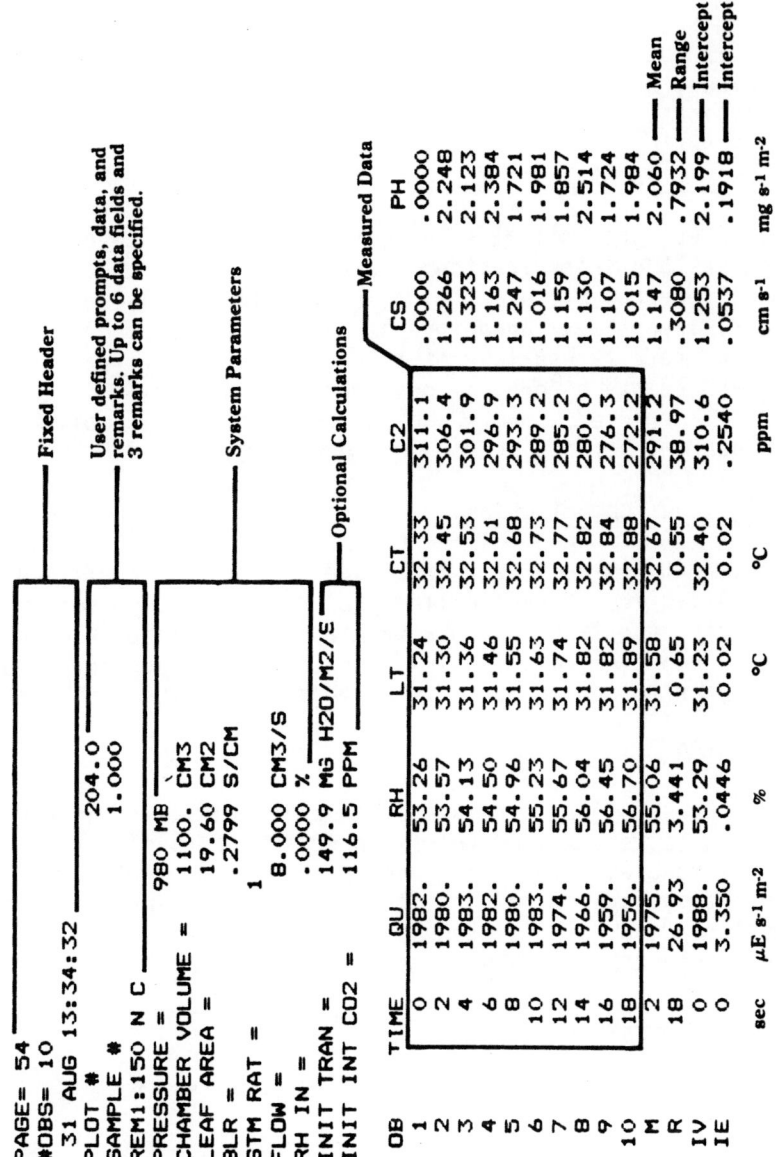

PAGE= 54
#OBS= 10
31 AUG 13:34:32 ——— Fixed Header

PLOT #   204.0
SAMPLE #   1.000   ——— User defined prompts, data, and remarks. Up to 6 data fields and 3 remarks can be specified.
REM1:150 N C

PRESSURE =   980 MB
CHAMBER VOLUME = 1100. CM3
LEAF AREA =   19.60 CM2
BLR =   -.2799 S/CM   ——— System Parameters
STM RAT =   1
FLOW =   8.000 CM3/S
RH IN =   -.0000 %
INIT TRAN =   149.9 MG H2O/M2/E
INIT INT CO2 =   116.5 PPM   ——— Optional Calculations

Measured Data

OB	TIME	QU	RH	LT	CT	C2	CS	PH	
1	0	1982.	53.26	31.24	32.33	311.1	.0000	.0000	
2	2	1980.	53.57	31.30	32.45	306.4	1.266	2.248	
3	4	1983.	54.13	31.36	32.53	301.9	1.323	2.123	
4	6	1982.	54.50	31.46	32.61	296.9	1.163	2.384	
5	8	1980.	54.96	31.55	32.68	293.3	1.247	1.721	
6	10	1983.	55.23	31.63	32.73	289.2	1.016	1.981	
7	12	1974.	55.67	31.74	32.77	285.2	1.159	1.857	
8	14	1966.	56.04	31.82	32.82	280.0	1.130	2.514	
9	16	1959.	56.45	31.82	32.84	276.3	1.107	1.724	
10	18	1956.	56.70	31.89	32.88	272.2	1.015	1.984	
M	2	1975.	55.06	31.58	32.67	291.2	1.147	2.060	——— Mean
R	18	26.93	3.441	0.65	0.55	38.97	.3080	.7932	——— Range
IV	0	1988.	53.29	31.23	32.40	310.6	1.253	2.199	——— Intercept Value
IE	0	3.350	.0446	0.02	0.02	.2540	.0537	.1918	——— Intercept Error
	sec	µE s-1 m-2	%	°C	°C	ppm	cm s-1	mg s-1 m-2	

Figure 8.

9.9 Internal CO_2 Concentration

Based on computed initial values of the relevant parameters (resistance, photosynthetic rate and CO_2 concentration), the internal CO_2 concentration is a good indicator of whether or not the resistance and photosynthetic rate are in balance or not. Usually one has some idea of what the internal CO_2 ought to be for a plant, at least under certain conditions. If the value computed by the instrument is out of line, then something is wrong with the measurement technique, or a sensor's calibration, or a parameter used in the calculations.

9.10 Transpiration Rate

Also based upon initial values, since the presence of the chamber causes the transpiration rate to decline over the course of the measurement. Because the computed value is also based upon the boundary layer experienced by the leaf in the chamber, the transpiration rate is usually not a true indication of what the leaf's water use was before the measurement was made. This computed value can be corrected, however, based upon an estimate of the leaf's boundary layer resistance in the canopy.

9.11 Storage and Communications

The LI-6000 has an RS-232 interface that can be configured in software. A variety of handshaking protocols are supported. The instrument also can function in a terminal mode whereby the console acts like a dumb terminal. This is useful for getting logged onto a host computer to set it up to receive data from the instrument.

10. CONCLUSION

In its present form, the system described in this paper has been found to be well suited for field measurements of gas exchange of broadleaf species. The onboard computer allows for on site data reduction and storage of results, greatly facilitating the entire measurement process. The system's development is an ongoing process, and future enhancements are expected.

11. LIST OF SYMBOLS

A $mg\ m^{-2}\ s^{-1}$ Apparent CO_2 assimilation rate.

A' $umole\ m^{-2}\ s^{-1}$ Apparent molar CO_2 assimilation rate.

a cm^2 Area of leaf (1 side) exposed in chamber.

C $ul\ l^{-1}$ or ppm   Volumetric CO_2 concentration.

E	$mg\ m^{-2}\ s^{-1}$	Transpiration rate.
E'	$mmole\ m^{-2}\ s^{-1}$	Molar transpiration rate.
f	$cm^3\ s^{-1}$	Flow of air returning to the chamber.
H		Relative humidity (q/q_2).
K		Ratio of stomatal resistances.
P_a	mb	Atmospheric pressure.
q	$mg\ m^{-3}$	Vapor density, or generic concentration.
q_c	$g\ m^{-3}$	CO_2 concentration.
q_s	$mg\ m^{-3}$	Saturation vapor density.
r_h	$s\ cm^{-1}$	Boundary layer resistance of 1 leaf side.
r_L	$s\ cm^{-1}$	Leaf resistance.
r_s	$s\ cm^{-1}$	Stomatal resistance (parallel of both sides).
r_{sb}	$s\ cm^{-1}$	Stomatal resistance of the bottom surface.
r_{st}	$s\ cm^{-1}$	Stomatal resistance of the top surface.
T_a	C	Temperature of the air within the chamber.
T_L	C	Temperature of the leaf surface.
t	s	Time.
v	cm^3	Chamber volume.
W	$ppm\ m^3\ g^{-1}$	Ratio of CO_2 to vapor density change.

REFERENCES

1. Page, C.H. and Vigoureux, P. (eds.) The International System of Units (SI). Nat. Bureau of Stand. Special Publ. 330, 3rd edition, U.S. Government Printing Office, Washington, DC, 1977.

2. CIE (Commission Internationale de l'Eclairage). International Lighting Vocabulary, 3rd edition, Bureau Central de la CIE, Paris, 1970.

38

3. McCree, K.J. Photosynthetically active radiation. In:
    Physiological Plant Ecology, Vol. 12A, Encyclopedia of Plant
    Physiology (new series)(O.L. Lange, P. Nobel, B. Osmond and
    H. Ziegler, eds.) Springer-Verlag, Berlin, Heidelberg, New
    York, 1981.

4. Incoll, L.D., Long, S.P. and Ashmore, M.A. SI units in publi-
    cations in plant science. In: _Commentaries in Plant Science_,
    Vol. 2, Pergamon, Oxford, pp. 83-96, 1981.

5. Shibles, R. Committee report: Terminology pertaining to
    photosynthesis. _Crop Sci._ 16:437-439, 1976.

6. McCree, K.J. Test of current definitions of photosynthetically
    active radiation against leaf photosynthesis data. _Agric._
    Meteorol. 10:443-453, 1972.

7. Illuminating Engineering Society of North America. Nomencla-
    ture and definitions for illuminating engineering. Publication
    RP-16, ANSI/IES RP-16-1980, New York, 1981.

8. Flowers, E.C. Copmparison of solar radiation sensors from
    various manufacturers. In: _1978 Annual Report from NOAA to_
    the DOE, 1978.

9. Kondratyev, K.Y. _Direct Solar Radiation. Radiation in the_
    Atmosphere, Academic Press, New York, 1969.

10. Combs, W.S., Jr. The Measurement and Prediction of Irradiances
    Available for Photosynthesis by Phytoplankton in Lakes. Ph.D.
    Thesis, University of Minnesota, Limnology, 1977.

11. Clegg, M.D., Sullivan, C.Y. and Eastin, J.D. A sensitive tech-
    nique for the rapid measurement of carbon dioxide concentra-
    tions. _Plant Physiol._ 62:924-926, 1978.

12. Jarman, P.D. The diffusion of carbon dioxide and water through
    stomata. _J. Exp. Biol._ 25:927-936, 1974.

13. McPherson, H.G., Green, A.E. and Rollison, P.L. The measurement,
    within seconds, of apparent photosynthetic rates using a
    portable instrument. _Photosynthetica_ 17, 1983.

14. von Caemmerer, S. and Rarquhar, G.D. Some relationships
    between the biochemistry of photosynthesis and the gas exchange
    of leaves. _Planta_ 153:376-387, 1981.

15. Williams, B.A., Gurner, P.J. and Austin, R.B. A new infrared
    gas analyzer and portable photosynthesis meter. _Photosynthesis_
    Res. 3:141-151, 1982.

INSTRUMENTATION FOR THE MEASUREMENT OF CO_2 ASSIMILATION BY CROP LEAVES

Steve P. Long

University of Essex
Colchester, UK

1. INTRODUCTION

1.1 The Role of CO_2 Exchange Measurements

Growth of crops, in terms of carbon, organic matter or dry-weight gain, has traditionally been measured by sampling, drying, weighing and chemical analysis of the dried material. While this technique is adequate for assessing long-term changes, its value is limited where interest centers either on short-term dry-matter gain, i.e. intervals of days, hours or minutes, or on contributions made by individual organs, e.g. the flag leaves of cereals. The shortest periods over which statistically significant change in plant dry-weight may be resolved, even by extrapolation with statistical regression techniques (1), usually exceeds one day. Since water content is highly variable, plants must be dried before weighing, thus making the method unavoidably destructive. Therefore, production may only be estimated by sampling from a population, so introducing a random error which will decrease sensitivity. Interest in the relation of photosynthetic activity to productivity centers on carbon gain. Direct measurement of CO_2 uptake provides an alternative method of measuring productivity with six important advantages over measurements of dry-weight change:

1) It is instantaneous, measuring production in vivo on the time-scale of both in vitro studies of sub-cellular photosynthetic processes and of in vivo slow chlorophyll fluorescence transients (2).

2) It is non-destructive, thus the same leaf may be measured throughout a treatment or throughout its life.

3) The immediate effects of sudden changes in microclimate or experimental treatments on photosynthetic productivity may be

determined, where change in dry-weight gain would require days
before an effect might be demonstrated. Transient effects of, for
example, environmental change or herbicide application on production
may be revealed by measurement of gas exchange, yet be of too short
a duration to be apparent in dry-weight changes.

4) It accounts for all photosynthetic C-gain, including the
large fraction of up to 30% of photosynthate which may be lost by
root exudation (3).

5) It allows separate investigation of individual leaves,
parts of leaves or other photosynthetic organs.

6) It allows separation of photosynthetic gain from respiratory losses of carbon.

1.2 Measures, Symbols and Units

A plethora of measures, terms and units may be and have been
applied in the study of crop gaseous exchanges. This paper has
adopted the measures and symbols described by von Caemmerer and
Farquhar (4). This system is attractive in its simplicity by comparison to those recommended earlier (5). It allows both simple
derivation and direct comparisons of fluxes, conductances and derived
terms such as quantum efficiency and carboxylation efficiency. The
units used are exclusively S.I. This is not simple pedantry; adher-
ence to S.I. and its conventions is invaluable in the calculation
of fluxes and derived terms.

In the system adopted here, the CO_2 content of air is described
as the mole fraction (c) which equals both the partial volume (cm^3
m^{-3}) and the ratio of the partial pressure of CO_2 with the total
pressure of the body of air. Many instruments indicate CO_2 content
as v.p.m. (volumes per million), this is directly proportional to the
mole fraction, where 1 v.p.m. = 1 cm^3 m^{-3} = 1 μmol mol^{-1}. Thus, over
most ambient conditions the mol. fraction will be given directly,
circumventing the need to correct for pressure and temperature; this
would have been necessary had v.p.m. been converted to mass per unit
volume. The assimilation rate (A) is expressed as amount of CO_2
assimilated per unit leaf area per second (mol m^{-2} s^{-1}) and may be
calculated as the product of the change in mole fraction of CO_2
across the leaf and the net mole flow of air, divided by the leaf
surface area. Amount rather than mass is used to express A since
the same units may be used for other fluxes, notably transpiration
(E) and photon flux (I_p). This simplifies calculation of quantum
efficiency (A/I_p) and efficiency of water use (A/E). The mass flux
is the product of A and the molecular weight of CO_2; i.e. 44.

A further advantage of this system is that not only fluxes,
but also conductances share common units. Conductances (g) are
given by the ratio of the flux and CO_2 gradient across the same
diffusion pathway. The CO_2 gradient will be described as mole
fractions which are dimensionless; the units of g will be those of
the numerator, i.e. A. To avoid the use of exponents, sub-multiples
of the S.I. base units may be used, but only in the numerator.

Thus, μmol m^{-2} s^{-1} would be correct for A, but mol dm^{-2} h^{-1} should not be used. Sub-multiples in the denominator add unnecessary complexity to the calculation of derived terms and complicate comparisons (6).

1.3 The Approach

Most CO_2 exchange studies have involved enclosure methods, i.e. a leaf, plant or stand of plants is enclosed in a transparent chamber. The rate of CO_2 assimilation by the material enclosed is then determined by measuring the drop in the CO_2 concentration of the air flowing across the chamber. Alternatively, CO_2 exchange of large areas of vegetation may be measured without enclosure, using micrometeorological techniques.

Micrometeorological techniques. In these methods CO_2 assimilation by a large area of crop (>0.5 ha) is determined through concurrent measurements of CO_2 concentrations and air movements above the crop. As the crop assimilates CO_2, so the CO_2 concentration in the air immediately above and surrounding the plants will decrease. The change in CO_2 concentration at different heights above the vegetation will be determined by the rate of CO_2 assimilation by the vegetation and air movements which replenish the CO_2. Two standard micrometeorological methods of measuring fluxes of CO_2 between vegetation and atmosphere have been developed (7). In the more direct method, eddy correlation, instruments are used to measure turbulent fluctuations in vertical wind speed, w' (m s^{-1}), and CO_2 concentration, c' (mol m^{-3}), associated with individual eddies of air. The molar flux density is given by the time-average of the product of these measures:

$$F = \overline{w'c'} \quad \ldots\ldots\ldots \text{(1.1)}$$

In the other method, flux gradient analysis, it is assumed by analogy with molecular diffusion that the flux density is proportional to the vertical gradient of mean CO_2 concentration, dc'/dz (averaged over several minutes).

$$F = -Kdc'/dz \quad \ldots\ldots\ldots \text{(1.2)}$$

The minus indicates that where concentration increases with height, the flux is towards the surface. Here K is derived for momentum from wind profiles and uses an empirical relationship between K for momentum and mass (7). A mean rate of CO_2 assimilation per unit leaf area A may then be calculated, if the leaf area index (LAI) is known:

$$\overline{A} = (R_{soil} - F)/LAI \quad \ldots\ldots\ldots \text{(1.3)}$$

Enclosure methods. The remainder of this paper will be limited to enclosure methods. The disadvantage of these, relative to the

micrometeorological methods, is that some alteration to the crop environment is inevitable when the material to be studied is enclosed.
Assimilation chambers used to enclose the plant material are usually
designed to minimize alteration of the environment and this is considered in section 4. The procedure for measuring CO_2 assimilation
by the enclosed material depends on the type of gas exchange system
employed. Three basic configurations of systems have been used and
these are considered below. In essence, all three make use of the
fact that the net rate of CO_2 assimilation by the plant material
enclosed is equal to the product of the net rate of air flow and
the change in CO_2 mole fraction across the chamber.

1.4 Gas Exchange Systems and Equations for Flux Determinations

Closed systems. In a closed system air is drawn from the
chamber enclosing the leaf or plant into an IRGA which will continuously record the CO_2 concentration of the system (Fig. 1).

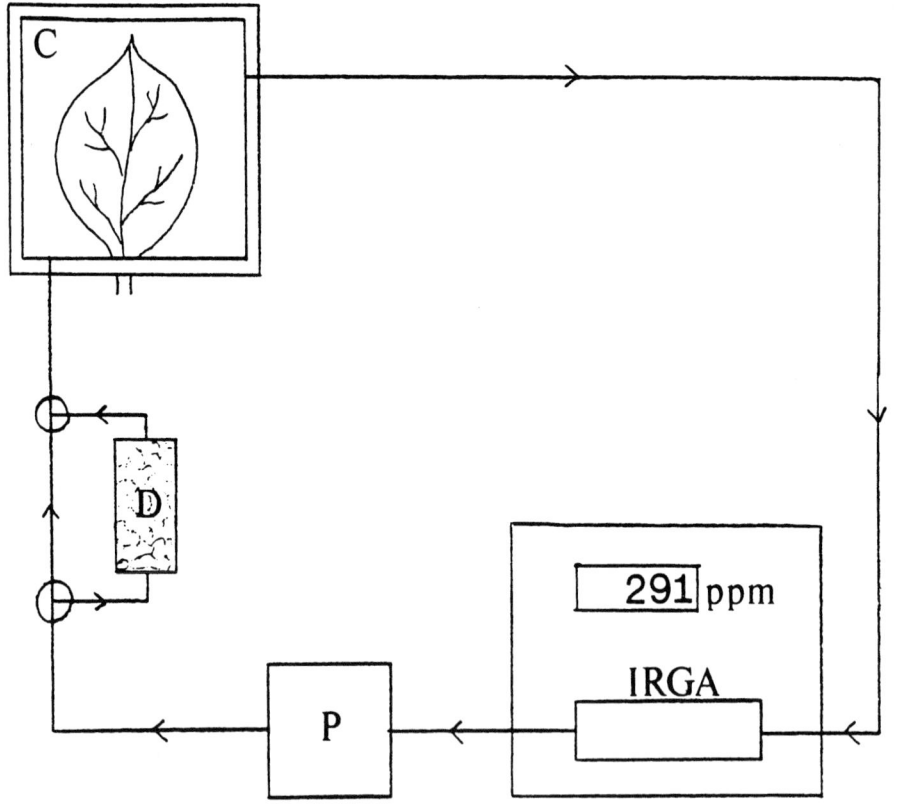

Figure 1. A closed system. Air is driven by a pump (P), partially
dried by a water vapor absorbent (D), to a leaf chamber (C) and
then onto an infra-red gas analyzer (IRGA) and back to the pump.

The air is then recycled from the IRGA back to the chamber. Thus
no air will leave the system and no air will enter it from outside.
If the leaf enclosed in the chamber is photosynthesizing, then the
CO_2 concentration in the system will decline, and continue to
decline until the CO_2 compensation point of photosynthesis (Γ) is
reached. In practice the CO_2 concentration is allowed to drop by
about 30 μmol mol^{-1} from the ambient level. In good light (i.e.
$I_p > 500$ μmol m^{-2} s^{-1}) A declines continuously with decrease in c
below the ambient level. A drop in c of more than 30 μmol mol^{-1}
would result in an unacceptably large underestimate of A. The rate
of CO_2 assimilation (A) is equal to the change in the amount of CO_2
in the system per unit time and may be determined by equation
(1.4).

$$A = \frac{c_1 - c_2}{(t_1 - t_2).10^3} \cdot \frac{v}{22.4} \cdot \frac{p}{1013.25} \cdot \frac{273.13}{T} \cdot \frac{1}{s} \quad \ldots\ldots\ldots (1.4)$$

Where: c_1 and c_2 are the partial pressure of CO_2 in the system at
times (t) 1 and 2 respectively (μmol mol^{-1}).
v is the total volume of the system (ml).
22.4 is the volume occupied by one mmol of CO_2 at STP in
(cm^3/nmol). STP is assumed to be p=101.325 kPa and
T=273.13K

If humidity is not controlled, transpiration will result in an
increase in water vapor concentration and a dilution of the mol.
fraction of all other gases, including CO_2. However, this will be
compensated for by an almost equivalent increase in pressure such
that number of moles per unit volume will remain constant. However,
if the system is not totally air-tight, this increase in pressure
will result in a loss of gas, including CO_2, thus a dilution will
result which must be accounted for in calculating A.
Closed systems, by comparison to other configurations, are the
simplest being the least demanding of the IRGA and require no
measure of flow. The IRGA need only be a single cell instrument
capable of resolving c to about 1 μmole mol^{-1} (=1 "ppm"). However,
such systems also have important disadvantages. Recirculation of
the air will result in a continuous rise in humidity. A humidity
trap cannot be used since this will produce a variable volume of
liquid water which would represent a sink for CO_2 and complicate
the determination of v. Alternatively, a portion of the recircu-
lated air may be passed through a drier, as in the LI-6000 (LI-COR)
photosynthesis meter. This, however, will necessitate measurement
of the flow and removes one of the advantages of a closed system.
A further disadvantage of closed systems is that errors resulting
from CO_2 absorption/desorption to and permeation through tubing and
chamber walls will be amplified by the continuous recirculation.
The estimate of A is in effect a mean for the period over which
change is measured.

The major theoretical objection to closed systems is that since the CO_2 concentration is changing, A cannot reach steady-state and the measured value may not be a true reflection of the rate which would be obtained under constant CO_2 conditions. Oscillations in CO_2 assimilation such as those produced by stomatal cycling at low humidities or those produced through feed-back effects in carbon metabolism may occur at a lower frequency than the period required for a measurement. Such cyclic variation in A, which would be apparent in a system capable of continuously monitoring A in a constant c_a, would appear as random noise in closed system measurements.

Although used extensively in the early studies of photosynthetic CO_2 assimilation, closed systems had lost popularity in recent years as a result of their theoretical and practical limitations (8). However, the use of this system configuration in one of the major portable gas-exchange systems, the LI-COR photosynthesis meter, has revived interest.

Semi-closed systems. These are a variation on the closed system which allows c_a to remain constant so that a steady-state A may be attained, thus overcoming the theoretical objections outlined for closed systems. In this system the IRGA is used as a null-point instrument which controls through an electronic circuit a flow of CO_2 into the system at a rate directly equivalent to the rate of uptake by the leaf (Fig. 2). In practice, when CO_2 is removed by the photosynthesizing leaf, a decrease in c_a sensed by the IRGA activates an input of CO_2 which is then maintained at a rate just sufficient to keep c_a constant.

$$A = \frac{f_c}{s} \quad \ldots\ldots\ldots \quad (1.5)$$

Where f_c = the mole flow of CO_2 into the system (μmol s^{-1})
s = the leaf surface area enclosed within the system (m^{-2})

To determine A in a semi-closed system, the requirements are that the IRGA is calibrated in absolute mode and that the rate of addition of CO_2 is known with great accuracy, since random error in the estimation of A will be directly proportional to and primarily dependent upon error in f_c. The humidity within the system must either be maintained at a constant level, or it must be monitored and c corrected to that of dry air, otherwise A would be overestimated since the mole fraction of CO_2 within the system would decline not only as a result of photosynthesis, but also as a result of transpiration, which through raising the humidity will decrease the mole fraction of CO_2. Thus a semi-closed system requires an IRGA of greater resolution. A further requirement of the IRGA, not essential to the closed system, is absence of zero drift, i.e. long-term stability, since any drift will produce a systematic error in the estimate of A.

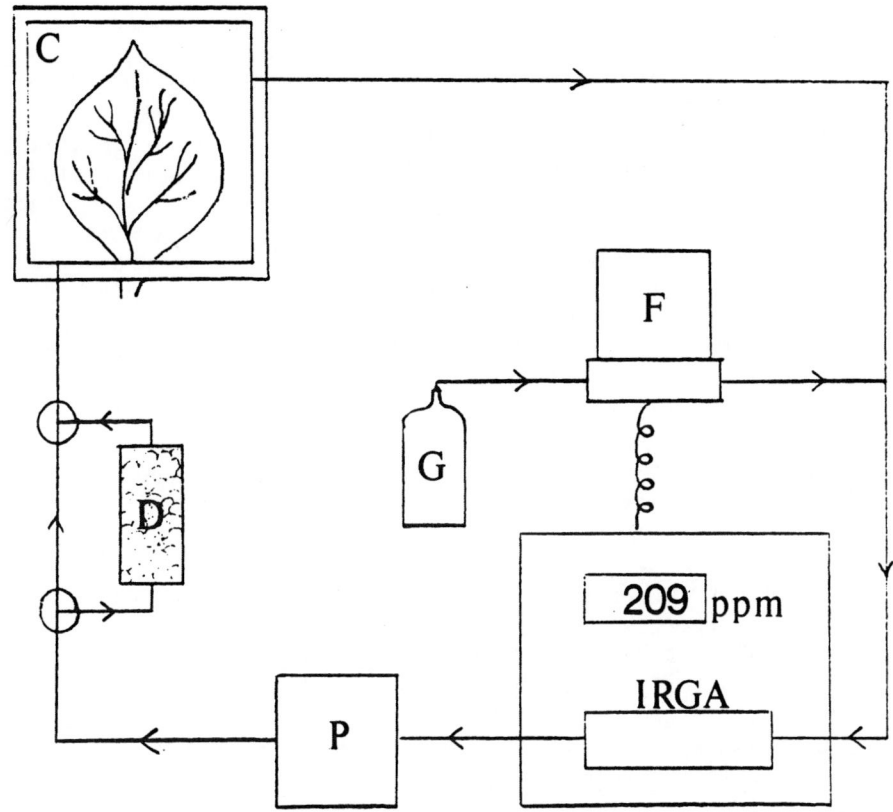

Figure 2. A semi-closed system. Air is driven by a pump (P), partially dried by a water vapor absorbent (D), to a leaf chamber (C) and then onto an infra-red gas analyzer (IRGA) and back to the pump. Decrease in system CO_2 concentration sensed at the IRGA, opens a controlled and measured flow (F) of CO_2 from a compressed gas cylinder (G).

Addition of CO_2 to semi-closed systems may be achieved with electronic dosing cocks (9) which add pulses of CO_2 into the system or with electronic flow controllers (10) which bleed CO_2 into the system at a constant rate.

The advantage of semi-closed systems over closed systems is that c_a is, by definition, maintained at a constant level and that steady-state A may be determined. The amplification of errors arising from adsorption/desorption of CO_2 from the system lining will be decreased, providing that the lining can rapidly come into equilibrium with c_a. Errors arising from permeation or leakage of CO_2 will be a constant, rather than an accumulating error as it is in a closed system. A further advantage is that A and, if humidity control and measurement is included, E may be studied at a range of values of c_a and e_a, simply by changing the set-point values in the system. A complex air-conditioning system would be required to achieve the same measurements in an open system.

Two practical disadvantages of semi-closed systems should be noted. First, only one leaf, plant or stand of plants may be monitored by one IRGA, thus such systems can be expensive relative to the number of measurements that may be made. Secondly, rapid transients in A, such as those arising from sudden changes in light level, cannot easily be resolved. If recirculation of air is slower than the transient, a damped oscillation will be set up in the system which is a function of the system and not the plant material. Similarly, the response times of the flow controllers may impose further oscillations in CO_2 concentration. To monitor transients, the volume of the system would need to be minimized and the air rapidly circulated. The controller would also need to be of the piezo-electric rather than electro-thermal type in order to give a response time faster than many transients in A.

Griffiths and Jarvis (10) provide a description of a transportable semi-closed system based on the Binos II IRGA.

Open systems. Open systems (Fig. 3) are characterized by a net flow of air through the system. The IRGA is used to monitor the change in CO_2 in the air as it passes through the system. Typically, such systems use a split beam IRGA capable of determining differences between two air streams. In the open system, the IRGA will be used to compare the inlet and exhaust air from the assimilation chamber. Where the air is dried prior to entering the IRGA, A will approximate to equation 1.6.

$$A = \frac{f.(c_e - c_0)}{s} \qquad \ldots\ldots\ldots (1.6)$$

Increase in water vapor content of the air as it passes over the leaf will affect calculation of A in the above equation in two ways. First, cross-sensitivity of CO_2 IRGAs for water vapor will increase the analyzer signal and cause an overestimate of c_0. This may be accounted for if both the response to water vapor and the increase in water vapor (i.e. $X_0 - X_e$) are known. Many CO_2 IRGAs now incorporate optical filters which minimize the response to water vapor and so remove the need for this correction. In the absence of such filters waver vapor may be reduced to a constant low quantity by passage through an ice bath or by passage through columns of calcium chloride and magnesium perchlorate in series (5); silica gel should be avoided in this application as it may exchange CO_2.

Secondly, increase in the water vapor mole fraction will cause a dilution of all the other gases, including CO_2. Thus, c_0 will be less than c_e simply as a result of transpiration. This could be corrected by reducing both gas streams to a constant humidity, e.g. by passage through an ice condenser prior to the IRGA.

47

Figure 3. An Open System

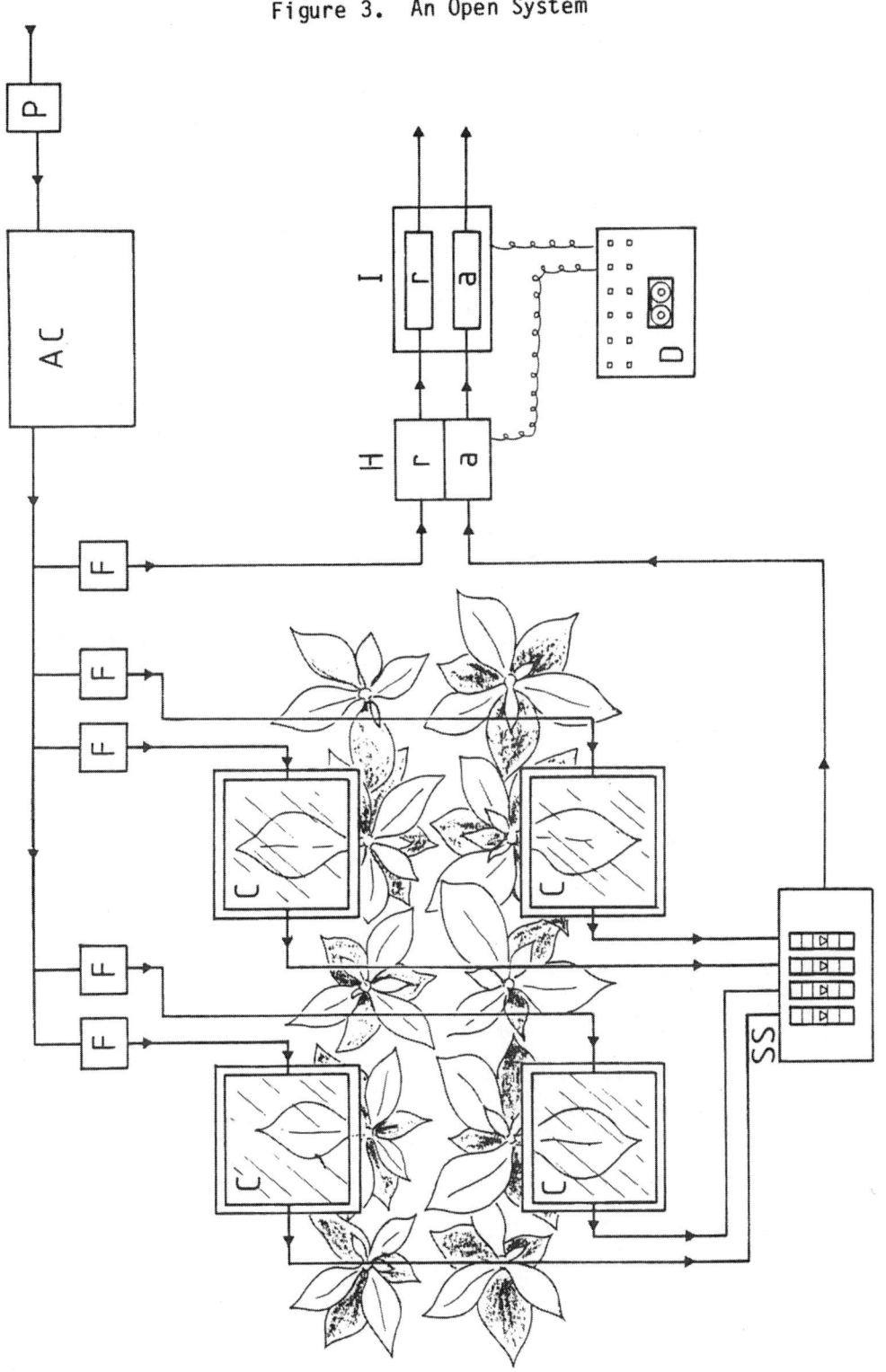

In Figure 3, outside air is drawn into the system by a pump (P) and then passed into an air conditioning system (AC) to control humidity and gas concentrations. Flow to a reference and several leaf chambers is controlled by individual mass flowmeters/controllers (F). Air from the leaf chambers (C) is passed to a sample selector (SS) which passes gas from each chamber in sequence to a differential hygrometer (H) and differential IRGA (I). Changes in water vapor and CO_2 concentrations across each leaf are determined by comparison to a reference (r) gas stream. Typically, the large amounts of data gathered by such a system would be sent to a data-logger (D).

Alternatively, if the change in χ is determined, then dilution of c may be accounted for in calculation of A:

$$ A = - \frac{f}{s} \left[\frac{(1 - \chi_i)}{(1 - \chi_0)} c_0 - c_e \right] \qquad \dots\dots\dots (1.7) $$

To determine A in an open system the requirements are: that the IRGA may be used and is calibrated in differential mode, that the change in both humidity and CO_2 across the leaf is known, that the flow rate of air through the leaf chamber is constant and accurately known, and that the leaf area is accurately determined. The main disadvantage of such a system is the initial expense, the requirement of an air-conditioning system, and the requirement of an IRGA which can accurately sense small differences in CO_2 mole fraction between two air streams, i.e. differences of the order of 1 $\mu mol \ mol^{-1}$. There are a number of advantages of such a system. Firstly, by use of a switching device, A can be simultaneously determined for a number of chambers. Secondly, the CO_2, O_2 and water vapor concentration around the leaf can be easily manipulated. Thirdly, by linking an H_2O IRGA or electrical humidity sensor in series with the CO_2-IRGA, transpiration and photosynthetic CO_2 assimilation can be measured simultneously for several chambers.

Two commercial open systems are available: H Waltz Co. based on the Binos II IRGA and described by Schultze et al (11); ADC Ltd based on the ADC LCA IRGA.

2.   INFRA-RED GAS ANALYSIS

2.1 Principle

CO_2 assimilation has been measured by a wide range of techniques, the most common being [14]CO_2-labeling, conductivity and IR spectroscopic analysis. The latter, infra-red gas analysis of CO_2, is the most widespread contemporary method of determining photosynthetic and respiratory CO_2 exchange in plants. This popularity stems from the reliability, accuracy and simplicity of this technique when compared to other available methods. To accurately determine CO_2 exchange from a leaf area of about 10 cm^2 in an open or semi-closed system, the instrument should be capable of resolving a CO_2 mole fraction of 0.1 - 1 µmol mol^{-1} (=.1 - 1 vpm) against the normal atmospheric concentration of about 340 µmol mol^{-1}. Although many IRGAs designed for laboratory operation will meet this specification, there are few truly portable instruments which are capable of this resolution. This section examines the principles of infra-red gas analysis with particular emphasis on developments for field operation.

Heteratomic gas molecules absorb radiation at specific infra-red wavebands in the sub-millimeter spectrum; each gas having a characteristic absorption spectrum. Gas molecules consisting of two identical atoms (e.g. O_2, N_2) do not absorb infra-red radiation (IR), and thus do not interfere with determination of the concentration of heteratomic molecules (12). IR gas analysis has been used for the measurement of a wide range of heteratomic gas molecules, including CO_2, H_2O, NH_3, CO, N_2O, NO and gaseous hydrocarbons (13). Thus IR gas analysis can be used not only for the accurate determination of CO_2 concentration but also H_2O vapor in transpiration studies.

The major absorption band of CO_2 is at λ = 4.25 µm with secondary peaks at λ = 2.66, 2.77 and 14.99 µm. The only heteratomic gas normally present in air with an absorption spectrum overlapping with that of CO_2 is water vapor (both molecules absorb IR in the 2.7 µm region)(13). Since water vapor is usually present in air at much higher concentrations than CO_2, this interference does present a significant problem. This is overcome either by drying the air that is to be examined or by filtering out all radiation at the wavelengths where absorption by the two gases coincides.

The absorption bands are in fact made up of a series of discrete lines, which correspond to rotational states of the molecules. The major absorption bands of CO_2 consist of between about 30 to 60 absorption lines (IUPAC 1960).

Absorptance of radiation by CO_2 at any one wavelength follows the Beer-Lambert law (14) and thus depends on the radiation path length through the measuring gas and the molar concentration of CO_2 (M, kmol m^{-3}).

$$\alpha_\lambda = 1 - e^{[-M.l.k_\lambda]} \qquad \dots\dots\dots 2.1$$

Where k_λ is the extinction coefficient at wavelength λ
l is radiation pathlength
M is molar concentration of CO_2 is air

However, most IRGAs for CO_2 will use broad band radiation, e.g. 4.1-4.3 µm, total absorptance will therefore be determined by integrating over all the c. 61 absorption lines of this band. Since k_λ will be different for each line, the spectral distribution of energy will change with passage of the broad-band radiation through the sample the more strongly absorbed wavelengths being depleted more rapidly than the weaker absorbed wavelengths.

2.2 Configurations

In its simplest form an IRGA will consist of three basic parts, an infra-red (IR) source, gas cell and detector (Fig. 4). The presence of CO_2 will decrease the quantity of radiation and this will be registered by the detector circuit. For true differential measurements two parallel chambers are needed, and the detector must be capable of measuring the difference in the amounts of radiation absorbed in both cells.

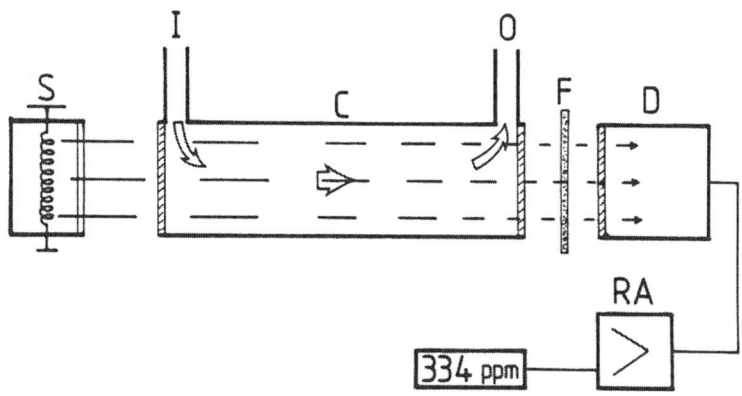

Figure 4. Lay-out of a simple infra-red gas analyzer. Infra-red radiation from a source (S) is passed through a gas cell (C), with an inlet (I) and outlet (O), allowing a continuous flow of the gas being analyzed. The infra-red radiation leaving the cell may be filtered (F) before reaching the detector (D). The detector signal will be rectified and amplified (RA) before display. Any increase in the concentration of the I.R. absorbing gas in the cell will result in a fall in the detector signal.

Source. Two basic groupings of IRGA may be recognized, i.e. dispersive (DIR) and non-dispersive (NDIR). In DIR the source radiation is passed through a monochromator; the selected narrow band of radiation is then passed through the cell. Thus the sample may be scanned and absorption by several heteratomic species measured. However, where concern centers on one molecular species, e.g. CO_2, this is unnecessary. Most instruments currently used in crop physiology are NDIR, i.e. use the broad-band radiation emitted by the source.

The IR source is typically a spiral of nichrome alloy or tungsten, heated to ca. 600-800 C (dull red glow) through a low voltage circuit. The coil may be coated with oxide to reduce sublimation which will otherwise contaminate windows and reflective surfaces (15). The delicate spiral of metal must be firmly mounted to minimize movement in response to vibration which would otherwise cause a random noise in the detector signal. Often the source will be embedded in a transparent ceramic material to prevent any movement. Here, care must be taken to check that the ceramic casing does not fracture.

In dual beam instruments, parallel beams of IR must be formed. This is achieved either by the use of two sources connected in series in the same circuit or by use of a single source split between the two parallel cells with reflectors (e.g. ADC 225/3 and Leybold-Heraeus Binos II). This latter method avoids the problem of differential aging between two sources.

The source radiation is chopped either mechanically with a rotating shutter or electronically by pulsing the electrical supply to the source. Mechanical choppers of older analyzers were belt driven (e.g. ADC 225/1) from a motor synchronized to mains frequency. This creates two problems, first a highly stable mains frequency is necessary so preventing operation from a simple field generator and secondly wear of the belt may result in slipping which will appear as a random noise at the detector (13). These problems are overcome in more recent instrument designs by use of direct drives and internal oscillators to maintain synchronous chopping independently of mains frequency (e.g. ADC 225/3 and Sieger 120). However, all mechanical shutters are inherently sensitive to vibration. This sensitivity and the resulting noise may be reduced, but not eliminated, by increased chopping speeds; for example, the Binos II operates at 150 Hz. This problem may be eliminated by solid-state chopping. Both the Liston-Edwards, as used in the LI-COR 6000, and the ADC LCA analyzers use solid-state chopping and so eliminate all moving parts from the optical bench of the IRGA.

Cells. Most instruments are dual beam, i.e. equal amounts of radiation pass into two parallel cells, termed the analysis and the reference cells (Figs. 5 & 6). The analysis cell will be a through-fall cell, i.e. there is a continuous flow of the sample gas through the cell. Reference cells may be factory sealed, e.g. with CO_2-free air, or may also be through-fall. The latter configuration ensures greater flexibility in the use of the instrument. Cells will therefore contain a gas inlet and outlet, windows of an IR transmitting material such as calcium fluoride and a highly reflective inner lining. To maximize transmission, cell inner surfaces are commonly gold plated. Need for this highly reflective coating, would be removed if imaging optics were used to produce a pencil of parallel radiation passing through the center of the cells. However, few instruments have imaging optics, exceptions include the ADC RF and IRI Series 700. By reference to equation 2.1, it may be seen that sensitivity will increase with increase in path length in the cell. This is normally achieved simply by increasing the physical length of the cell. In most current laboratory analyzers a cell length of about 250 mm is commonly employed to obtain 1 μmol mol^{-1} resolution. This is a limitation to miniaturization for field applications although it could be overcome by the use of folded optical paths.

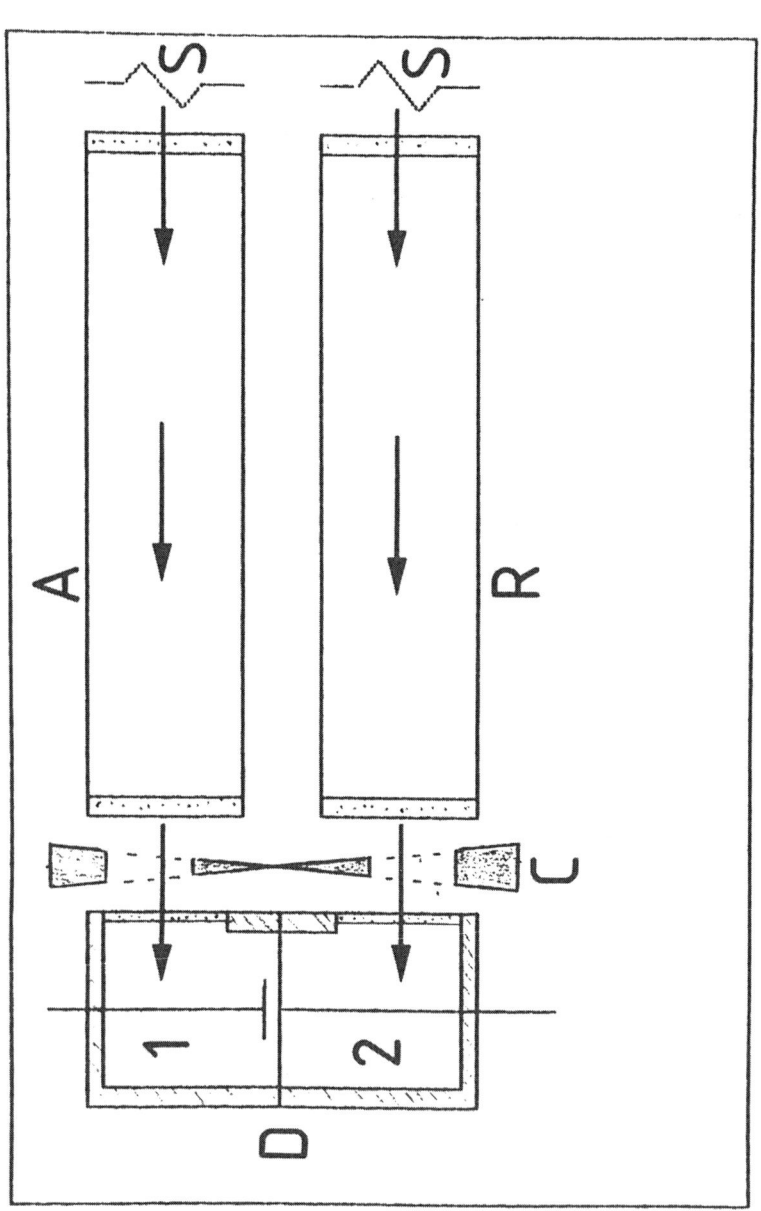

Figure 5. Double beam IRGA with Luft absorption cells of the detector (D) in parallel. Radiation from the sources (S) is simultaneously chopped so that radiation from both the analysis (A) and reference (R) cells reaches the detector simultaneously.

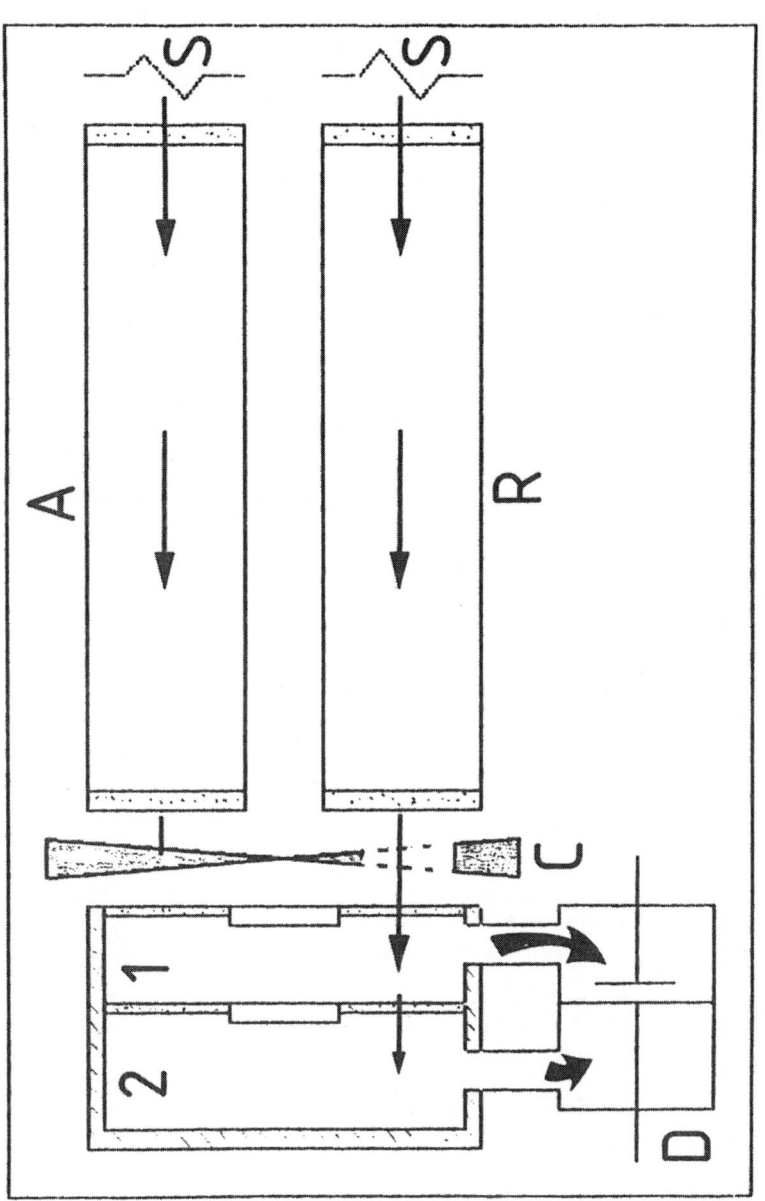

Figure 6. Double beam IRGA with Luft absorption cells (1 & 2) of the detector (D) in series. Radiation from the sources (S) is alternately chopped so that radiation from first the analysis (A) and the reference (R) cell reaches the detector separately.

Split cells, i.e. an analysis cell which is split by a window into two lengths in series, provide a simple means for varying cell length giving two broad ranges of measurement and sensitivity. In the ADC Series 225 instruments the analysis cell is split to provide a short and long cell, representing 5% and 95% of the total path length. Thus by passing the sample gas through the short cell only and CO_2-free air through the long cell, the instrument may be switched from a 0-50 µmol mol^{-1} range to a 0-1000 µmol mol^{-1} range, both of 1% precision.

Some of the recent developments in portable analyzers have been single cell instruments. In the ADC LCA a reference is provided by alternating the sample with CO_2 free air, a comparison is thus made in time rather than in space, as in dual beam instruments. In the Liston-Edwards, there are two sources in line which are heated alternately. Radiation from the first passes through the sample cell. Radiation from the second passes through a sealed cell of CO_2 and then through the sample cell (15). Radiation from the second cell will thus be stripped of CO_2 absorption bands on reaching the sample and can respond only to other gases in the sample, so providing a near ideal reference. A similar approach is used in the ADC RF where the chopper includes two gas cells, one filled with CO_2, the other with an inert gas (16). The only drawback of the technique used in the Liston-Edwards and ADC RF is that the CO_2 gas filters will not only remove response to CO_2 in the sample but will also decrease response to gases with absorption bands overlapping with those of CO_2 and thus the instruments will still respond to other gases, notably water vapor.

In open path instruments there are no cells. Radiation is emitted through the open air and that received by the detector at a fixed distance is recorded. A number of custom built open-path analyzers have been described primarily for eddy correlation work (15,17,18).

Detectors. The most common type of detector is known as the Luft-type which operates on the principle of positive filtration, that is, it absorbs IR in the CO_2 absorption bands. This is achieved by filling the detector with CO_2. The detector is divided into two chambers separated by a thin diaphragm of Copper-Berylium, Aluminium or Gold, which forms one electrode of a diaphragm condenser. The chambers may be arranged in parallel or series configurations (Fig. 5 and 6). The principle of detector operation will be illustrated by examination of parallel operation. In parallel configuration radiation passing through the reference cell enters one chamber and radiation passing through the analysis cell enters the other chamber. Both chambers will absorb radiation in the CO_2 absorbing bands, the amount available for absorption being inversely proportional to the amounts absorbed within the cells. The chopped radiation will thus cause periodical pressure changes in the detector with simultaneous vibration of the membrane. The amplitude of vibration is determined by the pressure difference between the two chambers which in turn is determined by the CO_2 concentration difference between analysis and reference cells. Change in the amplitude of vibration of the membrane produces a change in the condenser capacity which is inversely proportional to voltage change across the condenser (19). Where the detector chambers are arranged in series, radiation is chopped such that it passes alternately through reference and sample cells. The radiation passing through the cell then passes into the front and then the rear absorption chamber. The gas in the front chamber primarily absorbs radiation in the center of the wavebands, leaving the tails for absorption in the rear chamber. The rear chamber is made deeper such that pressure pulses will balance in the zero position, i.e. equal radiation transmission in both cells. This configuration, by comparison to parallel, leads to less cross-sensitivity to other gases and a more stable zero position (15,19). In the Leybold-Heraeus Binos IRGAs the second cell is isolated from all IR so that pressure fluctuations result purely from the chopped radiation reaching the first cell.

The Luft-type detector, although providing a very sensitive method for detecting small CO_2 differences, has two serious limitations when considering use in the field. First, the minimum practical size of these detectors is too great to allow the manufacture of miniaturized or even readily portable instruments. Secondly, the Luft detector suffers from "microphony" that is signal noise arising from spurious vibrations of the diaphragm. Thus, these instruments are very sensitive to vibration. "Microphony" may be reduced by incorporating a capillary between the two absorption chamber, the bore being selected to allow pressure equilibration between the chambers at frequencies below the chopper frequency so reducing sensitivity to low frequency vibration (13). In the Liston-Edwards narrow band width electronic filters are employed to remove noise resulting from low frequency vibration. Phase-selective rectification will also reduce sensitivity to vibration, but neither modification can completely eliminate it (15). Two relatively recent developments are helping to overcome these limitations in the Luft detector.

In the Binos II, the Luft principle of two absorption chambers filled with the absorbing gas, CO_2, is retained, but these are connected by a tube which incorporates a mass flow sensor, so that a measurement of flow replaces the diaphragm capacitance measurement. This eliminates the problem of vibration and allows the manufacture of smaller detectors. A development, potentially of even greater importance, is the solid-state detector. Generally, these are broad-band pyroelectric detectors. These are internally polarized and produce a voltage proportional to temperature change, e.g. when they receive a pulse of IR (19). These detectors are, therefore, very sensitive to ambient temperature change and could not simply replace a Luft detector. However, this limitation has been overcome in the ADC LCA and ADC RF instruments by "gas-chopping." That is, chopping only the CO_2 absorbing wavebands emitted by the source. This latter technique is used in the LCA IRGA which has a single

sample cell and a single solid-state radiation detector. In this
instrument energy from the source passes through sample gas in the
cell and out through a narrow band-pass thin-film filter to isolate
the 4.2 µm absorption band onto the detector. For absolute measurements the instrument pumps the sample gas through the cell and then
CO_2 free air alternatively for 2 second intervals (Fig. 7). The
signal from the amount of energy received in a half cycle is stored
and then compared with the amount received in the next half cycle.
The difference in the energy reaching the detector between half
cycles is thus directly proportional to the quantity of CO_2 in the
sample gas. This technique of gas alternation or gas chopping of
the radiation beam allows the manufacture of sensitive miniaturized
instruments, ideal for field use and for use in remote locations.
The optical bench of the ADC LCA is approximately 8 cm x 1 cm x 1
cm and yet it may resolve < 1 µmol mol^{-1} of CO_2. The most serious
practical limitation is that the instrument cannot be used in a
semi-closed system, since CO_2-free air will enter the system as the
analyzer switches between gas streams.

2.3 Calibration

Although the construction of the IRGAs described allows highly
sensitive and continuous monitoring of CO_2 concentration, most will
require frequent calibration. Instruments vary considerably in the
periodicity with which recalibration is required. Initially it is
therefore advisable to calibrate daily. If, however, no significant
shift in calibration settings are found, then a longer interval
could be employed. Regular calibration is also a useful diagnostic
tool. The development of chronic faults, such as a slow leak in a
Luft detector will be seen as a need to repeatedly increase amplifier gain in order to regain the set-points at each calibration.
The minimum requirement for reliable calibration is a source of CO_2
free air and a source of air containing a precisely known concentration of CO_2, in the range to be analyzed and contained preferably
in an aluminium cylinder (this should not absorb CO_2 into its walls
as a steel cylinder will). Alternatively, scale point or calibration
mixtures may be made by mixing known volumes of CO_2 and N_2 either
with precision gas mixing pumps (Wosthoff SA27 or M300 series) or
gas syringes (LI-6000-01, LICOR). However, care must be taken to
ensure isothermal conditions for the gas volumes to be mixed. A 1°C
difference in temperature between the two volumes will cause an error
of ± 1.2 µmol mol^{-1}. In the Wosthoff pumps, isothermal conditions
are achieved by the immersion of the two pump cylinders in a bathing
fluid. Some IRGAs provide a built-in "CO_2-free" air supply, which
pumps air through a column of CO_2 absorbent, e.g. CARBOSORB. It is
advisable to check the efficiency of these supplies by comparison to
high purity nitrogen passed through a column of carbosorb.

57

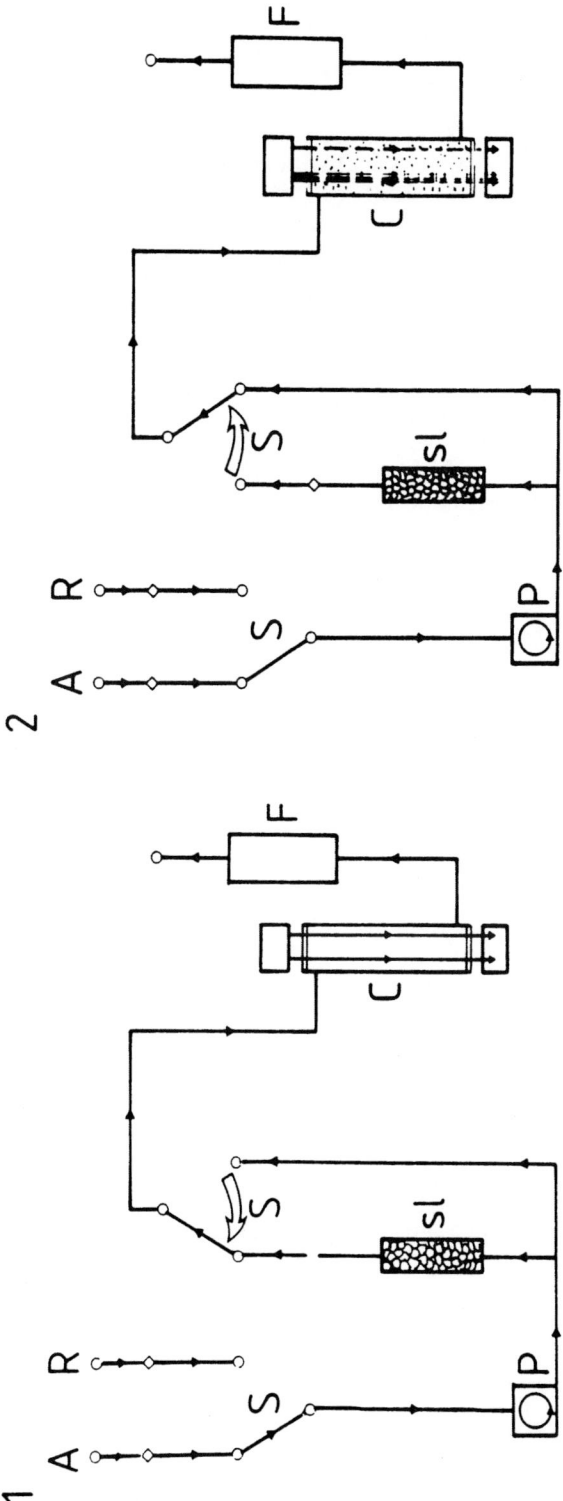

Figure 7. Gas pathway in a gas-chopped CO₂ IRGA. Analysis (A) or reference (R) air is drawn in via a pump (P) and scrubbed free of CO₂ by a soda lime column (sl) and then passed on to the measuring cell of the IRGA (C). In the absence of CO₂ a maximum signal will be obtained. At a regular frequency the switch (S) by-passes the soda lime column (diagram 2) and the air passes through the cell complete with its original CO₂ content. Decrease in detector signal on switching will be directly proportional to the CO₂ content of the air stream. (Based on gas circuit diagrams for the LCA IRGA of ADC Ltd.)

Principles. Calibration is the process of adjusting the IRGA output, either an electrical signal or meter deflection, to the quantities and units required for calculation of CO_2 exchange. The measure given on the output meters of most commercial IRGAs is the partial volume (synonymous with mixing ratio) with units of vpm (volumes per million), the SI equivalent being cm^3 m^{-3}. This measure is also directly equivalent to mole fraction in μmol mol^{-1}. However, spectrophotometric techniques do not measure the partial volume or mole fraction, but measure the molar concentration, i.e. amount per unit volume (mol m^{-3}). Essentially the instrument measures the quantity of absorbing molecules per unit volume irrespective of the amount of non-absorbing background matter. Because mixtures are commonly made by volumetric mixing, manufacturers of compressed gases normally provide scale-point gases, i.e. calibration mixtures, specified in terms of vpm, i.e. volume fraction. This has the advantage of independence of temperature and pressure. Volume fractions may be converted to molar concentrations for calibration:

$$M = (c/22.4).(273.13/T).(p/103.25).10^{-3} \quad \ldots\ldots\ldots 2.2$$

Where
M = the molar concentration of CO_2 (mol m^{-3})
c = the mole fraction (μmol mol^{-1})=volume fraction(cm^3m^{-3}=vpm)
p = the pressure of the gas (kPa)
T = the absolute temperature of the gas (K)

The IRGA could be calibrated in units of volume or mole fraction, but the calibration will only be valid for the temperature conditions of calibration. Ambient pressure changes will produce a small error in such a calibration, typically ± 2%. However, since the gas must flow through the cell there will be a pressure difference proportional to the rate of flow. It is therefore important to calibrate at the flow rate which will be used in taking measurements. A greater problem in field use will be the effect of ambient temperature variation on molar concentration. By reference to equation 2.2 it may be shown that an IRGA calibrated against a scale-point gas of 300 cm^3m^{-3} at 20°C will underestimate the true volume and mole fractions by 6.8% at 0°C and overestimate by 6.9% at 40°C.

Absolute calibration. Where the analyzer will be used to determine the exact CO_2 concentration of an air sample, it is calibrated in absolute mode, i.e. the sample is compared to CO_2 free air. To calibrate the analyzer, CO_2-free air is passed through both the analysis cells. Zero is set on the output meter by adjusting the zero shutter. Air samples of known CO_2 concentrations are then passed through the analysis tubes starting with the highest concentration. The output response to CO_2 concentration is not necessarily linear, and therefore a range of points spanning the range of mole fractions to be analyzed are needed. These may be obtained by using a single calibration cylinder, providing the highest scale

point and then obtaining lower scale points by dilution of the mixture with CO_2-free air using a precision gas mixing pump (e.g.
Wosthoff SA27) or through by-passing a portion of the gas through a
CO_2 absorber in a gas diluter (e.g. ADC GA600).

Differential calibration. Where the analyzer will be used to
determine a change in CO_2 concentration, for example, the difference
in CO_2 concentration in an air stream before and after it has passed
over a leaf, the analyzer is calibrated in differential mode. In
this mode it is possible to detect very small changes in CO_2 mole
fraction, down to 0.1 μmol mol^{-1} with several of the larger instruments. Precise calibration requires that the analysis and reference
cells are filled with air of known, but only slightly different, CO_2
concentrations. In practice this can only be achieved by a small
dilution of the calibration gas using a precision gas-mixing pump
or gas diluter. Here the calibration gas flow is split such that
one stream passes through the reference cell while the other is
precisely diluted with CO_2 free air and passed through the analysis
cell. However, an equally accurate differential calibration is possible, at no extra expense, on instruments in which the analysis
cell is split, as illustrated in Figure 8 (20). In the ADC Series
225, the analysis tube is split into two lengths, i.e. a long cell
representing 95% of the path length and a short cell representing
the remaining 5%. To zero the instrument, the air stream to be
used in the experiment and of previously determined absolute CO_2
concentration is passed through both the reference and analysis
tubes. A center zero is then set on the output meter. At atmospheric concentrations of CO_2, the amount of IR absorbed in the
analysis tube is only a small fraction of the total IR in the CO_2
absorbing bands which passes through the tube, thus removal of CO_2
from the short cell (by flushing with CO_2-free air) is optically
equivalent to reducing the CO_2 concentration throughout the length
of the analysis tube by 5%. The change in the output meter produced
by this operation represents the change that a 5% depletion in CO_2
concentration would produce, and thus provides a simple means of
calibration.

3. THE MEASUREMENT AND CONTROL OF GAS FLOWS

3.1 Flowmetering

Measurement of flow is fundamental to the measurement of CO_2
exchange between plants and the flowing atmosphere of an enclosing
chamber. In both open and semi-closed systems, the accuracy of flow
rate measurement is as important to the final estimate of A as the
measurement of CO_2 concentration difference. In an open gas exchange
system, A is proportional to the product of difference in CO_2 mole
fraction and the flow rate of air across the leaf (eq. 1.6). In
semi-closed systems a constant CO_2 mole fraction is maintained by
an inflow of CO_2 into the system, the rate of this flow being the
measure of CO_2 uptake by the plant (eq. 1.5).

60

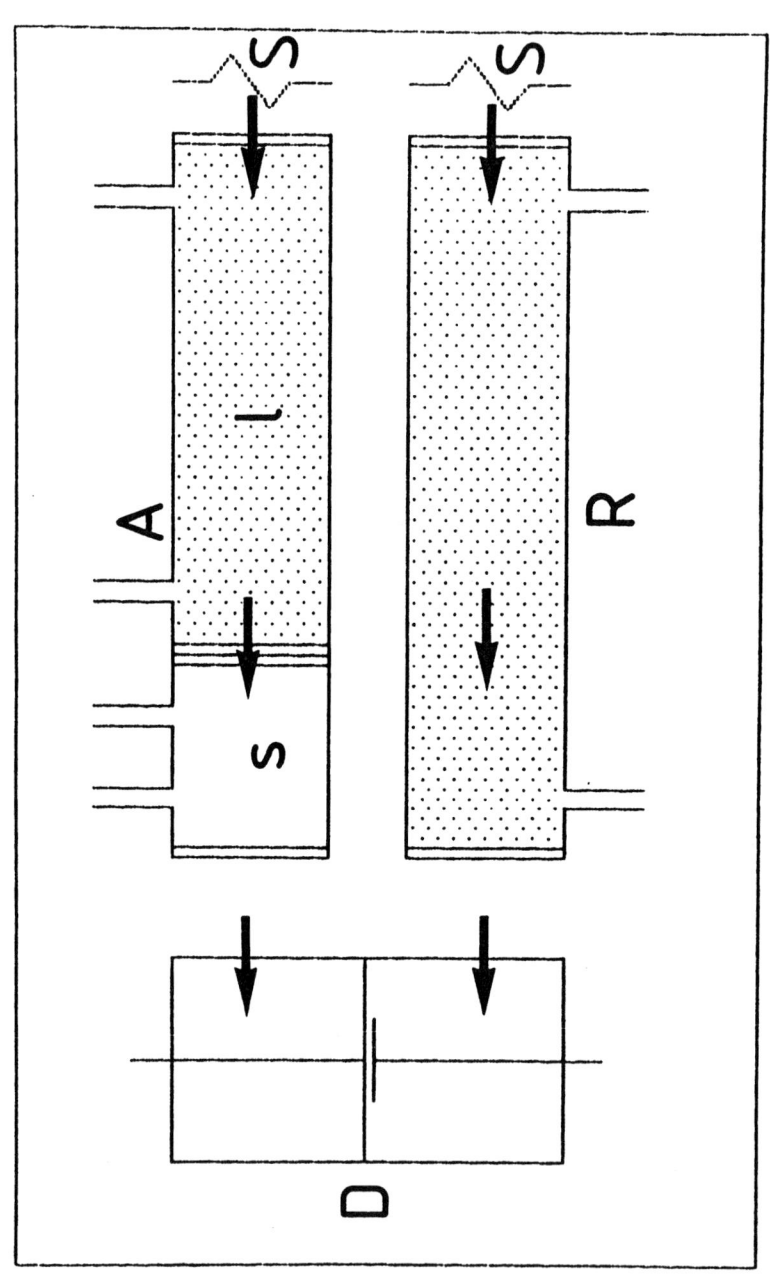

Figure 8. Double beam IRGA with a split analysis cell allowing the use of the split-tube differential calibration, in which the reference gas is passed through the reference cell (R) and long analysis cell (l), while CO_2-free air is passed through the short analysis cell (s).

Surprisingly then, measurement of flow rate has, by comparison to measurement of CO_2 concentration, received little attention.

As a result of the physical principles employed, most commercial flowmeters provide an estimate of either volumetric flow (f_v; $m^3 s^{-1}$) or mass flow (f_w; kg s^{-1}) and not the mole flow required for the direct calculation of A. Direct measurement of either f or f_w is clearly not possible since it is not practicable to collect and weigh gas (21). Direct mesurement of f_v is possible, however, and forms the basis of the calibration techniques considered later, but even these methods are unsuitable for use within a gas exchange system. Mole flow rates may be determined by equation 3.1 from volumetric flow rate and by equation 3.2 from mass flow rate.

$$f = (f_v/22.4).(273.13/T).(p/101.325) \qquad \ldots\ldots\ldots (3.1)$$

Where:
T = the temperature of the gas (K)
p = the mean pressure of the flowing gas (kPa)

$$f = f_v/m \qquad \ldots\ldots\ldots (3.2)$$

Where:
m = the molecular weight (g mol^{-1})
    (CO_2 = 44.01, H_2 = 18.02, dry air = 28.97)

To determine flow rate the experimenter must resort to the measurement of some physical effect arising from the motion of the gas in the tube. Three effects have been widely used: 1) consequent mechanical effects, such as the rate of rotation of a rotor mounted in the stream; 2) pressure changes; and 3) the rate of heat transfer from a heated body in the air stream. Four reviews list some 20 methods which utilize these effects for the measurement of flow rate of gases in pipes (21,22,23,24). This section is limited to the few techniques which have found application or appear to have potential application in the study of plant gas exchange of biological systems. The range of volume flow rates that it would be necessary to measure in biological gas exchange studies varies from 1 µmol s^{-1}, for CO_2 being fed into a semi-closed system for the measurement of photosynthesis by small leaves, to 100 mol s^{-1} ($f_v \approx$ 2 dm^3 s^{-1}), for air in open systems used for the measurement of CO_2 exchange by whole plants or swards in enclosures.

Variable-area flowmeters. These have been the most widely used instruments for flow rate measurement, but are now supplanted in many applications by thermal mass flow meters. However, the low cost, simplicity and the visual indication of flow rate suggest that variable-area flowmeters will continue to be used, at the least as a secondary flow rate indicator.

Variable-area flowmeters, developed from the ball-and-tube flow-meter of Ewing (25), consist of a transparent graduated tube with a slightly tapered bore, in which the diameter decreases downwards; and the gas flow to be measured passes upwards. A float of diameter slightly below the minimum bore of the tube is forced up the tube to the point where its weight is balanced by the force of gas flowing past it (Fig. 9).

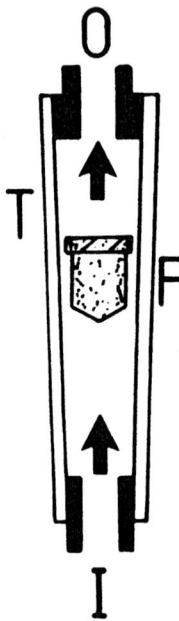

Figure 9. A variable area flow-meter. Air from the inlet (I) forces the float (F) to rise in the tapered glass tube (T) to an equilibrium position. The outlet (O) of these meters is necessarily at the top.

In a constant flow, the ball shape is said to be inherently liable to sudden fluctuations in position within the tube or may "chatter" against the side of the tube (21). One solution to this problem is the use of a conical float with angled grooves which cause rotation around the vertical axis, giving the float central stability (21,23). This type of float is used in several commercial instruments. Variable area flowmeters are available from a range of manufacturers for volume flow rate measurements in the range of 30 mm^3 s^{-1} to about 50 dm^3 s^{-1}, and the standard range of any one meter is usually 10:1 (e.g., 1,2,3, and 4; (22)). Precision depends on tube length, float shape and precision of the glass tube, but typical repeatability is ± 2%, and up to ± 0.5% may be obtained with very long precision tubes (22). Instruments are calibrated by the manufacturer for a given gas, at a specified pressure and temperature. These calibrations will only change if deposits of dirt are allowed to form on the tube or float, or if the tube or float become damaged or corroded. These instruments are particularly sensitive to ambient temperature and pressure fluctuations. For example, if a variable area flowmeter was factory calibrated at 15°C and 101.3 kPa, normal sea-level atmospheric pressure fluctuations from 88-108 kPa could produce an error of 7.3% to -3.3% in indicated flow rates. In the field, ambient temperature variations of 0°C-40°C could produce errors of 2.7% to -4% (24). Calibration of these flowmeters for the range of working temperatures is therefore necessary, unless the manufacturer provides temperature correction graphs. Since flowmeters in most gas analysis circuits would be inserted immediately upstream of the assimilation chamber, gas in the outlet of the flowmeter would be above the atmospheric pressure. Either the actual outlet pressure should be measured so that the equivalent flow rate at atmospheric pressure can be calculated or the flowmeter should be recalibrated in situ (24).

The major disadvantages of these instruments are that they must be mounted perfectly upright, that subjective errors in assessing the float position relative to the tube graduations are difficult to avoid, especially if the float position fluctuates, and that accuracy is inherently low except in the longest tubes (22). A further practical problem is that the slightest amount of moisture in the tube may cause the float to stick. The design is not well suited to the production of an electrical output. Alternative mechanical methods of flow measurement include turbine meters and pressure drop measurement across a tube constriction (21,23,24,26).

Thermal mass flowmeters. The application of thermal mass flowmeters to the measurement of photosynthetic gas fluxes greatly increases the potential accuracy of A determinations. These instruments utilize the thermodynamic principle that the heat carried by a gas flow is related to the heat capacity of the gas and the mass of gas flowing. They consist of a sensor tube which carries the flow and is precisely heated such that the temperature distribution is symmetrical about the mid-point (27). This may be achieved either by applying heat at the mid-point (Fig. 10), or by uniformly heating two points equidistant from the midpoint. Two temperature sensors, typically either platinum resistance thermometers or thermocouples, are situated one each side of and equidistant from the mid-point. With no gas flow the temperature at both sensors will be equal. Gas flow will transfer heat downstream, causing the temperature distribution to become asymmetric and the temperature at the sensor downstream (T_d) of the mid-point will then be higher than at the upstream sensor (T_u). The magnitude of the temperature difference will be a function of the flow rate (28):

$$\Delta T = T_d - T_u \equiv (W.f.C_p)/N \quad \ldots\ldots\ldots \text{ (3.1)}$$

Where
W = injected power
C_p = specific heat
N = a correction factor, dependent on the gas molecular structure
f = proportion of total flow passed over the sensors

64

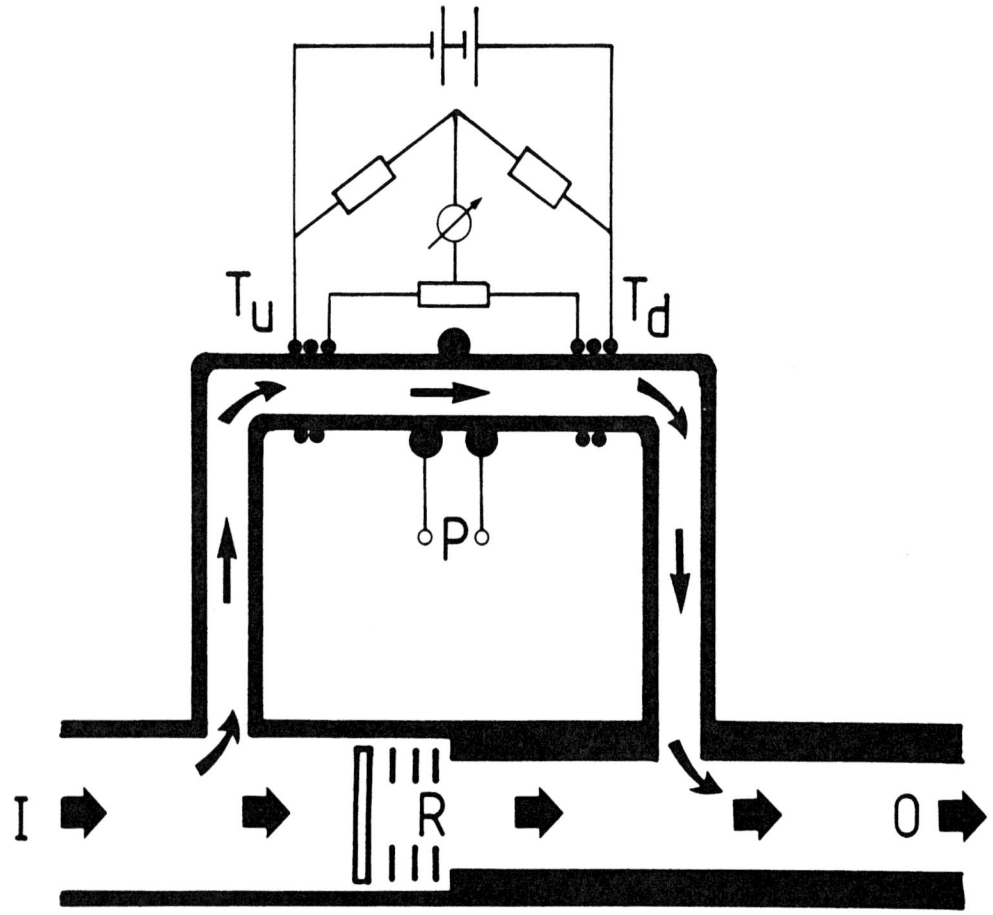

Figure 10. A calorimetric mass flowmeter with shunt arrangement. A portion of the inlet (I) flow is forced by a restrictor (R) through a shunt which is heated at its midpoint by a power source (P). Temperature is measured equidistant upstream (T_u) and downstream (T_d) of P. Flow in the tube is determined from T_d-T_u.

In many commercial units, e.g. Datametrics, MKS, Teledyne Hastings-Radyst and Tylan, ΔT is measured in a tube shunt which takes only a fraction (K) of the total flow. The relation between ΔT and f is complex, differing between instruments and flow conditions (28).

Thermal mass flowmeters are normally factory calibrated for one gas, but single correction factors (N), dependent on the molecular structure of the gas and its specific heat capacity, can be used to recalculate the flow rate of other gases; detailed conversion tables are given by Anon (29). The first commercial thermal mass flow meters were described by Thomas (27), more recent developments and calibration techniques were reviewed by Widmer et al.(28). As noted in the introduction, mass flow rate (f) measurement has fundamental advantages over volume flow rate (f_v) measurement in the study of gas exchange by living organisms. However application of the term mass flowmeter to these instruments is something of a misnomer. A true mass flowmeter measures the mass of gas flowing irrespective of the properties of the gas (24). By strict definition, thermal mass flowmeters are not true mass flowmeters, since they rely on the thermal properties of the flowing gas and thus meter response will change according to the molecular structure of the gas. However, because the heat transfer properties of a gas do not change markedly with changes in its density, they may approximate to mass flowmeters for a given group of gases over a moderate range of temperature and pressure (22). In practice performance will suffer less from fluctuations in pressure ($\pm$ 0.003% kPa^{-1}) than from fluctuations in temperature, which may be $\pm$ 0.1% K^{-1} between 5° C and 43° C (29). This source of error may be reduced by controlling the temperature of the flowmeter (22) or by including a compensating circuit, which measures the temperature of a duplicate or reference sensor tube with zero flow rate (e.g. some of the Hastings Mass Flowmeters). Both manufacturers and independent assessors (22) suggest accuracies of $\pm$ 0.5 - 1% (depending on design) of maximum flow rate for thermal mass flow meters, at a given temperature.

3.2 Flowmeter Calibration

It is frequently necessary to recalibrate a flowmeter for use with different gases or under different operating conditions. It is also advisable to recheck calibrations at regular intervals. For the flow rate ranges used in gas exchange systems, soap film flowmeters provide a simple means to calibration. Other calibration methods are reviewed by Ower and Pankhurst (21) and Hayward (22). Widmer et al.(28) have reviewed industrial methods for the calibration of thermal mass flowmeters.

Soap film meters, illustrated in Figure 11, were first suggested by Barr (30). To operate, the rubber bulb is squeezed until the level of soap solution in the reservoir rises to the air inlet where a soap film will form across the tube and will be forced up the vertical tube at the speed of the gas. The time taken for the soap film to travel between two points separated by a known volume is recorded and provides a direct measure of volume flow rate (f_v). Such a flowmeter may be simply constructed by adding a 'T' junction to the base of a high-quality burette. By using different diameter tubes a range of flow rates from about 10 mm^3 s^{-1} to 100 cm^3 s^{-1} may be measured with an accuracy of $\pm$ 0.25%, decreasing to $\pm$ 1% at 1 dm^3 s^{-1} (31). However, if the movement of the soap film is monitored by eye and timed with a stop watch, timing precision is unlikely to be better than $\pm$ 0.1 s. Thus to obtain the potential accuracy of this instrument, the combination of tube length and diameter should be such that the time required for the passage of the film between the

fixed points should exceed 20 s. Timing accuracy may be improved by
the use of photoelectric detectors which trigger an electric timer
(22). The slight curvature of the soap film moving up the tube may
also introduce some ambiguity into assessing the exact position of
the film. This may be overcome by replacing the soap film with a
low friction mercury piston. Kolk and Moulijn (32) have shown that
the use of such a piston in a high precision diameter glass tube
with an antistatic coating combined with photoreflection cells for
timing the passage of the piston allows the measurement of f_v from
50 mm^3 s^{-1} to 50 cm^{-3} s^{-1} with an accuracy and repeatability better
than ± 0.25%. For larger volume flow rates (>10 cm^3 s^{-1}), it is
more practicable and accurate to use a wet gas meter for calibration
(21,24).

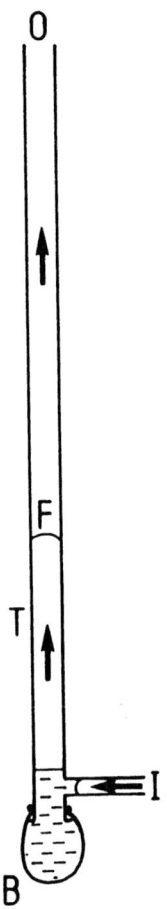

Figure 11. A soap-film flowmeter. From inlet (I) gas passes through
a soap solution contained in a compressible bulb (B). The films
(F) or bubbles generated rise up in the tube (T) to the outlet (O).
Time taken to rise through a known volume of tube provides an
accurate measure of flow.

3.3 Flow Control

Although its importance is less easily quantified, flow rate control is an essential consideration in gas exchange studies. Fluctuation in flow rate will decrease the resolution with which both flow rate and gas concentration difference may be measured, may cause fluctuations in gas composition if different gases are being mixed or humidified, and may introduce both random and systematic errors in flux calculations.

Critical flow orifices. A particular problem in mixing gases, either in the supply to gas exchange systems or for calibration of infra-red gas analyzers and other instruments measuring gas concentrations, is that the flow rate of one gas will be influenced by slight changes in downstream pressure produced by small changes in the pipework and connections or from change in flow rate of a second gas used in generating the mixtures. If a constriction is introduced into the gas stream such that the gas reaches the sonic velocity in the constriction, then that velocity and hence volumetric flow rate will be maintained within very close limits (22). Thus for any given constriction, once the sonic velocity is reached and provided that the critical ratio of upstream to downstream pressure remains sufficient to maintain sonic velocity, then volumetric flow rate will be insensitive to changes in downstream pressure. However, mass flow rate may be varied simply by altering the upstream pressure and hence density of the gas. Constrictions may be small nozzles, venturis or orifices (22,33). Parkinson and Day (34) have shown that a range of orifice sizes would provide a fixed range of constant volumetric flow rates provided that a constant upstream pressure is maintained. This principle is utilized in the ADC GD600 and WG600 instruments which can generate a wide range of CO_2 and water vapor concentrations, respectively, for the calibration of infra-red gas analyzers. These instruments also provide a simple, transportable and accurate method for providing known mixtures of CO_2 and water vapor at controlled flow rate to gas exchange systems.

Electronic flow control. An accurate control of flow rate may be obtained by linking the electrical output of a flowmeter to an electronic flow rate control valve, such that the flow rate is continually monitored and automatically adjusted to maintain a preset flow rate. A wide range of electronic flow rate controllers, many patented, are available commercially (e.g. Brooks, MKS and Tylan). The selection of a suitable valve will depend on the pressure drop, the range of flow rates and response time required.

The two established methods of electrical control are solenoid and servo-driven valves. In solenoid valves, the valve seat is typically connected to the armature and is lifted away from an orifice as increasing current is supplied to the solenoid. In servo-driven systems the valve is operated by a small stepping motor. Both of these methods can give good precision. More recent developments, now commercially available, are thermal expansion valves and piezoelectric crystal valves.

Thermal expansion valves, used for example in Tylan flow controllers, consist of a small thin-walled tube with a ball welded to one end which rests in the gas pipe. The tube contains a small resistance heating element and heat transfer fluid. When a voltage is applied to this element, the tube expands and moves the ball which thus controls the flow rate (29). This design has the advantage of having no moving seals, virtually no moving parts (total travel ca. 0.1 mm) and no friction and may thus be expected to be both precise and reliable. However, the device cannot completely stop the flow rate and thus a solenoid valve must be added to the line if it is necesssary to interrupt the flow during experiments. The slow response time of 6-10 s (Manufacturer's specification) would be a serious limitation in semi-closed systems if the experimental objectives are the study or record of non-steady state changes in gas exchange, as for example in kinetic studies of the effects of dark-light transitions on CO_2 assimilation.

Piezoelectric crystal valves, used for example in some MKS controllers, consist of a viton seal cemented to a piezoelectric crystal which has the property of flexing in response to an applied electrical potential. In the resting position the valve is closed but when voltage is applied, the flexing of the crystal opens the valve by an amount proportional to the voltage. The particular advantage of this design is a very fast response time, which according to the manufacturer is 2 ms (35). The low power requirements of both of these types of controllers mean that they may be operated via a digital-analogue computer (D-A) interface, so allowing control through a pre-programmed set of instructions. These instruments also have the advantage of being readily portable and suitable for use at remote field sites.

4.   CHAMBER CONDITIONS AND CONSTRUCTION

4.1 Principles

The environment of the enclosed leaf or plant will be influenced by the design of the chamber and the effectiveness of the air conditioning system which determines the composition of the air supplied to the chamber. Chamber design will be determined by the objectives of the study, the prevailing climatic conditions and the size and shape of the material which will be enclosed. Two broad groupings of objectives in the study of CO_2 assimilation by crops may be identified.

Measurement under prevailing field conditions. Where the objective is to determine A under the field conditions at that point in time, conditions within the chamber must be close to those oustide. Two approaches have been employed to achieve this objective: 1) to design the chamber such that its effect on the leaf environment is minimal; 2) to monitor the outside conditions and then control the chamber internal conditions such that they track those outside. The former approach is clearly the simpler, in that the chamber requires no ancillary control equipment and is thus more readily portable. The LI-COR LI6000 series and ADC Parkinson leaf chambers have been designed with this approach.

Measurement under controlled conditions, in the field. Where comparisons between treatments or genotypes in the field are required, for example screening of varieties in crop breeding, control of some aspects of the leaf chamber environment will be necessary. Since the natural leaf microclimate is continually varying, especially with respect to light and temperature, effects of genotype or treatment may be confounded with microclimate variation. It is therefore necessary to provide control of those aspects of microclimate which will strongly influence A so that valid comparisons are possible. This may be achieved, for example, by controlling the leaf temperature at a fixed value for all measurements and/or addition of saturating artificial light. Control of the chamber environment is also necessary if the study aims to establish the bases of genotype or treatment differences, through light and CO_2 response curves (Section 5).

4.2 Chamber Design

As discussed above, the objective of the field study will be a primary determinant of chamber design. Regardless of objective a further primary requirement will be that the environment within the chamber should be as homogeneous as is possible, i.e. gradients of temperature, CO_2 and water vapor across and along the leaf should be minimized so that A will be known for a well defined microclimate. The degree of temperature and gaseous homogeneity within the chamber will be determined by the boundary layer conductance to heat and gaseous transfers, which should be maximized. Chamber design must take this prerequisite into account with respect to the tissue under investigation, since a chamber designed for leaves, or a leaf of one species may not be suitable for many others.

Boundary layer conditions. The boundary layer conductance (g_b) will determine the homogeneity of the gaseous and thermal microenvironment of the leaf. It is maximized by obtaining a high rate of air movement around the leaf. This can be achieved either by vigorously stirring the air with a fan within the chamber or by rapidly recirculating the air with a pump outside the chamber. Even when a ventilation technique is used, care should still be taken in design to avoid the creation of pockets of still air which can occur where a leaf is in close proximity to the chamber wall and in the corners of a rectangular shaped chamber. In the absence of stirring or recirculation, g_b will depend on the net rate of flow of air through the chamber and its pattern of movement across the leaf. The velocity of air movement around the leaf may, however, be accelerated by reducing chamber volume and regular spacing of gas inlet and outlet ports. Unstirred chambers using these design considerations are described by Incoll and Wright (36) and Harris et al.(37). It is not possible to predict g_b exactly, and it will depend not only on chamber design, but on leaf size and shape. It is therefore advisable to determine g_b empirically for any new chamber or application.

Boundary layer conductance to water vapor transfer (g_b') may be determined by placing a wet filter paper replica of the leaf into the chamber and measuring the rate of water vapor efflux (E), leaf temperature (T_1) and the ambient humidity of the chamber (e_a), which will be e_0 in a stirred chamber and ca. $(e_e + e_0)/2$ in an unstirred chamber.

$$g_b' = E/(e_s/P - e_a/P) \dots\dots\dots (4.1)$$

Where
P = the atmospheric pressure (kPa)
e_s = the saturation water vapor pressure (kPa) at T_1

The temperature of the filter paper replica is normally deter-mined by a small wire thermocouple appressed to the lower leaf sur-face. The temperature measured may not be the mean for the whole replica and since a part of the junction is likely to be in the air stream, the observed temperature may be influenced by air temperature as well as the true surface temperature. Parkinson (38) provides an alternative approach in which mean replica temperature is determined from its energy balance so avoiding the uncertainties of thermocouple measurement. Alternatively, g_b may be determined from heat exchange either from a replica or from actual leaves (39).

If boundary layer conditions are to be similar to those commonly found for leaves in the open, then chamber design should be such that g_b' is > 5 mol m^{-2} s^{-1} (calculated from Jarvis, 39).

Temperature. From a physiological standpoint, it is the temper-ature of the leaf rather than the air that is of interest. Control of leaf temperature is greatly facilitated by minimizing the thermal radiation which reaches the leaf. The use of a long wave IR trans-mitting window in the chamber, such as polypropylene film (e.g. Pro-pafilm C, I.C.I.), is useful in preventing a "green house" effect within the chamber. Alternatively, a heat reflecting glass ("hot mirror") or a radiation shield above the chamber, as in the design of the ADC Parkinson chamber, will intercept much of the incoming IR.

Good chamber ventilation, as measured by g_b keeps leaf air temperature gradients to a minimum and minimizes temperature gradi-ents along and across the leaf. Leaf temperature is most commonly controlled by regulating the temperature of the ambient air in the chamber. This can be controlled either by jacketing the chamber so that coolant can be circulated over the chamber walls, by inserting cooling coils inside the chamber, or by building Peltier modules into the chamber walls (40). The thermal conductivity of aluminium is 205 W m^{-1} K^{-1} compared to ca. 0.2 W m^{-1} K^{-1} for acrylic plastics, i.e. 1000 x slower. An aluminium walled chamber will therefore be far less prone to heating above the ambient temperature. The speed of heat dissipation may be further enhanced by the addition of cooling fans on the chamber undersurface.

Photon flux density. Usually, the light required will be that which would have been incident on the leaf in the absence of the chamber. The chamber window must therefore be as transparent as possible. Few window materials are perfect transmitters. Acrylic plastic has a transmission of .92, when new. Scratches and smears produce surprising reductions in photon flux density (I_p) at points on the leaf surface. It is important, therefore, to have a supply of replacement windows. This is easier if thin film windows, such as "Propafilm" are used since they may be simply sealed with double-sided tape, making rapid replacement of a damaged window practicable in the field.

When a saturating light level is required, a supplementary light must be provided above the chamber. A criterion in selection of light sources used for controlled environment CO_2 exchange studies is similarity to natural daylight. Xenon-arc lamps provide provide a close match, but this includes a high heat output. A combination of high pressure Na and Hg lamps can also give a good spectral match to daylight without the same heat output. Photon flux density can be varied at the level of the chamber by placing neutral density filters or even sheets of muslin above the chamber. This is preferable to reducing the voltage supply to the lamps which will alter the spectral composition, as well as quantity, of emitted radiation.

Only by using a perfectly spherical chamber with reflective walls and light entering through an inserted optical pipe could the radiation supplied to the leaf be totally diffused. The advantage here is that they may be used to measure total light absorption by the leaf simultaneously with gas exchange and thus allow determination of the true quantum efficiency ($\emptyset$) (41,42). Such chambers have so far been limited to laboratory use, but there is no reason why they should not be used in the field. Most leaf chambers are designed to receive direct radiation on the upper surface. If the base of the chamber is painted with optically black paint, then the radiation conditions of the leaf can be precisely defined. Finally, the leaf should be held in the horizontal plane in such a chamber if all parts of the surface are to receive the same I_p. In large chambers this may be achieved by placing the leaf between two course meshes of fine transparent nylon. Chambers for field use are manufactured for example by ADC, LI-COR and H. Waltz Co.

4.3 Materials

Materials used in construction of chambers, gas connections between system components and the air conditioning system have a major influence on the effectiveness of the system and accuracy of determinations of gaseous fluxes (43). Permeation of CO_2 and water vapor between the surrounding air and that enclosed in the system will produce errors in determinations of A and E in all systems, although the error is likely to be most pronounced in a closed system since recirculation will cause an amplification of error in flux calculations. Adsorption and desorption of CO_2 and water vapor from internal surfaces will also produce errors in calculations of steady-state A and E in closed systems. Strictly, this problem is avoided in open systems since here steady-state fluxes cannot be achieved until the whole system is in equilibrium and thus adsorption must be balanced by desorption. Materials with a high adsorptive capacity for CO_2 or water vapor will greatly affect the apparent response time of an open system imposing a long lag on the responses of A or E to changes in microclimate. In addition, the character of non-steady state changes, e.g. the induction of CO_2 assimilation following a dark-light transition, will be altered since the system will impose an additional lag in response on to any biological lag. Ideally then all systems should be constructed from materials which neither adsorb, absorb or allow permeation of either CO_2 or water vapor. Many materials not only adsorb, but also absorb water vapor. Since CO_2 is soluble in water (36.5 mol m^{-3} at 20°C) any material which absorbs water vapor will also adsorb CO_2. The amount of CO_2 adsorbed will be strongly dependent on temperature and humidity. Any sudden change in temperature, for example a sun-fleck falling on the tubing, could suddenly raise the CO_2 level and produce a spurious flux. A slow change in temperature, e.g. gradual warming through the day may produce a systematic error which may well go undetected.

Metals. Metals, unless of micron thicknesses, have near-zero permeability to gases. However, if the metal is reactive then it may adsorb or absorb gases. Non-stainless steels will of course absorb water vapor. Rusting produces the additional problem of surface roughness. A rough surface will contain many microcavities trapping still air and greatly increasing the time taken for the surfaces to come into equilibrium with air in the system. The problem of surface reactivity is avoided in good quality stainless steels, though even here the material should not be assumed to be perfect. Contamination of surfaces with greases and oils is common and if the surfaces have been roughly worked, they will contain surface microcavities. Stainless steel surfaces should be washed in a degreasing agent and preferably polished to improve surface smoothness. A practical disadvantage in chamber construction is that stainless steels are difficult to work. Copper, aluminium and alloys based on either of these are easier to work, but are more likely to be reactive. In particular copper may absorb and adsorb CO_2 strongly. This problem may be overcome by chrome plating which will provide a smooth and largely unreactive surface (Table 1). Aluminium is easily worked and provides heat conduction properties only slightly inferior to those of copper. However, it will rapidly oxidise and absorb water because the oxides possess surface hydroxyl groups which form hydrogen bonds with water (44). This problem is removed in the more inert alloys, i.e. duralinium. The ADC Parkinson chambers are constructed from duralinium (H(E)30) which has been found to have negligible adsorptivities (K.J. Parkinson, personal communication).

Table 1. Water adsorption by surfaces[1]

Material	Water adsorption (mmol m^{-2})
Acrylic plastic	550
Aluminium (oxidized)	40
Brass (tarnished)	60
Brass (chrome-plated .01mm)	21
Brass (nickel-plated .01mm)	3
Glass	3
Stainless steel	3

[1] Calculated from Table 2 of Bloom et al (1980)

Stainless steel tubing is difficult to fit, lacking flexibility, and is therefore only convenient for permanent connections. Copper is easier to work, but should be avoided because of its reactivity both with water and CO_2. Semi-rigid butyl rubber tubing with a thin stainless steel or polyethylene coated aluminium liner provides in many ways an ideal tubing (e.g. Dekaron tubing, Eaton Corp.). This combines the flexibility of plastics with the good surface and permeability properties of metals.

Plastics. Although few plastics approach the excellent low permeability and adsorptivity properties of metals, their use in some parts of the system is unavoidable and often more convenient. Since at least some part of the assimilation chamber must be transparent, a plastic or glass window is essential. Plastics are generally easier to work and the use of adhesives means that mechanical workshop facilities are not essential for chamber construction. There is a wide variety of types of flexible plastic tubing and connectors, which may make system construction simpler. Properties of some plastics and rubbers are listed in Table 2. These are intended only as a guide as properties vary considerably depending on density and method of manufacture.

Acrylic plastics ("Perspex" or "plexiglass") have been used widely in the construction of chambers. They uniformly transmit 92% of light in the 400-800 μm wavelengths (40). They are also light, easy to cut and simple to bond. However, water adsorption is high and thermal conductivity low. This second property makes the temperature of an acrylic plastic chamber difficult to control, unless a large heat exchanger is added. Water and CO_2 adsorption represent a very significant problem. Bloom et al.(43) show that rates of CO_2 and water vapor adsorption in such a chamber are large enough to create errors in excess of 50% for estimates of both A and E; significant wall fluxes of both CO_2 and water vapor were apparent even four hours after changing the chamber temperature.

Absorption by acrylic plastics may be reduced by coating the internal surfaces with a material of lower water permeability. Self-adhesive transparent PTFE tape provides one simple solution to this problem.

For tubing, reference to Table 2 shows that PTFE has excellent properties with respect to both water absorption and CO_2 permeability. PTFE has a very low water absorptivity while nylon 12 has a low CO_2 permeability. It must be appreciated that values given in Table 2 are only mid-points on ranges. Plastics vary considerably depending in particular on the quality of the tubing. Poorly manufactured tubing may have small holes making properties such as water absorption irrelevant. Temperature will also have a marked effect on permeability. In PTFE, this increases fivefold between 23°C and 35°C (45). The density of the tubing also has an important effect, the higher density polyethylenes and nylons have lower permeabilities and absorptivities. Soft polyvinyl chlorides which are used widely in gas exchange systems adsorb significant amounts of water vapor (0.25% of their weight) and consequently CO_2. They have the further disadvantage that volatile plasticizers are slowly released and these not only support microbial growths but have infra-red absorption spectra which coincide with CO_2 (43).

4.4 Air Conditioning

An essential part of any system is a means of controlling the concentration of gases entering the chamber, particularly CO_2, O_2 and water vapor.

CO_2 mole fraction. Since CO_2 will often limit A, its precise control is essential. A more·practical problem is that rapid fluctuations in the CO_2 mole fraction of air supplied to the chamber will be apparent as random noise in the determination of Δc with an IRGA and so decrease the accuracy of determination of A. In a closed system, CO_2 cannot be controlled while in the semi-closed system CO_2 concentration is controlled by definition. However, in open systems control of CO_2 concentration can become far more complex. Where only the atmospheric CO_2 mole fraction is required (340 $\mu mol\ mol^{-1}$ of CO_2 in air) air from outside the system could be used. However, the inlet must be distant from any source of CO_2, i.e. chimneys, combustion engines and people. In practice a height of 4 m above the ground will usually prove adequate. Even when distant from obvious sources of CO_2 pollution small fluctuations in atmospheric CO_2 concentration occur. These can be dampened by passing the air intake through two or three large containers linked in series, the actual volume required depends on the flow rate into the system. Most of the units that have been described previously are large and transportable only within a mobile laboratory (47). The air supply unit available from ADC does provide a portable alternative, although its maximum delivery flow of about 10 $cm^3\ s^{-1}$ limits its use to a single leaf or small plant chambers. It consists of a light weight telescopic mast, extendable to 4m.

Table 2. Properties of plastics and rubbers used for some common types of tubing gas exchange systems [1]

Material	Trade Names	Permeability[2] $(nmol.mm.s^{-1}.m^{-2}.Pa^{-1})$		Water Absorption $(mmol\ kg^{-1}\ d^{-1})$
		CO_2	H_2O	
PLASTICS				
Polyethylene (low density)		1.5	42.0	<5
Polyethylene (high density)		0.6	5.4	0
Polypropylene		0.6	22.2	<2
Polyvinyl chloride (soft)	Tygon	0.7	----	128
Polyvinyl chloride (hard)		0.04	55.5	---
Polyamide 6	Nylon 6	0.03	560.0	5300
Polyamide 12	Nylon 12	0	72.2	140
Polycarbonate		2.2	560.0	---
Polytetrafluoroethylene	Teflon PTFE	0.3	7.0	<10
Fluorinated ethylpropylene	Teflon FEP	1.7	10.0	<2
Polytrifluorochloroethylene	Plaskon CTFE	0.02	----	0
RUBBERS				
Polyisoprene	Natural rubber	23.0	-----	270
Polychloroprene	Neoprene	4.5	-----	270
Poly(dimethylbutadiene)	Methyl rubber	1.3	-----	170

1 Permeabilities of plastics calculated from Oberbach (46), excepting Teflons and Rubbers which were calculated from DuPont (45) and Bloom et al.(43), respectively. Data on absorption of water vapor after Bloom et al.(43).

2 Permeability expressed as the product of net amount diffusing and wall thickness per unit time, surface area and applied pressure.

Air is drawn down the mast by a pump capable of continuous operation for ca. 12 hours using its internal 12v rechargeable battery. All or a portion of the air stream may be passed through two columns. Normally, these are fitted with silica gel so that the air may be partially dried. However, if these are filled with a CO_2 absorber, e.g. soda-lime, then the unit may be used to generate a range of CO_2 concentrations.

An alternative method of controlling CO_2 concentration in the field is to provide the leaf with air from a light weight compressed gas cylinder. A range of CO_2 concentrations could be generated from the cylinder by use of critical flow orifices, to divert a precise portion of the flow through CO_2 absorbent or to mix gas from a second cylinder of CO_2 free air (24,34).

<u>Humidity control</u>. Water vapor concentration has an important influence on stomatal opening. The water vapor pressure deficits (VPD) that can be withstood vary from species to species. In many mesophytes stomatal closure begins at a VPD of > 1.0 kPa. Thus, care should be taken that VPD does not become inhibitory. In the laboratory humidity may be controlled by bubbling the air through water at a known temperature. A more efficient system is to pass the air first through water well above the required dewpoint and then bubble the air through water at the required temperature. This second bubbler acts as a condenser. Alternatively, water may be replaced by hydrated salt crystals with a high equilibrium water vapor pressure (e.g. ferrous sulphate crystals). Columns of such salts may be simply incorporated into air conditioning systems, and by combining this with the critical orifice technique an accurate range of humidities may be generated in the field (48); this is the basic method of the ADC WG600 water vapor generator. In many instances in the field, the humidity required will be that of the ambient air, thus if there are no materials within the system which will absorb water vapor, the humidity provided to the chamber will be that required. However, transpiration by the enclosed leaf will raise the actual chamber humidity and in a well stirred chamber the increase could be very significant. It may therefore be necessary to partially dry the air before it enters the chamber, so that the humidity rise resulting from enclosure may be compensated for. Silica gel is commonly used for drying and has the advantage of easy regeneration. However, it will adsorb CO_2 and is not appropriate where CO_2 concentration is being varied. Magnesium perchlorate or zinc chloride provide effective driers which will equilibrate more rapidly with any change in CO_2 mole fraction (13).

5. ANALYSIS OF GAS EXCHANGE MEASUREMENTS

Measurement of CO_2 exchange provides not only an instantaneous measure of productivity, but also an <u>in vivo</u> probe of limitations to photosynthetic C-assimilation, allowing a quantitative assessment of the effects of environmental variables on different steps in the vapor fluxes, it is possible to separate stomatal limitations from those within the mesophyll and to separate effects on the light and CO_2 limiting phases of photosynthesis.

5.1 Resistance Analogues

Resistance analogues have been applied extensively to gas exchange measurements in analyses of the effects of environmental variables on A and on the fluxes of other gases from leaves (39,49, 50). Gaastra (51) was the first to apply resistance analogues in the analysis of limitations to gas exchange in leaves. CO_2 enters the leaf in photosynthesis because a diffusion gradient exists between the atmosphere and the sites of photosynthetic CO_2 assimilation within the mesophyll. By applying Fick's Law of Diffusion in an integrated form, it may be shown that the net flux of a gas in a one-dimensional diffusion pathway is equivalent to the ratio of the concentration difference across that gradient with the resistance to physical diffusion within the gradient (52). Thus, in photosynthetic CO_2 assimilation and transpiration:

$$A = \Delta c/\Sigma r \quad \ldots\ldots\ldots \quad (5.1)$$
$$E = \Delta\chi/\Sigma r' \quad \ldots\ldots\ldots \quad (5.2)$$

Where:
Δc is the CO_2 gradient between the atmosphere and the site of assimilation.
$\Delta\chi$ is the water vapor gradient between the site of evaporation and the atmosphere.
Σr and $\Sigma r'$ are the total resistances to transfer of CO_2 and water vapor, respectively, across these gradients.

These equations are analogous to Ohm's low where Δc and $\Delta\chi$ are analogous to the potential difference and A and E are analogous to the current.

The diffusion pathway into the leaf may be divided into a number of discrete stages, each analogous to a resistor in an electrical circuit. Most recent studies have preferred the term conductance, i.e. reciprocal of resistance (g). The major advantage of conductance, as an expression of limitations to gas exchange, is that it is directly proportional to flux, i.e. A and E, thus its interpretation is simpler. In older analyses conductance was expressed as $m\ s^{-1}$. However, these units are determined by those used to express the flux and concentration gradient. In the older literature these concentration gradients have been described in amount of gas per unit volume. Cowan (53) suggested an alternative method of expressing conductance, resulting from expression of the gradient as the difference in mole fraction of the gas. This will equal the ratio of the difference in partial pressure to total air pressure. The units of concentration gradient are therefore dimensionless, since moles are the units of numerator and denominator, the units of conductance will be those of flux, i.e. A or E.

78

5.2 The Gaseous Diffusion Pathway

Gas phase conductance and transpiration. To reach the leaf,
CO_2 must first diffuse through the boundary layer; g_b being a function of the aerodynamic properties of the leaf and leaf chamber,
windspeed and turbulence (54). The empirical determination of g_b'
was described in Section 4.2. Under field conditions the boundary
layer conductance for many crops and mesophytes will be an order of
magnitude, or more, greater than the maximum stomatal conductance;
although it has been suggested to be of greater importance in some
desert annuals (50).

Since the stomata are considered to be the dominant limitation
to diffusion of CO_2 in the gas phase, this conductance is often referred to as the stomatal conductance, however g_s phase conductance
(g_g') provides a less ambiguous term. By assuming that the trans-
piratory efflux of water vapor diffuses through the pathway by which
CO_2 enters the leaf, g_g may be derived from g_g', the equivalent
conductance to the diffusion of water vapor. By measuring the flux
of water vapor and the atmospheric humidity, and by assuming that
the humidity of air at the mesophyll/internal air space interface
is saturated at the leaf temperature, g_g' may be calculated:

$$g_g' = (E.P)/(e_s - e_a) \quad \ldots\ldots\ldots (5.3)$$

The conductance of CO_2 through this pathway (g_g) has then been
assumed to be g_g' divided by the ratio of the binary diffusivities
of water vapor/air and CO_2/air; where the accepted value for this
ratio is 1.6 (50). The ratio in the boundary layer will be lower
because molecular transfer will be by both diffusion, dependent on
molecular size, and turbulent transfer, independent of molecular
size. A lower ratio of 1.37 has been suggested to be more appropriate to the boundary layer (39). The gas phase conductance to CO_2
will therefore be given by:

$$1/g_g = 1.61/g_s' + 1.37/g_b' \quad \ldots\ldots\ldots (5.4)$$

Where

$$1/g_s' = 1/g_g' - 1/g_b' \quad \ldots\ldots\ldots (5.5)$$

From considerations of diffusion alone it may be assumed that mean
CO_2 mole fraction at the mesophyll cell wall/internal air space
interface (c_i) will be:

$$c_i = c_a - A/g_g \quad \ldots\ldots\ldots (5.6)$$

The volumetric efflux of gas, predominantly water, from a
photosynthesizing leaf will normally exceed the total influx so that
a pressure gradient will exist driving a mass flow of gases including
CO_2 out of the leaf and so depressing c_i, relative to c_a. This will

occur even in the absence of any consumption of CO_2 within the meso-phyll (55,56). Thus a correction to equation 5.6 which takes account of the transpiration rate is necessary (4):

$$c_i = [(g_g - E/2).c_a - A] / [g_g + E/2] \quad \text{........ (5.7)}$$

The calculation of g_g from g_g' assumes that water vapor and CO_2 follow the same diffusion pathway between the atmosphere and meso-phyll/internal air space interface, i.e. the cell surfaces that are most actively assimilating CO_2 are also those which are most actively transpiring. Experimental evidence and an electrical analogue (53) suggest that most evaporation takes place from sites close to the stomata rather than from the mesophyll cell surfaces. However, Sharkey et al.(57) have shown good agreement in c_i estimated from equation 5.7 with a direct measurement of c_i for the same leaf, suggesting that any error introduced by additional sites of trans-piration do not significantly alter the estimate of g_g.

Feedback control of stomatal conductance by CO_2 assimilation.
Many studies in the 1960's and 70's used the resistance analogue approach to determine the relative importance of stomatal limita-tions. By assuming c at the site of carboxylation to be zero, Σr could be assumed to represent the total diffusion resistance to CO_2 assimilation. Thus, the relative limitation imposed by the stomata would be $r_s/\Sigma r$. To be correct the response of A to c at all points in the diffusion pathway must be linear. All evidence suggests that this condition would rarely be satisfied. The response of A to c shows a hyperbolic or asymptotic response and thus if A is approach-ing saturation, its value will be largely independent of r_s. The resistance analogue also ignores the fact that stomatal aperture and the capacity of the mesophyll for CO_2 assimilation may be closely linked. Stomata serve to balance the need of the leaf to allow the entry of CO_2 for photosynthesis while limiting the trans-piratory loss of water vapor (53). This compromise is effective if conditions favoring rapid CO_2 assimilation tend to raise g_s and those favoring rapid transpiration to lower g_s, such that dE/dA remains constant (58). Wong et al.(59) hypothesized that rather than g_g influencing the rate of CO_2 assimilation, the capacity of the mesophyll to fix CO_2 influences g_g. This is supported by two independent observations (60). An alternative method of assessing stomatal limitation is therefore necessary and is provided by the response of A to c_i.

5.3 The A/c_i Response

To remove stomatal influence, A is plotted against CO_2 concen-tration or mole fraction at the mesophyll cell surface (c_i); c_i determined according to equation 5.7. The plot illustrates the response of A to c_i in the absence of stomatal limitations. Although loosely referred to in many recent studies as the leaf internal CO_2, the measure determined by equation 5.7 is strictly the CO_2 concentration at the mesophyll cell surface, since this is the start of the water vapor gradient out of the leaf. Recent analyses have suggested that c_i is similar to the CO_2 concentration within the chloroplasts, and thus the response of A to c_i may be considered to approximate to the response of A to substrate concentration at the site of carboxylation (61).

The response has had two important applications. First, as an alternative method of separating stomatal from mesophyll limitations and secondly, in separating in vivo carboxylation and electron transport limitations within the mesophyll.

Separation of stomatal and mesophyll limitations. Farquhar and Sharkey (50) have developed a simple method of separating stomatal and mesophyll limitations using this response. A, measured at the normal atmospheric CO_2 concentration is subtracted from A_0, the rate which would occur if there was no stomatal limitation, i.e. the value of A interpolated from the response curve at $c_i = 340$ µmol mol^{-1}. The relative limitation (l) which the stomata impose may then be calculated:

$$l = (A_0 - A) / A_0 \qquad \text{....... (5.8)}$$

Thus, l is the proportionate decrease in A that may be attributed to the stomata and other gas phase limitations. The method has the great advantage, when calculated graphically, that it makes no assumptions on the shape of the response of A to c_i. Examination of many A/c_i response curves shows that c_i is often maintained, in a normal atmosphere, at a point close to the inflexion in the response (Fig. 12). Thus, elevation of c_i by removal of the gas phase limitation will have only a small effect on A and thus l will be small.

The internal CO_2 concentration also provides a means of assessing the dynamics of changes in the relative importances of stomatal and mesophyll processes in limiting A, following a change in the environment, e.g. imposition of water or temperature stress. If an increase in stomatal limitations is the dominant cause of a reduction in A, then c_i must decrease; if on the other hand an increase in limitations within the mesophyll dominated the reduction in A, then an increase in c_i might be expected.

Separation of carboxylation and electron transport limitations. CO_2 assimilation(A) responds in a characteristic manner to increase in c_i. Farquhar et al.(62) and von Caemmerer and Farquhar (4) suggested from a steady-state model of photosynthetic carbon metabolism that this response should consist of two phases, an initial linear response where the efficiency of carboxylation, i.e. amount of active Rubisco, determines the slope dA/dc_i, followed by an inflexion to a slower rise or plateau where dA/dc_i approaches zero because A is limited by the supply of substrate (RubP) for carboxylation. Many recent descriptions of the response of A to c_i fulfill this expectation (4).

81

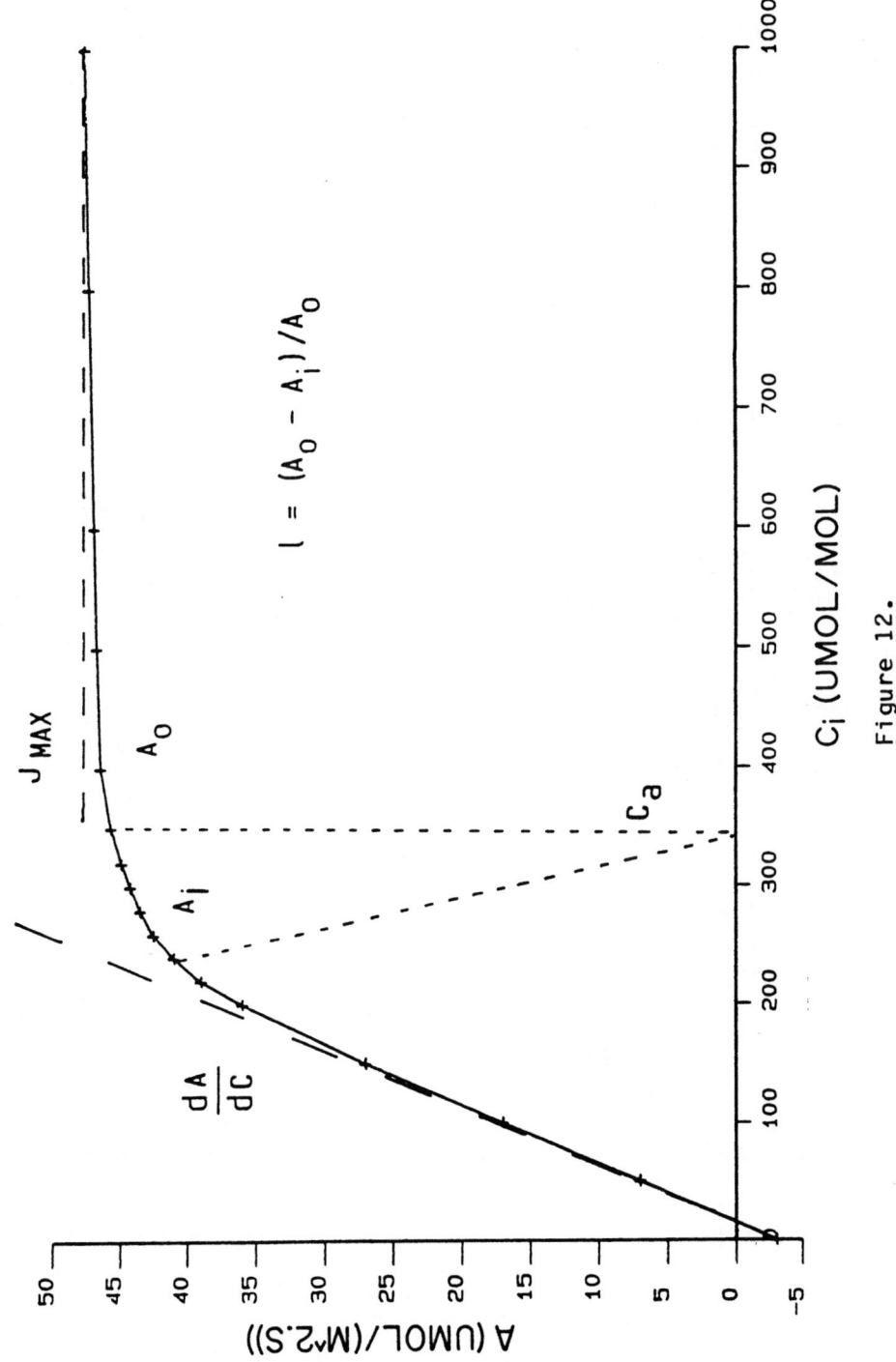

Figure 12.

Seeman and Berry (61) found a very close agreement between the initial slope of the A/c_i curve, predicted by Farquhar et al.(62) to reflect Rubisco levels and extractable Rubisco activity.

Thus when
   c_i approaches 0
   $dA/dc_i = v_{RubPc}$
   Where:
   v_{RubPc} is the velocity of RubP carboxylation (mol $m^{-2}s^{-1}$)

and when
   A approaches A_{max}
   $A = J_{max}$
   Where:
   J_{max} is the maximum rate of RubP regeneration (assumed to
      equal the maximum rate of coupled photosynthetic electron
      transport) (μmol $m^{-2}s^{-1}$)

The A/c_i response curve, measured in relation to other environmental parameters, provides a method of separating effects on carboxylation efficiency from those on RubP regeneration, generally assumed to be photosynthetic electron transport (4,50).

5.4 Light Response

The response of A to photon flux I_p describes a curve or curvilinear progression consisting of three phases: 1) an initial linear phase of increase in A with I_p through the light compensation point; 2) a progressive decrease in the slope of the curve (dA/dI_p) with increase in I_p to a plateau, the light saturated A, or A_{max}; 3) with further increase in I_p the gradient (dA/dI_p) remains at zero, but as a result of photoinhibition, will eventually become negative with still further increase in I_p. Normally, such photoinhibition will not occur at naturally occurring photon flux densities, except in shade adapted leaves or in leaves suffering from other environmental stresses (63,64).

Quantum yield (∅). The initial slope of the light response curve (α) may be described as the apparent maximum quantum yield. The qualification, apparent, is used since the estimate is based on incident and not absorbed photon fluence. If account is taken of reflected and transmitted light, then the true maximum quantum yield (∅) is obtained.

The parameter of efficiency of light utilization by photosynthesis is the quantum yield (∅), the moles CO_2 fixed per mole quanta absorbed by a leaf. Since light becomes of less importance as a factor limiting photosynthesis with increasing photon fluence rate (I_p), the maximum quantum yield can only be measured at low I_p, when photosynthesis is strictly light limited and proportional to I_p.

The majority of gas exchange studies have concerned light-saturated rates of CO_2 assimilation, perhaps because in most C_3 crops A is saturated by light levels well below full sunlight. However, in the field situation and particularly in canopies with a large LAI, a large proportion of leaves may experience light-limiting conditions, since even in full sunlight a large proportion of leaves will be shaded by others. The significance of this may be seen in measurements of CO_2 assimilation by whole stands. Unlike single leaves, stands of plants will show a much longer phase in which CO_2 assimilation responds linearly to increase in I_p (Fig.13). There is now much evidence that CO_2 assimilation by mature crop canopies is determined more by ϕ than by the light saturated rate of A (60).

Light saturated rate of CO_2 assimilation (A_{max}). Net photosynthesis responds hyperbolically to quantum fluence rate as light becomes of decreasing importance as a limiting factor. Individual leaves of many C_3 plants are unable to use additional light above about I_p= 500 μmol m^{-2}s^{-1}, roughly 25% of full sunlight, but this is not true of C_4 plants which in general fail to saturate even at full sunlight. The light saturated A (A_{max}) may be considered as a measure of the photosynthetic capacity of the leaf. In contrast to ϕ, A_{max} varies markedly both within and between C_4 and C_3 species. Maximum rates of photosynthesis of C_4 plants exceed those of C_3 plants; those of C_4 grasses (32 - 66 μmol m^{-2}s^{-1}) are the highest recorded (49). A_{max} varies with almost all environmental variables which influence photosynthesis and with pre-conditioning, leaf age and ontogeny.

Considerable variation in A_{max} may be found within one genotype or even within the same plant.

6. CONCLUSION

Measurement of CO_2 assimilation provides not only a sensitive probe for determining the short-term growth of crops, but also a method for separating the bases of effects of the environment or genotype differences on photosynthetic capacity. Recent developments in miniaturization, especially with respect to infra-red gas analyzers means that this technique, once confined to the laboratory, may now be applied routinely to crops in the field.

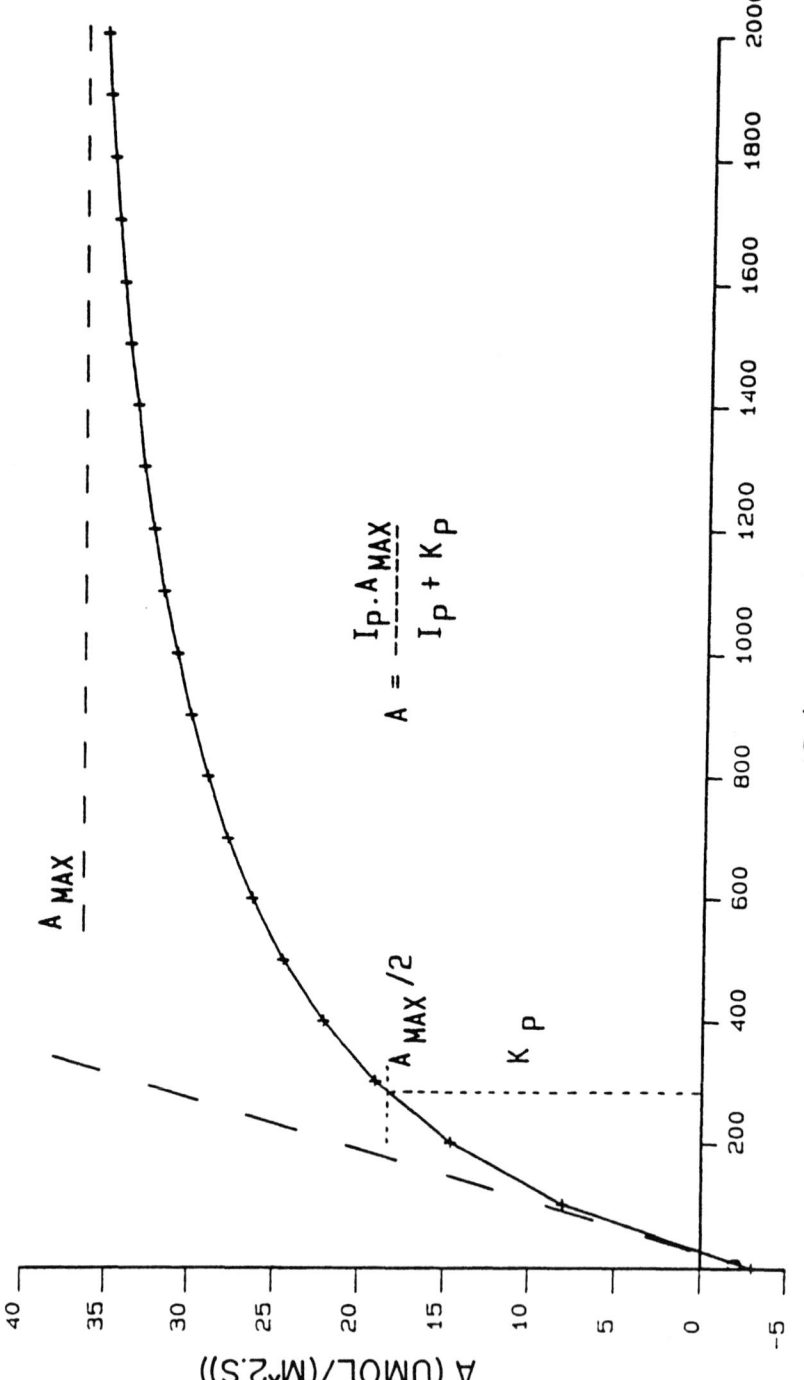

IP (UMOL/(M^2.S))

Figure 13.

APPENDIX

List of manufacturers of equipment utilized in crop gas exchange studies. Inclusion of a manufacturer in this list is not a recommendation by the author, nor a guarantee of the suitability of this equipment. The author has neither been an employee nor paid consultant of any of the listed companies.

Aluminium cylinders for compressed gases:
Cliff Impact Div., 33800 Lakeland Blvd., Eastlake, OH 44094 USA
P.K. Morgan Ltd., 4 Bloors Lane, Rainham, M3I 7ED, UK.

Calibration gases for infra-red gas analyzers:
Matheson Gas Products Div., P.O. Box 85, East Rutherford, NJ 07073 USA.
P.K. Morgan Ltd., 4 Bloors Lane, Rainham, M3I 7ED, UK.

Complete gas exchange systems:
Analytical Development Co Ltd (ADC), Pindar Rd, Hoddesdon, EN11 0AQ, UK
LI-COR
H. Waltz Co., Eichenring 10-14, D-8521, Effeltrich, FRG

Flowmeters (Mass flow):
Brooks Instruments, Emerson Electric Co., Hatfield, PA 19440 USA
Datametrics Instrument Division, Willmington, MA 01887 USA
MKS Instruments Inc., 34 Third Ave., Burlington, MA 01803 USA
Sierra Instruments Inc., P.O. Box 909, Sierra Bldg, Carmel Valley, CA 93924 USA
Teledyne Hastings-Raydist, Hampton, VA 23661 USA
Tylan Corp., 23301 S. Willmington Ave., Carson, CA 90745 USA

Flowmeters (Variable-area):
Brooks Instruments, Emerson Electric Co., Hatfield, PA 19440 USA
Fischer and Porter GmbH, 1 Gibraltar Plaza Bldg, Horsham, PA 19044 USA
Fisher Controls Ltd., Brenchley House, Week St. Maidstone, ME14 1UQ, UK
Krohne Messtechnik GmbH, Ludwig-Krohne-Strasse, Postfach 100970, D-4100 Duisberg 1, FRG
Nixon Instrumentation Ltd., Charlton Kings Ind. Est., Cirencester Rd., Cheltenham, GL53 8DZ, UK
G.A. Platon Ltd., Wella Rd, Basingstoke, RG22 4AQ, UK

Gas blending systems:
Analytical Development Co Ltd (ADC), Pindar Rd, Hoddesdon, EN11 0AQ UK

Signal Instrument Co., St. Mary's Works, Krooner Rd, Camberley,
Surrey, GU15 2QP
(See also: Brooks, MKS, Teledyne..., Tylan under Flowmeters
Mass flow).

Gas mixing pumps:
H. Wosthoff oHG, Hagenstrasse 30, D-463 Bochum, FRG

Infra-red gas analyzers:
Analytical Development Co Ltd (ADC), Pindar Rd, Hoddesdon,
EN11 0AQ UK
Beckman Instruments Inc., 2500 Harbour Blvd, Fullerton, CA
92634 USA
Foxboro Co., Bristol Park, Foxboro, MA 02035 USA
Hartmann & Braun Ag., Postfach 900507, D-600 Frankfurt 90, FRG
Horiba Ltd, Miyanohigashi, Kisshoin, Minami-Ku, Kyoto, Japan
Infra-red Industries (IRI), PO Box 989, Santa Barbara, CA
93102 USA
Leybold-Heraeus GmbH, Wilhelm-Rohn-Strasse 25, Postfach 1555,
D-6450 Hanau 1 FRG
Liston Edwards, Inc., 3800 Campus Dr., Newport Beach, CA 92660
USA
MSA (Britain) Ltd., Instrument Division, East Shawhead,
Coatbridge, ML5 4TD UK
Sieger Ltd., 31 Nuffied Estate, Poole, BH17 7RZ UK

Leaf chambers
See manufacturers of complete systems.

Parkinson Grass Leaf Chamber, ADC Ltd. (see Figure A1). This
is a commercially available example of a hand-held leaf chamber for
use in field gas-exchange systems. The chamber is constructed from
duralinium with a polymethyl pentene (PMP) window; both materials
which have very low water absorptivities. A plate of acrylic
plastic ("perspex") above the window removes part of the incoming
infra-red radiation so reducing the heat load on the chamber. The
chamber is stirred by an integral polyethylene paddle. The chamber
is opened by a scissor action and is sealed on closing by two layers
of closed cell foam, which will readily mould around leaves placed
into the chamber. Two ports in the base of the chamber hold a
capacitance humidity sensor and a thermistor, while a cosine
corrected quantum sensor ("PAR" sensor) is incorporated into the
chamber top. Electronic circuits for all of these sensors are
incorporated into the handle.

87

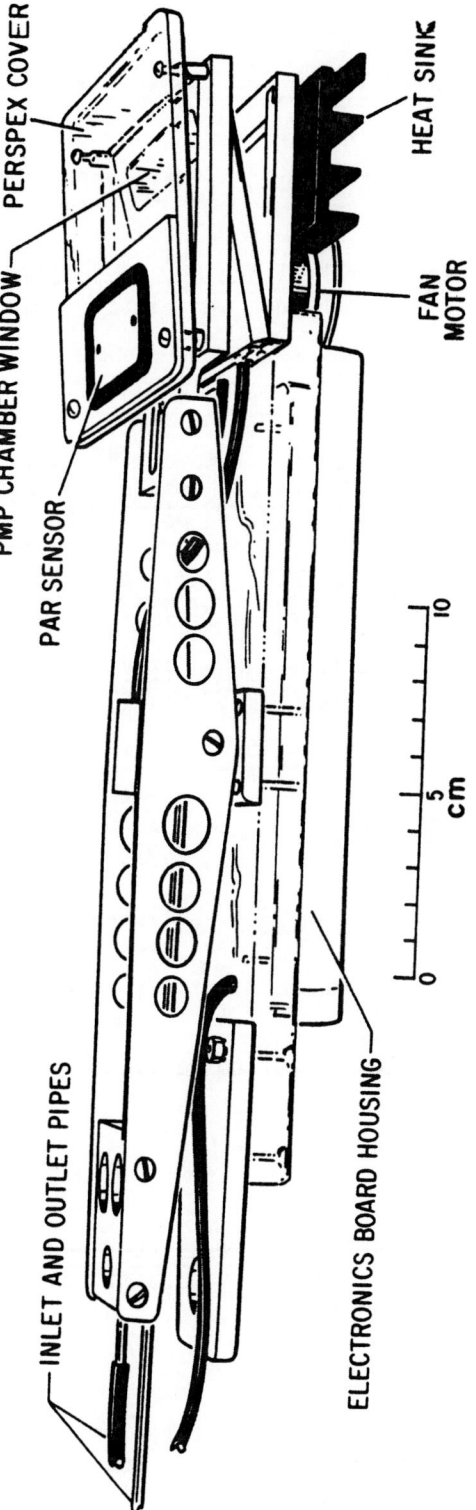

Figure A1.   Parkinson Grass Leaf Chamber, ADC Ltd.

REFERENCES

1. Hunt, R. Plant Growth Curves, pp. 248, Edward Arnold, London, 1983.

2. Ireland, C.R., Long, S.P. and Baker, N.R. Planta 160:550-558, 1984.

3. Bowen, G.D. In: Contemporary Microbial Ecology (E.C. Ellwood, J.H. Hedger, M.J. Latham, J.M. Lynch and J.H. Slater, eds.), pp. 283-304, 1980.

4. von Caemmerer, S. and Farquhar, G.D. Planta 153:376-387, 1981.

5. Catsky, J., Lake, J.V., Begg, J.E. and Voznesenskii, V.L. In: Plant Photosynthetic Production. Manual of Methods (Z. Sestak, J. Catsky and P.G. Jarvis, eds.), The Hague, pp. 198-237, 1971.

6. Incoll, L.D., Long, S.P. and Asmore, M.R. Curr. Adv. Plant Sci. 27:331-343, 1977.

7. Unsworth, M.H. In: Plants and Their Atmospheric Environment (J. Grace, E.D. Ford and P.G. Jarvis, eds.), Glackwell, Oxford, pp. 111-138, 1981.

8. Sestak, Z., Jarvis, P.G. and Catsky, J. In: Plant Photosynthetic Production. Manual of Methods (Z. Sestak, J. Catsky and P.G. Jarvis, eds.) pp. 1-48, The Hague, 1971.

9. Jones, M.B. and Milburn, T.R. J. Exp. Bot. 25:595-597, 1974.

10. Griffiths, J.H. and Jarvis, P.G. J. Exp. Bot. 32:1157-1168, 1981.

11. Schultze, E.D., Hall, A.E., Lange, O.L. and Walz, H. Oecologia 53:141-145, 1982.

12. Banwell, C.N. Fundamentals of Molecular Spectroscopy, McGraw-Hill, London, 1966.

13. Janac, J., Catsky, J. and Jarvis, P.G. In: Plant Photosynthetic Production. Manual of Methods (Z. Sestak, J. Catsky and P.G. Jarvis, eds.) pp. 111-192, The Hague, 1971.

14. Janac, J. Photosynthetica 4:302-308, 1970.

15. Jarvis, P.G. and Sandford, A.P. In: Instrumentation for Environmkental Physiology (B. Marshall and F.I. Woodward, eds.), Cambridge University Press, London, in press, 1984.

16. Coombes, R.G. and Stroud, D.J. International Environment and Safety, June 1982.

17. Ohtaki, E.1and Matsui, M. Boundary-Layer Meteorol. 24:109-119, 1983.

18. Desjardins, R.L., brach, E.J., Alvo, P. and Schuepp, P.H. Science 216:733-735, 1982.

19. Hill, D.W. and Powell, T. Non-dispersive Infra-red Gas Analysis in Science, Medicine and Industry, Adam Hilger Ltd, London, pp. 222, 1968.

20. Parkinson, K.J. and Legg, B.J. J. Phys. E. Sci. Instrum. 4:598-600, 1971.

21. Ower, E. and Pankhurst, R.C. The Measurement of Air Flow, 5th edition (in SI/Metric Units), Pergamon Press, Oxford, 1977.

22. Hayward, A.T.J. Flowmeters. A Basic Guide and Source Book for Users, Macmillan, London, 1979.

23. Brain, T.J.S. and Scott, R.W.W. J. Phys. E. Sci. Instrum. 15:967-980, 1982.

24. Long, S.P. and Ireland, C.R. In: Instrumentation for Environmental Physiology (B. Marshall and F.I. Woodward, eds.), Cambridge University Press, London, in press, 1985.

25. Ewing, J.A. Roc. Roy. Soc. Edin. 45:308-321, 1924.

26. Studman, C.J. and Compton, S.E. J. Phys. E. Sci. Instrum. 16:190-192, 1983.

27. Thomas, C.C. J. Franklin Instit. 61:411-460, 1911.

28. Widmer, A.E., Fehlmann, R. and Rehwald, W. J. Phys. E. Sci. Instrum. 15:213-220, 1982.

29. Anon. Mass Flowmeters. Mass Flow Controllers, Tylan Corporation, Carson, 1981.

30. Barr, G. J. Sci. Instrum. 11:321-324, 1934.

31. Levy, A. J. Sci. Instrum. 41:449-453, 1961.

32. Kolk, J.F.M. and Moulijn, J.A. J. Phys. E. Sci. Instrum. 11:259-261, 1978.

33. Brain, T.J.S. and Reid, J. Performance of Small Diameter Cylindrical Critical-flow Nozzles. Report #546, National Engineering Laboratory, Glasgow, 1973.

34. Parkinson, K.J. and Day, W. J. Appl. Ecol. 16:623-632, 1979.

35. Anon. Vacuum-Pressure-Flow Control Systems. Bulletin PFC-11/79, MKS Instruments, Burlington, 1979.

36. Incoll, L.D. and Wright, W.H. Spec. Soils. Bull. Conn. Agric. Exp. Stn. Number 30, 1969.

37. Harris, G.C., Cheesebrough, J.K. and Walker, D.A. Plant Physiol. 71:102-107, 1983.

38. Parkinson, K.J. Plant Cell Env., in press, 1985.

39. Jarvis, P.G. In: Plant Photosynthetic Production. Manual of Methods (Z. Sestak, J. Catsky and P.G. Jarvis, eds.) pp. 566-622, The Hague, 1971.

40. Jarvis, P.G., Catsky, J., Eckardt, F.E., Koch, W. and Koller, D. In: Plant Photosynthetic Production. Manual of Methods (Z. Sestak, J. Catsky and P.G. Jarvis, eds.) pp. 49-110, The Hague, 1971.

41. Oquist, G., Hallgren, J.E. and Brunes, L. Plant Cell Env. 1:21-27, 1978.

42. Idle, D.B. and Proctor, C.W. Plant Cell Env. 6:437-440, 1983.

43. Bloom, A., Mooney, H.A., Bjorkman, O. and Berry, J. Plant Cell Env. 3:371-376, 1980.

44. Zettlemoyer, F., Micale, J. and Klien, K. In: Water. A Comprehensive Treatise, Vol. 5 (F. Franks, ed.), Plenum Press, pp. 241-291, 1975.

45. DuPont. Teflon Fluorocarbon Resins, reprint 125-D, pp. 6, DuPont Plastics Div., Geneva, 1970.

46. Oberbach, K. von. Kunststoff-Kennwerte fur Kunstrukteure, Carl Hanser, Munchen, p. 171, 1975.

47. Long, S.P. and Woolhouse, H.W. J. Exp. Bot. 29:567-577, 1978.

48. Parkinson, K.J. and Day, W. J. Exp. Bot. 32:411-418, 1981.

49. Korner, C., Scheel, J.A. and Bauer, H. Photosynthetica 13:45-82, 1979.

50. Farquhar, G.D. and Sharkey, T.D. <u>Ann. Rev. Plant Physiol.</u> <u>33</u>:317-345, 1982.

51. Gaastra, P. <u>Meded. Landb. Hogesch. Wageningen</u> <u>59</u>:1-68, 1959.

52. Monteith, J.L. (ed.) <u>Vegetation and the Atmosphere</u>, 2 vol., Academic Press, London, 1975.

53. Cowan, I.R. <u>Adv. Bot. Res.</u> <u>4</u>:117-228, 1977.

54. Grace, 1983

55. Parkinson, K.J. and Penman, H.L. <u>J. Exp. Bot.</u> <u>21</u>:405-409, 1970.

56. Leuning, R. <u>Plant Cell Env.</u> <u>6</u>:181-194, 1983.

57. Sharkey, T.D., Imai, K., Farquhar, G.D. and Cowan, I.R. <u>Plant Physiol.</u> <u>69</u>:657-659, 1982.

58. Farquhar, G.D., Schultze, E.D. and Kuppers, M. <u>Aust. J. Plant Physiol.</u> 7:315-327, 1980.

59. Wong, S.C., Cowan, I.R. and Farquhar, G.D. <u>Nature</u> <u>282</u>:424-426, 1979.

60. Long, S.P. In: <u>Topics in Photosynthesis</u>, Vol. 6 (J. Barber and N.R. Baker, eds.) Elsevier, Amsterdam, in press, 1984.

61. Seeman, J.R. and Berry, J.A. <u>Carnegie Inst. Wash. Yr. Bk.</u> <u>81</u>:78-83, 1982.

62. Farquhar, G.D., von Caemmerer, S. and Berry, J.A. <u>Planta</u> <u>149</u>:78-90, 1980.

63. Osmond, C.B., Winter, K. and Ziegler, H. In: <u>Encyclopaedia of Plant Physiology</u>, Vol. 12B (O.L. Lange, P.S. Nobel, C.B. Osmond and H. Ziegler, eds.) pp. 497-547, Springer-Verlag, Berlin, 1982.

64. Powles, S.B., Berry, J.A. and Bjorkman, O. <u>Plant Cell Env.</u> <u>6</u>:117-124, 1983.

NON-INVASIVE MEASUREMENTS OF PHOTOSYSTEM II REACTIONS IN THE FIELD USING FLASH FLUORESCENCE

Howard Robinson

Department of Physiology and Biophysics
University of Illinois
Urbana, Illinois 61801

This chapter discusses a new technique for making measurements on the photosynthetic apparatus in leaves of plants in the field. A device for this purpose which can be used under field conditions employs flashlamps to non-invasively measure the kinetics of the reactions of the quinone acceptor complex of photosystem II (PSII, see abbreviations below). Studies in the laboratory on the isolated chloroplasts have led to a detailed understanding of the functioning of the quinone acceptor complex of PSII. This understanding is the basis for the interpretation of the measurements made on the plants in the field.

Experiments on isolated chloroplasts show that in plants that are resistant to the herbicide atrazine there has been an alteration in the functioning of the quinone acceptor complex. This alteration is demonstrated by measuring changes in the kinetics of the reactions associated with this complex. The kinetics are determined using precisely timed flashes of light to measure changes in the level of fluorescence emitted from the pigment bed of PSII. The field flash fluorescence apparatus uses this same technique to measure the functioning of the quinone acceptor complex in whole leaves in the field. The characteristic altered kinetics of the reactions of the quinone acceptor complex which have been demonstrated in the laboratory experiments on isolated membranes are clearly discernable in the leaves of the atrazine-resistant plants using the field fluorescence apparatus.

Abbreviations

DCMU	3-(3,4-Dichlorophenol)-1,1-dimethylurea
F, F_0, F_v, F_{max}	Fluorescence intensity, initial, variable, maximum
PSI(II)	Photosystem I(II)
$Q_A(^-)$	primary quinone acceptor of PSII (semiquinone)
$Q_B(^-)(^=)$	secondary quinone acceptor of PSII (semiquinone)(quinol)

1. INTRODUCTION

The photosynthetic process of higher plants converts carbon dioxide, water, and light quanta into reduced sugars. The enzymes which catalyze this conversion are contained within the chloroplasts and can be thought of as split into two distinct groups. In one group are the enzymes of the "light" reactions, which harvest the light, oxidize water and generate two energy-rich intermediates, ATP and NADPH. These two intermediates plus carbon dioxide are the substrates for the second group, the "dark" reactions or Calvin cycle, which produce the reduced sugars.

The chloroplast is completely bounded by an outer membrane. There is a separate, highly convoluted membrane system, the thylakoid system, enclosed within the chloroplast. The enzymes of the light reactions are physically associated with the thylakoid membranes, while the Calvin cycle enzymes are free-floating within the stroma, or soluble part of the chloroplast (Fig. 1).

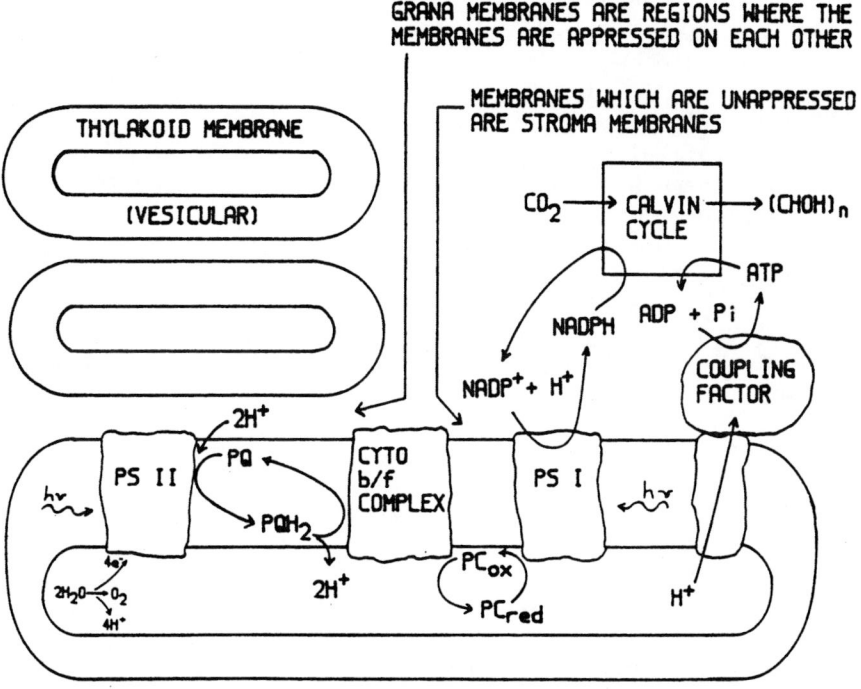

Figure 1. Schematic representation of the photosynthetic process in plant chloroplasts. PC_{ox} and PC_{red} are the oxidized and reduced forms of plastocyanin. PQ and PQH_2 are plastoquinone and plastoquinol. For explanation, see text.

The familiar "Z-scheme" of photosynthetic electron transport is shown in Figure 1, with three enzyme complexes embedded within the

thylakoid membrane. The complexes are linked by the mobile oxidation-reduction cofactors, plastoquinone/plastoquinol and plasto-
cyanan$_{ox}$/plastocyanin$_{red}$. The synthesis of ATP from ADP and
inorganic phosphate occurs at the coupling factor enzyme. The
synthesis is driven by the proton gradient generated across the
membrane by the vectorial movement of charged species, and by the
release of protons within the lumen space of the thylakoid membrane.
The NADPH and ATP produced by the light reactions are released to
the stroma and are thus accessible to the enzymes of the Calvin
cycle.

2. FLUORESCENCE FROM PHOTOSYSTEM II

Light quanta are collected by the light-harvesting pigment
complexes in the vicinity of the two photosystems by chlorophylls
and other accessory pigments. Light harvested in the vicinity of
PSI does not give rise to a measurable amount of fluorescence at
room temperature, so fluorescence reflects the behavior of PSII.
The pigments are anchored within the membranes to proteins which
order them so that the absorbed light energy can migrate within the
pigment bed by a process of excitation transfer. The PSII pigment
beds are connected, so that the migrating excitons can wander to any
of the PSII reaction centers where a photochemically induced charge
separation can take place. Alternatively, the migrating excitons
can deactivate through thermal processes, or re-radiation (fluorescence). The fluorescence will be from the lowest excited singlet
state of the chlorophylls in the PSII pigment bed (685 nm, $E = hc/\lambda$).
The probabilities of these alternatives can be determined by experimentation. The probabilities of thermal decay and fluorescence are
small compared to the probability of capturing the excitation energy
as photochemical work at the reaction center, thus fluorescence is
usually low.

3. FLUORESCENCE AND [Q_A^-]

Change separation occurs at the reaction center. In this process
a specialized pigment donates an energized electron through several
intermediates to the primary quinone acceptor, Q_A, to form Q_A^-.
The oxidized specialized pigment (P^+_{680}) becomes re-reduced by an
electron from the donor side of the reaction center. (These electrons on the donor side come from the oxidation of water via a
complex mechanism with several intermediate steps.) The arrival of
the electron at Q_A^- (in less than 1 ns) and the re-reduction of
P^+_{680} (in less than 500 ns) occur rapidly. In this state a reaction
center cannot accept another exciton and is thus closed.

When all the reaction centers are open, fluorescence is unlikely.
If a few of the reaction centers become closed, the probability of
fluorescence remains small compared to the probability of the exciton finding an open center. Although fluorescence is slightly more
likely, the increased probability is smaller than the increased ratio

of closed to total reaction centers, because of the possibility of exciton migration to a neighboring open center. If nearly all of the reaction centers are closed, then the probability of fluorescence becomes large compared to the probability of the exciton finding a reaction center, and the fluorescence probability is changing faster than the ratio of closed to total reaction centers. As a consequence, the relationship between the variable fluorescence intensity (F_v, where $F_v = F - F_0$ and F_0 is the fluorescence intensity when all the reaction centers are open), and the ratio of closed to total reaction centers ($Q_A^-/(Q_A^- + Q_A)$), is not quite linear; however, a correction can be applied to account for this property (3,5,12).

4. THE MODEL FOR THE QUINONE ACCEPTORS OF PSII, Q_A AND Q_B

In the model proposed by Bouges-Bocquet (2) and Velthuys and Amesz (13) and revised by Velthuys (14,15) shown in Figure 2, there are two quinone binding sites. At the Q_A site there is a tightly bound plastoquinone that can only be either oxidized (Q_A) or singly reduced (Q_A^-). A neighboring iron atom electronically interacts with the unpaired electron on Q_A^- or Q_B^- but its function is not clearly understood. The secondary quinone binding site (Q_B-site) can reversibly bind either plastoquinone or plastoquinol from a pool in the lipid bilayer (5-10 per PSII). Each reaction center operates independently in the oxidation of water and reduction of plastoquinone. At an individual reaction center, Q_B becomes reduced to Q_B^- by a first electron from Q_A^-, and then by a second electron from Q_A^-, following two separate single electron reductions of Q_A at the center (Fig. 3).

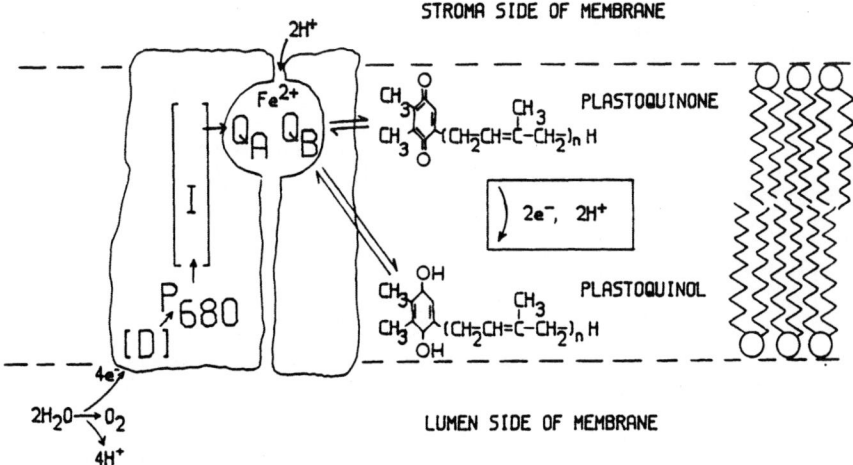

Figure 2. Schematic model for the functioning of PSII in thylakoid membranes. The symbol D represents a collection of intermediate steps between the reduction of oxidized P_{680} (the primary pigment donor at the reaction center of PSII) and the oxidation of water. The symbol I represents a series of pigments which undergo rapid oxidation-reduction reactions leading to the reduction of Q_A.

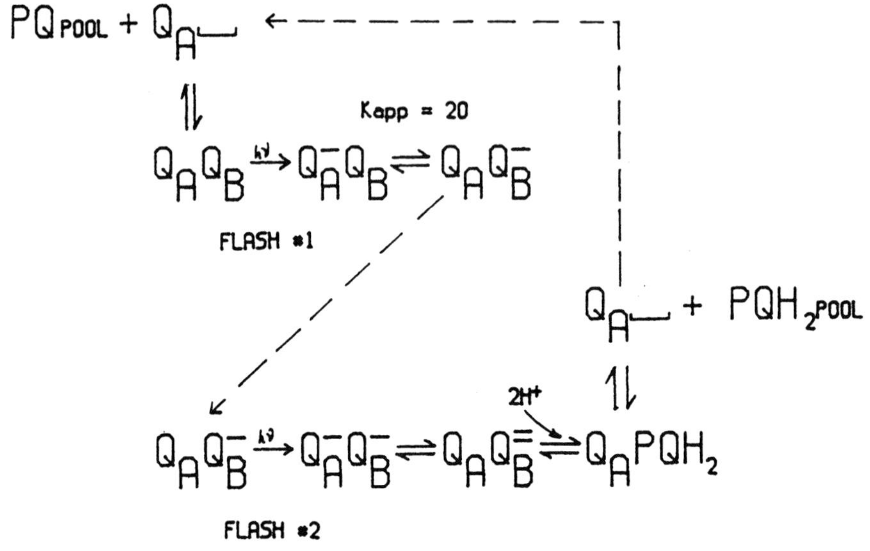

$$PQ_{POOL} + Q_A \text{---}$$

$$\Updownarrow \qquad K_{app} = 20$$

$$Q_A Q_B \xrightarrow{h\nu} Q_A^- Q_B \rightleftharpoons Q_A Q_B^-$$

FLASH #1

$$Q_A \text{---} + PQH_{2POOL}$$

$$\Updownarrow$$

$$Q_A Q_B^- \xrightarrow{h\nu} Q_A^- Q_B^- \rightleftharpoons Q_A Q_B^= \overset{2H^+}{\rightleftharpoons} Q_A PQH_2$$

FLASH #2

Figure 3. Kinetic model of the oxidation-reduction reactions of the quinone acceptor complex of PSII. The symbol ⌐___⌐ represents the unoccupied secondary quinone binding site where plastoquinone, plastoquinol, and some photosynthetic herbicides compete for occupancy based upon the concentration and binding affinity of the particular molecule.

Two protons become associated with Q_B^- to form plastoquinol which is then released from the Q_B-site. When there is a single electron at an individual center's quinone acceptor complex, plasto-semiquinone cannot leave the Q_B-site to either the water phase (because of the quinone phytol tail) or the non-polar membrane lipid phase (because the lipid is not able to stabilize the charge of the unpaired electron on the head group of the quinone).

5. Q_A^- OXIDATION AFTER A SATURATING FLASH

In isolated thylakoid membranes that have been dark-adapted under mildly oxidizing conditions with benzoquinone (10,11), all the PSII quinone acceptor complexes are fully oxidized. A single saturating flash will cause charge separation in 95% of the centers within 2 µs. All of the centers are then closed, so a weak measuring flash (sampling less than 2% of the centers) shortly after the actinic flash will yield a high fluorescence intensity (nearly F_{max}), compared to the fluorescence intensity measured just prior to the saturation flash (F_0). If the intensity of fluorescence is measured after a delay, the fluorescence intensity is lower than the level measured immediately after the actinic flash. This is because a certain number of the centers have reopened due to the oxidation of Q_A^- by Q_B. If this experiment is repeated with fresh dark-adapted samples, with a series of delays between the actinic flash and the measuring flash, the data shown in Figure 4, flash #1, are obtained. The decrease in the fluorescence intensity due to oxidation of Q_A^- by Q_B shows a half-time of 150-250 μs. Q_A^- will not be oxidized in all of the centers, even after a long delay (1 s). This is because the equilibrium constant for sharing the one electron between the states $Q_A^-Q_B$ and $Q_AQ_B^-$ is about 20, so that Q_A^- will be present in about 5% of the centers at equilibrium (10, 11).

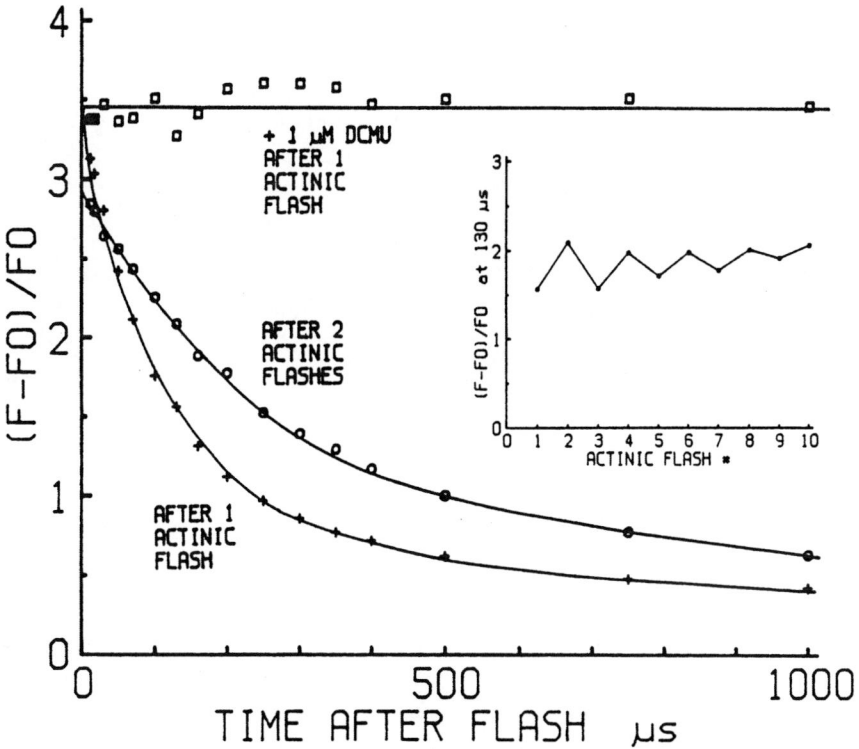

Figure 4. Decay kinetics of variable fluorescence after one or two saturating flashes in isolated thylakoid membranes from pea seedlings. Each data point was obtained with a fresh dark-adapted sample supplied by a flow-cuvette system. For a description of the instrument, details of the preparation of thylakoid membranes and the suspension solution, see Robinson and Crofts (10,11). The inset is the level of variable fluorescence measured at 130 μs after each of a series of saturating flashes.

After equilibrium has been reached (100 ms - 1 s), a second saturating flash again causes all the centers to become closed (Q_A^-). This time Q_A^- is oxidized by Q_B^- in most of the centers, and after protonation of Q_B^-, plastoquinol will dissociate from the Q_B-site. As shown in Figure 4, this process is slower (250-350 µs half-time) than the oxidation after the first flash. The acceptor complex then returns to the state prior to the first saturating flash, and a third actinic flash will have decay kinetics similar to those after the first flash. The inset to Figure 4 shows that the level of variable fluorescence, measured at a fixed delay after each of a series of actinic flashes, oscillates, reflecting the faster and then slower rate of Q_A^- oxidation after alternating actinic flashes. Thus, there is a two-electron gating mechanism (see Fig. 3) which connects the single-electron events at the reaction center to the two-electron, two-proton reduction of plastoquinone.

6. HERBICIDES WHICH BLOCK ELECTRON TRANSPORT AT PSII

Herbicides such as atrazine or DCMU function by interfering with the operation of the two-electron gating mechanism of PSII. Figure 4 shows that when 1 µM DCMU is present, no oxidation of Q_A^- occurs. DCMU (or atrazine) and plastoquinone are known to compete at the Q_B-site. DCMU has a very high affinity for binding at the Q_B-site, thus plastoquinone is displaced and no oxidation of Q_A^- occurs after an actinic flash. The electron on Q_A^- will then eventually either return to the donor side of PSII (possibly accompanied by emission of a photon--luminescence) or the electron may react with molecular oxygen, generating destructive oxygen radicals.

7. ATRAZINE-RESISTANT WEEDS

Over the last twenty years, atrazine has been extensively applied to fields throughout North America and Europe. A number of weed biotypes resistant to atrazine have appeared (1). However, this resistance is costly to the plant in that there is a lower quantum yield for CO_2 reduction (7) and probably there is an increased like-lihood for damage caused by oxygen radicals formed at the Q_B-site. These plants have a greatly lowered association constant for binding atrazine at the Q_B-site (9). The decay of variable fluorescence after one actinic flash is shown in Figure 5, for isolated thylakoid membranes from resistant and susceptible plants of Amaranthus hybridus. Although the initial rates of decay are iden-tical for both types, the equilibrium constant for the sharing of the one electron between Q_A and Q_B has changed from about 20 for the susceptible thylakoids to about 1 to 2 for the resistant thylakoids. These equilibria are sensitive to pH and the concentration of plas-toquinone in the membrane (Robinson & Crofts, in preparation). The initial rate of electron transport from Q_A^- to Q_B is the same in both cases (see Fig. 5), indicating that the forward rate constant is unchanged. However, in contrast to the susceptible plants, the reverse rate constant from Q_B^- to Q_A in the resistant plant thyla-koids must be similar to that for the forward reaction, since the equilibrium is close to one. From this it seems clear that in these atrazine-resistant weeds, not only does atrazine bind less tightly, but also the semiquinone form of Q_B (Q_B^-) is much less stable than in the susceptible plants. This decreased stability will have two primary effects on the resistant plants.

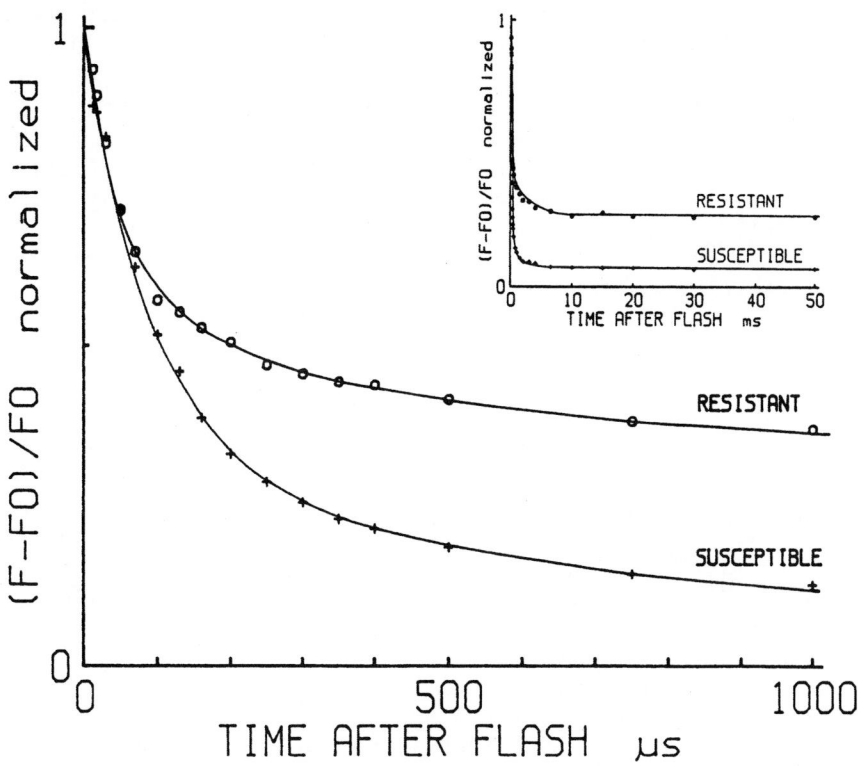

Figure 5. Decay of variable fluorescence after a saturating flash in thylakoid membranes isolated from an atrazine-resistant and an atrazine-susceptible plant of <u>Amaranthus hybridus</u>. The thylakoid membranes were isolated by the procedure of Ort, et al.(7) except that ascorbate was omitted from the procedure. The membranes were subjected to the incubation described in Robinson and Crofts (10, 11) to oxidize centers with residual Q_B^-. The inset shows results of the same experiment over a longer time interval.

First, in the one-electron reduced state, since the backreaction is a first-order process occurring with a rate proportional to the Q_A state, back-reactions from the quinone acceptor complex to the donor side of PSII will occur at least 10 times more rapidly in the resistant plants (this will cause a lowered efficiency especially at low light intensities). The back-reaction half-times will change from 20 to 30 seconds in susceptible plants to 2 to 3 seconds in resistant plant. Secondly, in the one-electron reduced state of the quinone acceptor complex, about 50% of the centers will be closed (Q_A) in the resistant plant. This increases the probability that a migrating exciton will not find an open center before being lost through a dissipative pathway (thermal deactivation or fluorescence). Although an individual center in this state will be closed 50% of the time, this does not cause a 50% reduction in efficiency, due to the possibility of an exciton migrating between centers, as discussed above. However, both these problems will cause a higher proportion of absorbed photons which do not lead to photochemical work. Ort et al. (7) have shown that the probability that an absorbed photon will lead to photochemical work (quantum yield of CO_2 reduction) is 25% lower in the atrazine-resistant versus susceptible biotypes of Amaranthus hybridus. Additionally, it seems possible that the atrazine-resistant plants could incur more damage from oxygen radicals, since the more unstable electron present in the quinone acceptor complex would be more likely to reduce molecular oxygen.

8. MEASUREMENTS ON LEAVES OF ATRAZINE-RESISTANT WEEDS

Measurements of the same reactions described above can also be made on whole leaves. A battery-operated portable device for making these measurements is described in the appendix. This machine performs essentially the same measurements on whole leaves as those performed on isolated thylakoid membranes discussed above. One major difference is that dark-adaptation becomes a problem with leaves. Instead of trying to thoroughly dark-adapt the leaves to try to return all the centers to the fully oxidized state, the protocol adopted with this machine was to give a number of pre-flashes (saturating) to scramble the two-electron gate mechanism of the quinone acceptors of PSII. Thus after scrambling, in about half of the centers there will be an odd electron remaining in the quinone acceptor complex, while the complexes in the remaining centers will be completely oxidized. After scrambling, a series of saturating and measuring flash pairs is given. The delay between the two flashes is progressively increased, and the fluorescence intensity is measured during the weak second flash and plotted graphically as a function of the delay interval. The resulting kinetic profile is the average of the decays seen after one or two actinic flashes when the acceptor complexes are all oxidized prior to the first flash. Figure 6 shows data from leaves of biotypes of Chenopodium album resistant and susceptible to atrazine (the results were essentially identical with Amaranthus hybridus biotypes). Both curves of Figure 6 have been normalized (the resistant plants generally have a slightly elevated F_0 level). The data show that in the resistant plants the oxidation kinetics of Q_A^- are quite different from the susceptible plants. This characteristic makes it possible to distinguish these two biotypes on the basis of a well understood biochemical difference which can be readily measured in the field without disturbing the plant.

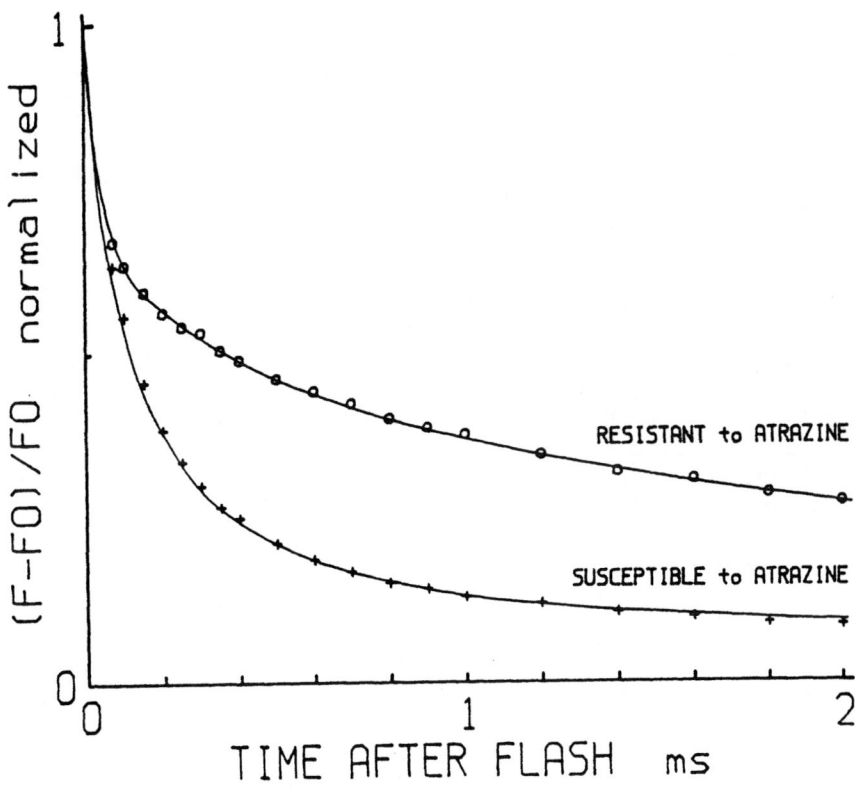

Figure 6. Decay of variable fluorescence with time after a series of saturating flashes from a leaf of <u>Chenopodium album</u> resistant to at-razine and a leaf susceptible to atrazine. Six saturating flashes were given prior to the start of the experiment to scramble the two electron gate of the quinone acceptor complex of PSII. After that one saturating flash was given, followed by a delay, and then the measurement was taken. This sequence was repeated at the same spot on the leaf (20 measuring flashes at 4 Hz). Details of the apparatus are in the appendix.

An attempt was made to discover weed biotypes which were resist-ant to atrazine but did not have an altered Q_B^- stability (Robinson, Kyle and Arntzen, in preparation). Seeds of seven independent atra-zine-resistant weed biotypes were collected and grown in the green-house. The plants were then screened with the flash fluorescence device, and the variable fluorescence decay recorded. Although there were some variations in amplitude of the signals, all of the atrazine-resistant plants (as determined by subsequent inhibitor studies on the isolated thylakoids of the plants) showed the same characteristic altered decay of variable fluorescence, compared to the non-resistant biotypes. This suggests that all the atrazine-resistant plants have the same genetic alteration. Hirschberg et al. (4) have shown that in the atrazine-resistant biotypes of both Amaranthus hybridus and Solanum nigrum, there has been a single base substitution in the chloroplast genome which leads to the substitution of a glycine for a serine in the primary amino acid structure of the Q_B-protein (6). The flash fluorescence studies suggest that all the biotypes examined have an equivalent substitution, which causes the plants to be resistant to atrazine.

9. OTHER POSSIBLE APPLICATIONS OF THE FLASH FLUORESCENCE DEVICE

The relative novelty of the flash fluorescence device for leaves has meant that other applications for this device have not yet been explored in any detail. There are several areas of study where this machine is likely to be of use: 1) water stress; 2) chilling stress; 3) leaf damage from environmental factors such as pollutants, agri-chemicals, disease and insects; 4) photosynthetic processes where isolation of chloroplasts is not possible; 5) measurement of photo-inhibition in situ; 6) photosynthesis in canopies.

In stressed plants it is extremely difficult to pinpoint the precise source of limited plant growth and productivity, a point discussed at length by Ort and Boyer (8). They point out that neither measurements of gross photosynthetic properties (such as measuring leaf gas exchange), or studies in vitro with isolated systems (such as PSII electron transport), are likely alone to reveal the causes of the stress-induced decrease in plant productivity. The application of the leaf flash fluorescence machine to these problems should be of great use, because a single well-defined step in a complex process can be monitored in isolated membranes, where the system can be easily manipulated biochemically, and also in leaves in the field under the in vivo stressed condition.

It should also be possible to study the extent of damage and recovery of leaves with the flash fluorescence apparatus. These studies could include damage from pests or diseases or photoinhibition, and also recovery from herbicide application. Acid rain may also damage leaves, possibly at the level of PSII. A rather different application is in the study of plants where chloroplasts are rather difficult to obtain, either due to the mechanical strength of the tissue or due to the release of detrimental compounds from the tissue during the isolation (e.g. pine needles).

10. CONCLUSION

The operational parameters of a particular set of steps in the process of photosynthesis, the two-electron gate of PSII have been examined. The operation of these steps can be measured by using flashlamps to follow the kinetics of changes in the field of variable fluorescence in the sub-millisecond range. This technique can be applied to isolated membrane preparations where the individual steps can be examined in detail. The technique has also been applied to allow non-invasive measurements of PSII functioning in leaves of whole plants in the field. Although the application to whole leaves does not allow ready measurements of the separate steps of the reactions of the quinone acceptor complex of PSII, complementary measurements in isolated preparations provide a powerful tool for interpreting the results obtained from the whole leaf measurements.

APPENDIX: NON-INVASIVE FLASH APPARATUS FOR LEAVES

Constructing an apparatus which is useful for field measurement of plants requires that the machine be battery powered, relatively small and maneuverable, and hardened to mechanical shock, dust and water. Additionally, the machine must take into account the particular geometry of the tissue being measured, and also the requirements of the specific measurement being made. The machine described here precisely times the firing of two xenon flashlamps aimed at the surface of a leaf to measure the decay kinetics of variable fluorescence from PSII. The measurement can be performed in ambient field light if the leaf is shaded by a black cloth. The kinetic profile obtained from this measurement can be interpreted in terms of the model developed for the functioning of PSII quinone acceptor complex by studies with a similar apparatus designed for use with isolated thylakoid membranes.

The decay of variable fluorescence indicates the oxidation of Q_A^-. The time range of this oxidation can be measured from 40 μs to several seconds after a flash. Thus, formation of Q_A^- in all the centers that might later be sampled during the measurement of F_v should occur within a window of $\pm$ 1 μs. The interval over which F_v is measured should also be similarly short, while at the same time disturbing a minimum number of centers (less than 2%). Two identical xenon flash lamps have been used for this purpose (lamp FX-201 with trigger pack FY-602, EG&G Salem, MA, recently superseded by 12B3 and FY-903). Figure 7 shows the placement of the flashlamps and plexiglass lenses such that a spot of nearly uniform density blue light (glass filter CS4-96, Corning) will be formed at the surface of the masking hole at the front end of the device. The beam-splitter (microscope slide) is placed to give an identical illumination geometry for both flashlamps, while greatly attenuating the measuring flashlamp. Fine-tuning of the illumination intensity of both lamps can be accomplished by adjusting the voltage of the discharge capacitor (using a voltage controlled DC/DC converter, PS-350, EG&G). The fluorescence is detected by a silicon photodiode (S1223-01, Hamamatsu) masked by a 690 nm interference filter (Corion, Holliston, MA) and a Wratten 70 gel filter (Eastman Kodak). The light from the leaf is routed to the detector by a small plexiglass rod positioned near the surface of the leaf. The photocurrent from the diode is integrated for 3 μs during the measuring flash, and fed to an 8-bit analog-to-digital converter (ADC0844, National Semiconductor).

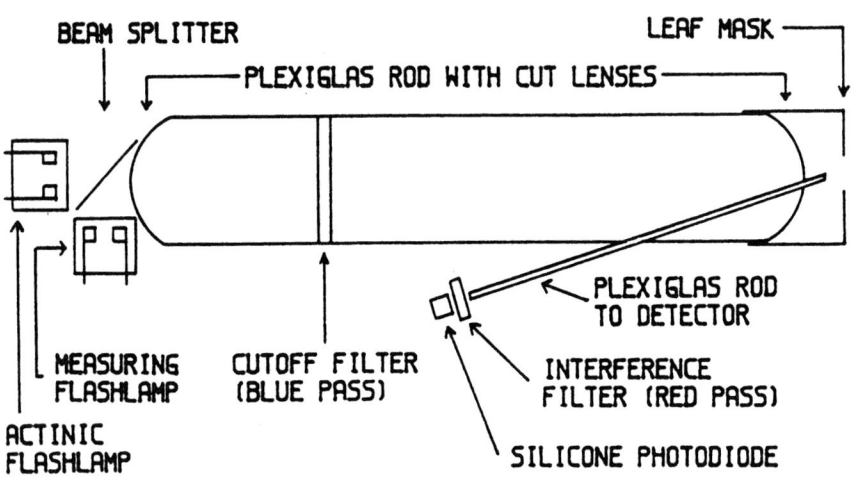

Figure 7. Detail of the arrangement of the optical components of the Non-Invasive Flash Apparatus for Leaves (NIFAL). The instrument is used for the measurement of the decay of fluorescence after a saturating flash given to a leaf.

The level of variable fluorescence is calculated and displayed on a graphic liquid crystal display screen (LM-213, Hitachi). The computer and memory (on STD bus cards, MSI-C800 and MSI-C764, Microcomputer Systems, Baton Rouge, LA) use complementary-metaloxide-semiconductor chips (the microcomputer chip is a NSC-800, National Semiconductor). The memory has 64K bytes and has a small battery which allows the information to be maintained when the power to the computer is turned off. Power for the computer components, display screen, external communications port (RS232-C), and analog circuits is provided by one 4-cell gel-type lead-acid battery (EP 626-26, 2.6 ampere-hours, Technacell Elpower, Santa Ana, CA) and suitble voltage regulators. Power for the flashlamps (1500 VDC and 200 VDC) is provided by one 8-cell geltype lead-acid battery (EP 1218-26, 1.8 ampere-hour, Technacell). These batteries are quite rugged and can be recharged overnight or swapped with a fresh battery for extended field use in the absence of AC power.

This unit contains no moving parts and is controlled from a small membrane key-switch key pad and a menu-driven control program. Programming was accomplished by cross-compiling (MVP-FORTH Cross Compiler, Mountain View Press, Mountain View, CA) source code written in the programming language Forth. The compiled code is loaded into the unit via the serial communications link or programmed into a programmable-read-only-memory chip.

The entire unit weighs less than 9 kgrams and can be carried with a strap for field use. The memory has room to store about 800 experiments, and programs and all data are maintained by the battery back-up for the memory circuits while the computer is turned off. The data can be either plotted, to a graphics device, or printed through the serial communications port, or fed to another computer, either directly or via a modem over phone lines. Thus, data can be collected over an extended period of time and later fed to another computer for further analysis.

The device needs only a planar photosynthetic surface of 0.6 cm^2. Most leaves (even succulants) will suffice, and pine needles (with 6-8 needles bunched) have also been used as experimental material. Other photosynthetic material such as isolated chloroplasts, or (with suitable filters) suspensions of photosynthetic bacteria or algae can be measured if they are placed in front of the detecting area in a thin flat cell (EPR flat-cell).

This device can easily be taken into the field to measure the functioning of the acceptors of PSII through the kinetics of decay of variable fluorescence after a saturating flash. The device is capable of performing these measurements on a broad variety of photosynthetic organisms under a variety of physiological and environmental conditions.

ACKNOWLEDGEMENTS

The technical assistance by Ms. Mary Hadden in the preparation of biological materials is gratefully acknowledged. This work was supported by a grant to Dr. Antony Crofts from the USDOE (DOE DEACO2 80ER10701). The invaluable advice and encouragmeent of Dr. Crofts is acknowledged. The experiments on Chenopodium leaves in Figure 6 were performed in the laboratory of Dr. C.J. Arntzen with Dr. D.J. Kyle.

REFERENCES

1. Arntzen, C.J., Pfister, K. and Steinback, K.E. The mechanism of chloroplast triazine resistance: alteration in the site of herbicide action. In: Herbicide Resistance in Plants (H.M. LeBaron and J. Gressel, eds.), pp. 185-224, New York: Wiley, 1982.

2. Bouges-Bocquet, B. Electron transfer between the two photosystems in spinach chloroplasts. Biochim. Biophys. Acta 314: 250-256, 1973.

3. Butler, W.L. and Kitajima, M. Fluorescence quenching in photosystem II of chloroplasts. Biochim. Biophys. Acta 376: 116-125, 1975.

4. Hirshberg, J., Bleeker, A., Kyle, D.J., McIntosh, L. & Arntzen C.J. Molecular basis of triazine herbicide resistance in higher plant chloroplasts. Z. Naturforsch. 39c:412-420, 1984.

5. Joliot, A. and Joliot, P. Etude cinetique de la reaction photochimique liberant l'oxygene au cours de la photosynthese. C.R. Acad. Sc. Paris 143:4622-4625, 1964.

6. Kyle, D.J. The 32,000 dalton Q_B protein of photosystem II. Photochem. Photobiol., in press, 1984.

7. Ort, D.R., Ahrens, W.H., Martin, B. and Stoller, E.W. Comparison of photosynthetic performance in triazine-resistant and susceptible biotypes of Amaranthus hybridus. Plant Physiol. 72:925-930, 1983.

8. Ort, D.R. and Boyer, J.S. Plant productivity, photosynthesis and environmental stress. In: Changes in Gene Expression in Response to Environmental Stress (Atkinson and Walden, eds.) in press, New York: Academic Press, 1984.

9. Pfister, K. and Arntzen, C.J. The mode of action of photosystem II-specific inhibitors in herbicide resistant weed biotypes. Z Naturforsch. 34c:996-1009, 1979.

10. Robinson, H.H. and Crofts, A.R. Kinetics of the oxidation-reduction reactions of the photosystem II quinone acceptor complex and the pathway for deactivation. FEBS Lett. 153:221-226, 1983.

11. Robinson, H.H. and Crofts, A.R. Kinetics of proton uptake and the oxidation-reduction reactions of the quinone acceptor complex of PSII from pea chloroplasts. In: Advances in Photosynthesis Research, Vol. 1 (C. Sybesma, ed.), pp. 447-480, The Hague: Nijhoff/Junk Publ., 1984.

12. Van Gorkom, H.J., Pulles, M.P.J. and Etienne, A.L. Fluorescence and absorption changes in tris washed chloroplasts. In: Photosynthetic Oxygen Evolution (H. Metzner, ed.) pp. 135-145, London: Academic Press, 1980.

13. Velthuys, B.R. and Amesz, J. Charge accumulation at the reducing side of system 2 of photosynthesis. Biochim. Biophys. Acta 333:85-94, 1974.

14. Velthuys, B.R. Electron-dependent competition between plastoquinone and inhibitors for binding to photosystem II. FEBS Lett. 126:227-281, 1981.

15. Velthuys, B.R. The function of plastoquinone in electron transport. In: Function of Quinones in Energy Conserving Systems (B.L. Trumpower, ed.) pp. 401-408, New York: Academic Press, 1983.

FLUORESCENCE AND ABSORBANCE MEASUREMENTS IN LEAVES: SENSORS OF PHOTOSYNTHETIC PERFORMANCE

Willem J. Vredenberg

Laboratory of Plant Physiological Research
Gen. Foulkesweg 72, 6703 BW WAGENINGEN, the Netherlands

1. INTRODUCTION

Application of absorption and fluorescence difference spectro-photometry has since long been proven to be an extremely useful measuring technique to study energy transfer and energy conversion in photosynthetic systems varying from isolated membrane fragments (chromatophores and stiripped chloroplasts) to intact leaves.

This contribution aims to illustrate the principals and features of these techniques. Special emphasis will be laid on applications to intact leaves or leaf discs with relatively simple instrumentation.

Growth and maintenance of plants depends on photosynthetic energy conversion which takes place in the chloroplasts of the plant cells. Light, absorbed by the photosynthetic pigments (chlorophylls a and b, carotenoids) of two different pigment systems PS I and PS II, promotes a charge separation in the reaction centers of each PS and electron and proton transfer across the inner-chloroplast (thylakoid) membrane. An oriented membrane-localized electron transport chain with electron and hydrogen carriers serves this purpose (Fig. 1). Thus, in the light the membrane becomes energized, resulting in electrochemical potential gradients amongst other of H^+ ($\Delta\mu_H+$) and production of reducing power, in the form of NADPH. A membrane-bound ATP-ase (CF_0/CF_1) catalyzes the production of ATP at the expense of the generated $\Delta\mu_H+$ • ($\Delta pH + \Delta\psi$). ATP and NADPH are required and consumed in the CO_2-reduction cycle (Calvin cycle) serving the incorporation of CO_2 into carbohydrates. This process takes place in the chloroplast stroma. Rates of leaf photosynthesis can be measured as rates of oxygen evolution or of CO_2 uptake.

108

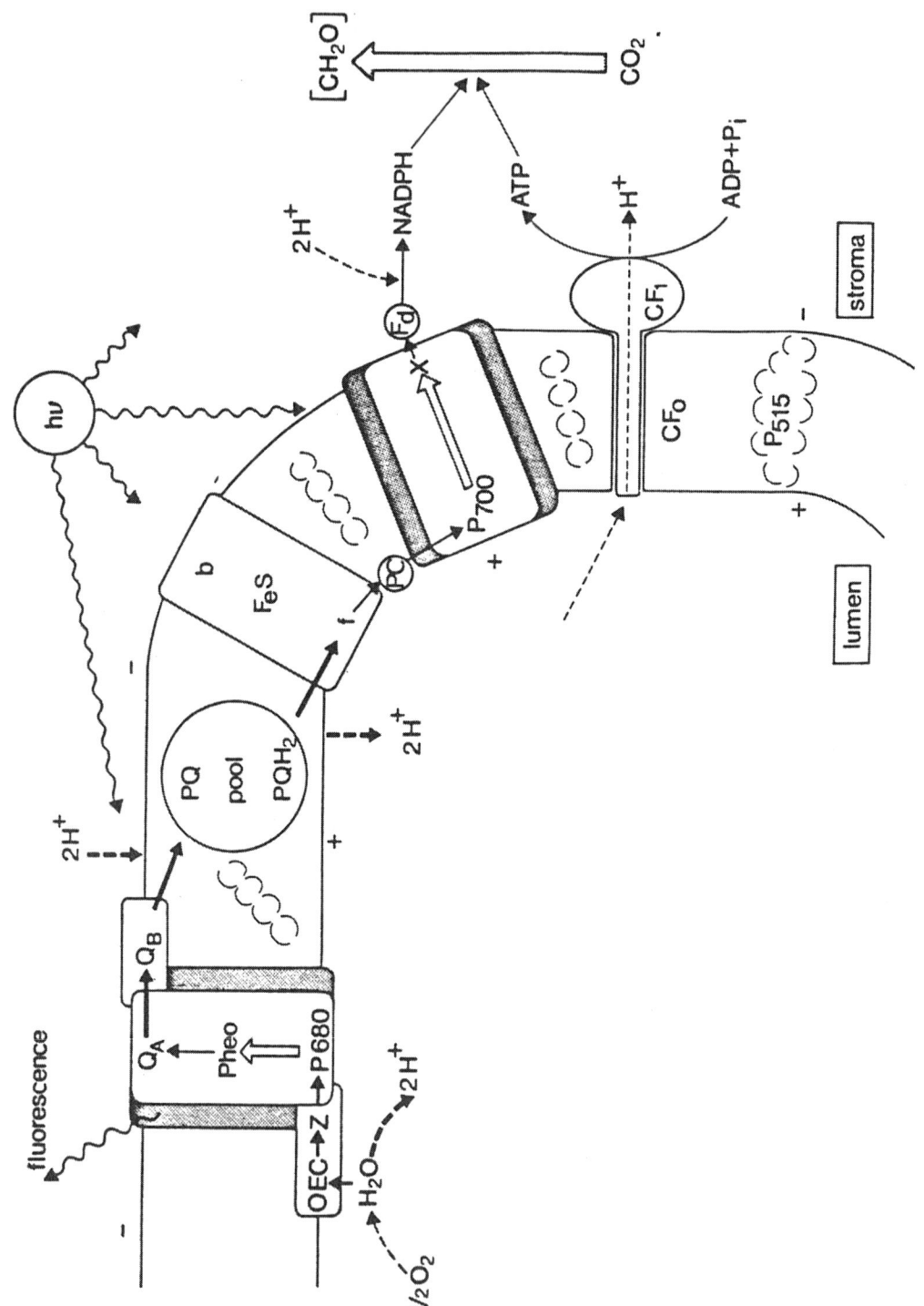

Figure 1.

Schematic representation of the thylakoid membrane with an organization pattern of the components of the electronproton carrying transport chain serving the oxidation of H_2O, reduction of NADPH, and the generation of $\Delta\mu_H^+$ (Δ pH plus $\Delta\psi$, positive inside). Two photo-reactions are initiated by light ($h\nu$) absorbed in PS I and PS II. Each PS consists of an antenna complex surrounded by a light harvesting pigment complex LHC (indicated by shaded areas). P680 and P700 are the reaction chlorophylls a of PS II and PS I, respectively.

PS II: OEC is the oxygen evolution complex.
Z: primary electron donor (probably a plastoquinol).
Pheo: pheophytin, primary electron acceptor of P680 after charge separation (open arrow).
Q_A: special quinone, secondary electron acceptor.
Q_B: plastoquinone, bound to a 32 kD protein and 2-electron acceptor of Q_A.
PQ: plastoquinone.
PS I: FeS/b/f (Rieske Fe-cyt$b$/f protein complex).
PC: plastocyanin.
X: primary electron acceptor of PS I reaction center.
Fd: ferredoxin.
CF_o/CF_1: ATP synthetase activated by $\Delta\mu_H^+$ (synthesis) and supporting $\Delta\mu_H^+$ under proper condition
          (ATP hydrolysis).

[CH_2O] stands for carbohydrate generated by the photosynthetic flux]CO_2 (shaded open arrow).
P515: electric field-sensitive carotenoid-chlorophyll b pigment moiety.

As indicated at room temperature fluorescence comes from PS II antennas and LHC.

For maximum rate of photosynthesis to occur all reactions underlying this process have to occur at optimal rate. Any hindrance somewhere in the sequence of energy transferring, -transducing and -accumulating reactions will lead to a change in energy accumulation in the other parts, and ultimately become expressed as a change in the energized state of the thylakoid membrane. Variations in this state can be monitored in intact leaves by measuring the red fluorescence of chlorphyll a and/or the absorbance change of a chlorophyll b-carotenoid pigment moiety (P515) with a specific absorbance bandshift at 515 nm.

An apparatus with which measurements of changes in fluorescence of chlorophyll a and of changes in absorbance of P515 in intact leaf discs (diameter 5 cm) is schematically depicted in Figure 2. For the description of an apparatus in which the rate of O_2 evolution and fluorescence can be measured in leaf discs, the reader is referred to references 5 and 35.

Chlorophyll a fluorescence; induction patterns

The basis of using the fluorescence yield of chlorophyll a as a monitoring probe of variations in the rate of photosynthesis rests upon the following facts and principles (for reviews, refs. 17,24):

(i) Light energy absorbed by the photosynthetic competent pigments is distributed between the two pigment systems and transferred to the long wavelength absorbing antennas of these pigment systems. It is finally captured by the photochemically active chlorophyll a molecules P700 and P680 of the respective reaction centers of photosystem I and II (PS I and PS II). Part of this energy is re-emitted as fluorescence.

(ii) Room temperature chlorophyll fluorescence mainly originates from the chl a/chl b light harvesting complex LHC and the antennae chlorophylls of PS II (chl a_2). The fluorescence yield of chl a_1 at room temperature is comparatively low.

(iii) The fluorescence yield of chl a_2 rises (i.e. becomes de-quenched) when the primary acceptor (Q_A) of PS 2, collecting the electron involved in the charge separation in the reaction center of PS 2, becomes reduced (8). This occurs in the so-called initial induction phase OP, where the fluorescence rises from a low level F_0 to a maximal level at P.

(iv) The fluorescence yield of chlorophyll a becomes lower, as a consequence of increased quenching, when $\Delta\mu_H^+$ across the thylakoid membrane increases, irrespective of the redox state of Q_A, or when Q_A becomes re-oxidized (13). This mainly occurs in a second induction phase PSMT, where the fluorescence decreases from P to a lower level S and often rises again toward a level M in between P and S and again decreases towards a lower T level.

111

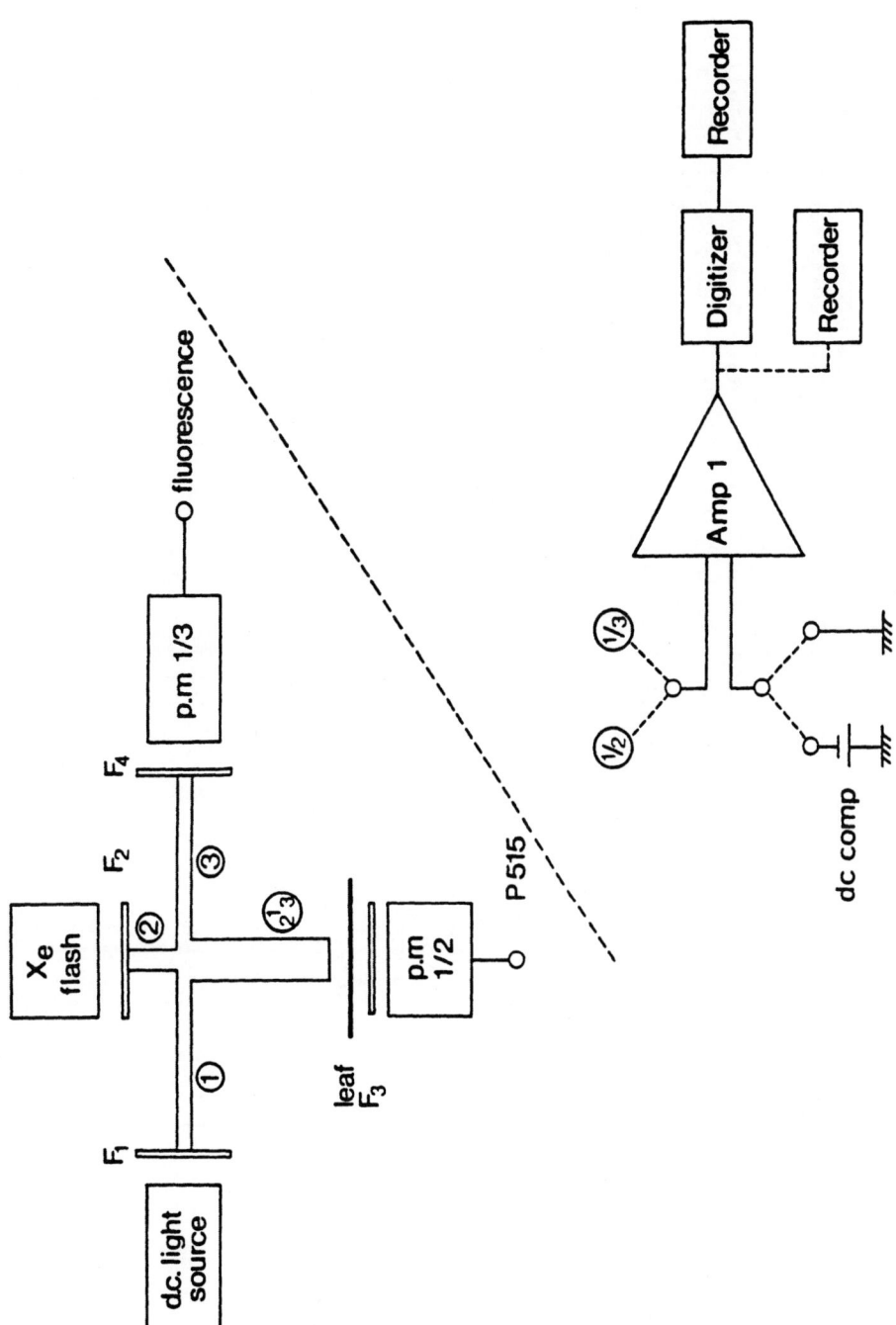

Figure 2. Schematic drawing of experimental set up for measuring fluorescence emission and induction, and absorbance change of P515 in intact leaves. A multi-branched optical fibre system (3 branches ar used here) is placed in front of the leaf in an air-conditioned cuvette. Absorbance measuring light and fluorescence excitation light comes from dc light source (24V, 300W) and is passed through branch 1. Actinic light from a Xe flash lamp (GEC/FT 230) is passed through branch 2, and fluorescence emission from the leaf is transmitted through branch 3. p.m. are multipliers (EMI 9558) detecting transmitted light of 518 nm (p.m. 1/2) or fluorescence (p.m. 1/3). Fluorescence and P515 measurements are done separately. Anode signals from p.m. are amplified (in P515 measurements after d.c. compensation, needed for amplifying the small absorbance difference signal with respect to the "dark" signal). The output of the amplifier is fed into a digitizer (Tektronix 5 D 10 waveforum digitizer) which i read-out by a recorder. The output of the amplifier can also be fed directly to the recorder if no high time resolution is needed. P515 measurements:

F1:   Interference filter 518 nm (Balzer);
F2:   Schott filter R6 630, 4 mm.
F3:   Colorfilter BG39 (Schott 4mm) plus CS-4-96 (Wratten, 2mm).
Fluorescence F1: see F3 above;
F4:   interference filter 685nm (Schott) plus RG 665 2mm (Schott).

The fluorescence set up in principle is similar to a commercial Plant Productivity Fluorometer (Brancher Model SF-10) which, as recently published (Norrish, et al. Photosynthesis Res. 4:213-227, 1983), can be linked to a Multitech Industrial Corporation Microprocessor Microcomputer for accurate measurements of rapid and slow fluorescence induction simultaneously.

As a consequence of these principal facts, the fluorescence yield of chlorophyll changes in response to any process which results in a change in the redox state of Q_A and/or a change in energization. The principles of fluorescence will be discussed briefly in relation to the rate of photochemistry. Relevant aspects of induction phenomena during subsequent phases associated with electron transfer and membrane energization will also receive attention.

Fluorescence, principles

Fluorescence intensity is at any time proportional to the concentration of excited molecules of the fluorescing species. This concentration is the expression of a quasi-steady state in which light absorption (excitation) is at balance with a number of energy dissipation (de-excitation) processes, namely fluorescence(F), energy transfer(T), non-radiative de-excitation(D), utilization in photochemical conversion(P). Under quasi-steady state conditions the yield ϕ_j of a de-excitation process (j = F,T,D,P) is defined as the ratio of its rate constant k_j to the sum of all de-excitation rate constants, i.e.

$$\phi_F = k_F / (k_F + k_P + k_D + k_T) \tag{1}$$

The fluorescence intensity F is at all instances F= ϕ_F*I, where I is the absorbed light intensity. At constant I changes in F are due to changes in ϕ_F.

In a functional photosynthetic unit excitation energy is rapidly spread by intra-molecular transfer among the antennas and finally the mobile "exciton" is efficiently trapped in a specialized chlorophyll molecule P of the reaction center DPA, where it is used for photochemistry via the reaction DPA+hυ $\rightarrow$ DP*A $\rightarrow$ DP$^+$A$^-$ $\rightarrow$ D$^+$PA$^-$. D and A are electron-donor and acceptor molecules of the reaction center. Trapping will be impossible (i.e. Kp=0) when the reaction center is "closed," i.e. unable to perform photochemistry. This occurs, for instance, when it is in either one of the two latter states. Thus fluorescence is high(er) in a unit with a closed trap (closed unit).

The absorption (exciton) cross-section of a trap will become increased when exciton transfer from a closed unit to neighboring open unit occurs. This has consequences for the kinetics of the change in fluorescence emission (yield) in relation to the change in rate of photochemistry (i.e. photosynthesis) during the initial induction period OP. Energy transfer between units of different photochemical systems, ie. PS II and PS I, may occur (spill-over), and will affect the fluorescence yield, which at 20°C mainly originates from PS II (17,24). Change in spill-over will result in a change in ϕ_F. Finally, change in energy transfer between units may occur due to lateral rearrangements of the units within the membrane matrix. Change in spill-over and in rearrangements can collectively be represented by changes in K_T and according to Eq. 1 show up as changes in F (ϕ_F). In addition, the formation of ΔpH across the thylakoid membrane in the light associated with an increase in H$^+$-concentration in inner thylakoid space (lumen) also results in a lowering of the fluorescence. This has been suggested to be due to an increase k_D (14). Changes in k_T and k_D become apparent in the PSMT induction phase. Changes in spill-over result in a higher fluorescence emission of antennas of PS I, which is low at 20°C, but is observed at 77°K as a band around 730 nm. The fluorescence emission of PS II at 77°K shows separate bands with maxima at 695 and 685 nm. This is attributed to antennas of PS II and the light harvesting complex LHC, respectively.

Fluorescence induction in relation to rate of photochemistry

With respect to the exciton concentration (E) in a photosynthetic unit, the following relation holds:

$$dE/dt = -[\Sigma k_i] E + I \tag{2}$$

Under quasi-steady state conditions $dE/dt = 0$. With definitions for the rate of photochemistry $V_P(= k_p[E])$, fluorescence $F(= k_F[E])$ and $\beta = (k_F + k_D + k_T)/k_F$ one obtains

$$\beta F + V_P = I \tag{3}$$

This important relation expresses that under constant excitation I photochemical rate V_P and fluorescence F are complimentarily related, and even linear when β is constant.

As the chlorophyll fluorescence in plant and algae cells at room temperature mainly originates from PS II, we will briefly discuss the fluorescence-induction during the OP phase observed at the onset of illumination, in relation to the rate of photochemistry in PS II. This rate depends on the redox state of the primary quinone acceptor Q_A of the reaction center complex DPQ_A. The rate is maximal when the acceptor is oxidized (Q_A: open center) and zero when the acceptor reduced (Q_A^-: closed center). Thus for a collective set of PS II units, the rate of the photochemical reaction in PS II is dependent on the fraction of "open" centers $q = Q_A/Q_0$ with $Q_0 = Q_A + Q_A^-$, and can be formalized by $k_p = f(q)$. For a unit with an open trap $V_P = V_{max}$ and $F = F_0$ with

$$\beta F_0 + V_{max} = I \tag{4}$$

For a unit with a closed trap $V_P = 0$ and $F = F_m$ ($=I*1/\beta$) with

$$\beta F_m = I \tag{5}$$

Combination of Eqs. 3, 4 and 5 results in a simple complimentary relation between V_P and the variable fluorescence $V_F = (F-F_0)/(F_m-F_0)$ with $o < V_F < 1$

$$V_P = V_{max} (1 - V_F) \tag{6}$$

When fluorescence F is taken with reference to the dark-adapted
(all units open) low fluorescence state with $F=F_0$, the relative
fluorescence yield $\phi = F/F_0$ is an often used parameter. The variable (relative) fluorescence yield is then designed by $\Delta\phi = \phi-1$
where $\Delta\phi_m$ ($=(F_m-F_0)/F_0$) is the maximal (relative) fluorescence
yield. If expressed in terms of fluorescence yield, the complimentary relation becomes

$$V_P = V_{max} (1 - \Delta\phi/\Delta\phi_m) \qquad (6a)$$

Thus, the fluorescence induction curve measured at the onset of
illumination gives us information on the rate of the photochemical
reaction of PS II. In a healthy leaf $\Delta\phi_m$ is 2-4.
    Another important source of information on the fraction of
closed traps is contained in the area above the induction curve
bordered by F_0 ($\phi=1$) and F_m ($\phi=\phi_m$). This may be exemplified by
the following.
    The rate of photochemistry V_P is equal to the rate of conver-
sion of units with open traps Q.

$$V_P = dQ/dt \qquad (7)$$

Thus the number of traps converted by an energy dose I*t after a
time t, i.e., the concentration of produced closed traps can be
derived by integrating Eq. 7.

$$Q_0 - Q = - \int_0^t (dQ/dt)dt = V_{max} \int_0^t (1\ \Delta\phi/\Delta\phi_m)dt = V_{max}S(t) \qquad (8)$$

$$Q_0 = V_{max} \int_0^\infty (1 - \Delta\phi/\Delta\phi_m)\ dt = V_{max} * S(\infty)$$

in which S(t) and S(∞) are the areas above the fluorescence induction curve before and after the end, respectively, of its completion. Thus at any time the ratio $S = S(t)/S(\infty)$ ($= 1-Q/Q_0 = 1-q$)
is equal to the fraction of closed traps, or in other words, equals
the number of redox-capable equivalents dumped in a pool which is
intermediary in the photosynthetic transport chain.
    It should be stressed that the equality between S and 1-q
holds, irrespective of all, none, or partial grouping of units
within the photochemical system. Differences in assemblies of units
in which interunit exciton transfer does (grouped units) or does not
(separate package) occur become apparent in the kinetics of the
fluorescence induction. The latter further influences the kinetics
of the ingrowth of the area above the induction curve. It can
easily be shown (for a thorough overview see refs. 22,30) that in a
separate package model (isolated, non-communicating units) the ingrowth of the area and the rise in fluorescence yield ϕ/ϕ_m occurs
exponentially. In this case, the photochemical rate V_P is linearly

dependent on q and $V_F = \Delta\phi/\Delta\phi_m = 1-q$. Then, according to Eqs. 6 and 7:

$$q = 1 - S(t)/S(\infty) = e^{-\omega t} \tag{9}$$

with $\omega = V_{max}/Q_0$

When dealing with exciton-communicating units, the fluorescence induction curve is sygmoidal. This is a direct consequence of the fact that an exciton can migrate from a closed unit to any other unit. If this occurs with a probability p the rate of photochemical reaction was predicted (10) to be

$$V_P = V_{max} (1 - V_F) = V_{max} Q/[1-p(1-q)] \tag{10}$$

By plotting V_F versus S (= 1-q) this can be tested (see ref. 30). Thus by analysis of the ingrowth curve of the area above the fluorescence induction curve, one is able to obtain information concerning:

1) the relative number of traps collecting redox-capable equivalents;
2) the abundance, relative size and exciton-transfer behavior of units operative in the photochemical system.

Pool size and area

First, it should be stressed that according to Eqs. 1 and 8 the following relation holds

$$Q_0 = S.V_{max} = S.k_p.E = S.k_p.(I*t) \tag{11}$$

This means that when dealing with a constant number of traps and when kp is constant, S should be inversely related to I*t. In an uninhibited electron transport chain, this is only fulfilled if I is sufficiently high such that during the time t at which the pool $Q_0 = [Q_A]+[Q_B]+[PQ]$ is filled, there is negligible transmission of equivalents (electrons) toward the next redox carrier in the chain. As the transfer time of electrons going from PQH_2 to the Q-b/f complex is about 20 ms one will approach the ideal situation at intensities at which the completion of the induction takes less than 20 ms. No problem arises in the presence of the plastoquinone antagonist DBMIB which inhibits the reoxidation of PQH_2. It will be clear now that in the presence of an inhibitor like DCMU which blocks the electron transfer between Q_A and Q_B, only the Q_A pool is filled. As such, the areas S in the absence and presence of DCMU should be in a ratio equal to $[Q_A]+[Q_B]+[PQ]$ and $[Q_A]$. In chloroplasts, this ratio has been found to be approximately equal to 8 (20). An illustrative experiment is illustrated in Figure 3. In chloroplasts isolated from spinach leaves HCO_3^- depletion results in an inhibition of electron transport between Q_B and PQ, without affecting the number of active PS II centers Q_A.

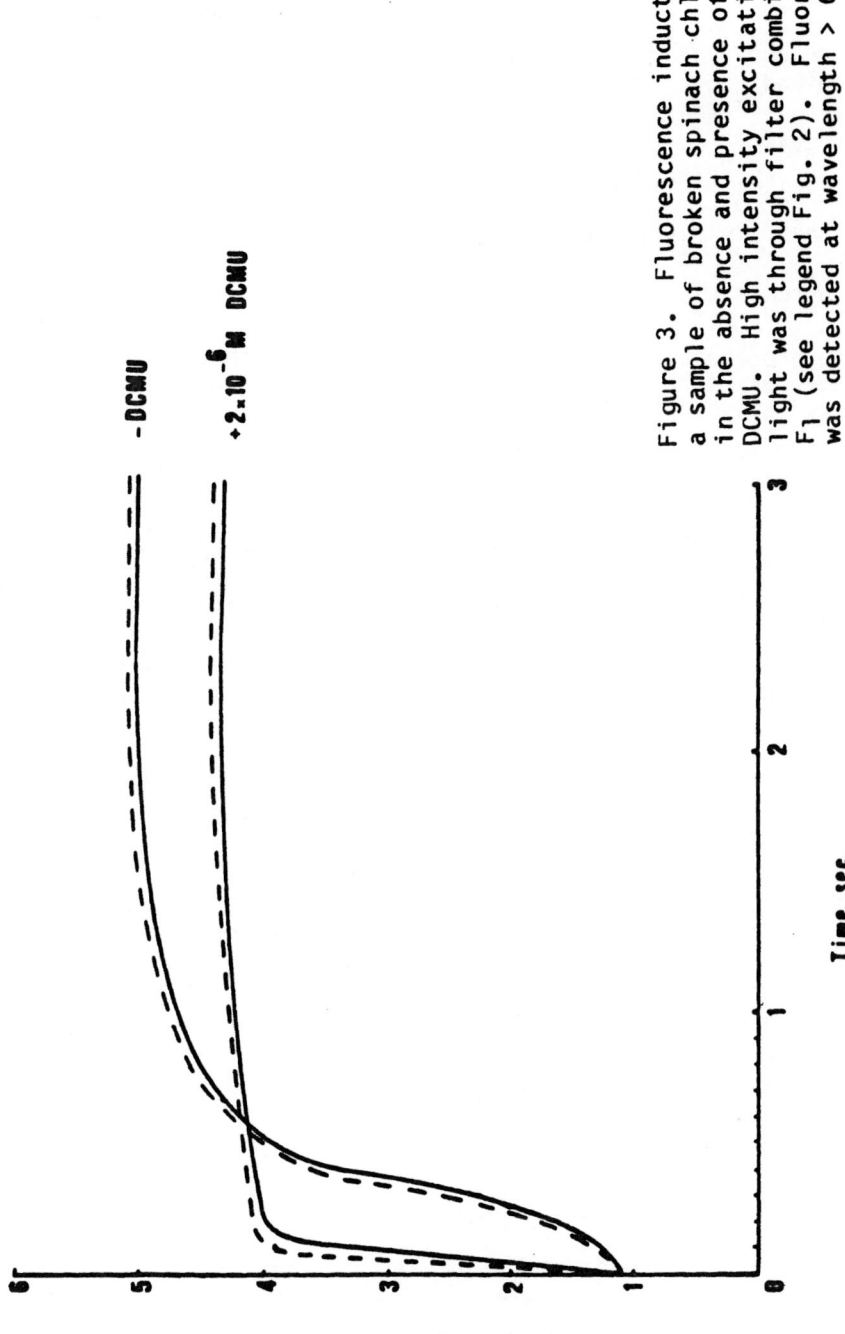

Figure 3. Fluorescence induction in a sample of broken spinach chloroplast in the absence and presence of 2 μM DCMU. High intensity excitation light was through filter combination F₁ (see legend Fig. 2). Fluorescence was detected at wavelength > 670 nm.

Thus, in chloroplasts in the absence HCO_3^- the pool size of $[Q_A]+[Q_B]$ is measured. As Q_B is a 2 electron acceptor and all Q_A and Q_B would be in the oxidized state before illumination, an area ratio of 3 would be found in the absence and presence of DCMU. A ratio of 2.2 has been reported (31).

The above examples illustrate that approximate information can be obtained about the pool size of redox catalysts involved in the hotosynthetic reaction chain between PS II and PS I. Any change therein caused by either internal physiological factors during growth or by external conditions (a.e. herbicide treatment, temperature and/or water stress) influence the relatively easily measurable area above the fluorescence induction curve.

Area ingrowth kinetics

Heterogeneity between relative size and distribution of different units within PS II influence the kinetics of the area ingrowth above the fluorescence induction curve, as briefly outlined before. Some potential features and prospects are worth mentioning, in view of the organization of the units within the membrane.

Separate units which are associated with the LHC, i.e. those present in the stacked membranes have a high absorption cross-section (exciton density) and consequently turn over rapidly. This becomes manifest in a high value of ω (see Eq. 9). Those with a lower cross-section (i.e. with less or no LHC) turn over with a low value of ω. It has been found that two populations of units exist by plotting log (1-S) (=-ωt) versus t. Those with high ω are called α and with low ω are called β centers (22).

Quantitative information on the abundance and conversion rate (at a given intensity) of α and β units, which are determinants of the rate at which engery passes through PS II, is obtained from this type of analysis. So far these studies have mainly been done with isolated chloroplasts, but can also be done with intact leaves (see ref. 30). This is also true for the detection and characterization of PS II units with interunit exciton transfer. When the fluorescence induction curve is corrected for (by subtraction) the contribution of α and β units by proper means, one is left with the curve representative of the so-called grouped units. For these it has been shown that, according to theoretical reasoning, the relation between relative variable fluorescence yield $\Delta\phi/\phi m$ and area above the fluorescence induction curve S = 1-q (i.e. normalized to $S(\infty) = 1$) has the following hyperbolic form

$$V = \Delta\phi/\Delta\phi m = S/[1 + k(1 - S)] \qquad (12)$$

in which k is related to the probability p (see Eq. 10, ref. 10) for inter-unit exciton transfer, with k = p/(1-p). As shown by Strasser (30) relative abundance of these groups and their interaction, i.e. values of k or p can be determined by estimating the relative size of S and making an Eddy Hofstee plot of S/V versus

S, respectively. For isolated chloroplasts it has been shown that after closer packing of units, induced by the addition of Mg-ions, which causes more stacking the units communicate more strongly. This is reflected by an increase in p (or k).

Thus, in principle, pertinent data on the rate of energy channeling through the photochemical systems, and the organization and behavior of flexible assemblies of units can be obtained in leaves by the non-evasive fluorescence method. Any condition (c.f. light quantity or quality, temperature, gas atmosphere) which inter-acts with the organization and assembly of the photosynthetic membrane apparatus during growth of a leaf becomes manifest in the kinetic pattern of the fluorescence induction curve.

Long term fluorescence induction (PSMT phase)

We will now turn our attention to the fluorescence induction observed in prolonged illumination. The general pattern as shown here (Fig. 4) for an intact spinach leaf shows, after the initial rise OP which is completed within 1 s, a slower decline of fluores-cence emission from the P to the State S. The S level frequently is followed by a slower S-M-T fluorescence decline, which can be considered as an increase in fluorescence quenching with no, or relatively small, change in F_0. This may have different causes, which dependent on individual species and/or conditions, can effec-tively become manifest especially in isolated chloroplasts. This decline has found different interpretations and for intact leaves it is still a matter of debate. We will briefly discuss the separate effects which lead to a decrease in fluorescence yield (i.e. in-crease in fluorescence quenching).

1.   Partial reoxidation of Q (initially reduced during the OP phase) by withdrawal of electrons by PSI.

2.   Redistribution of excitation energy in favor of PSI due to increased energy spill-over from PS II to PS I. This increase in spill-over causes a transition from a high fluorescence state (state 2) to a low fluorescence state (state 1). The concept of State 1 - State 2 changes was first introduced by Bonaventura and Myers (4). It is the reflection of a regulatory process which is governed by a subtle phosphorylation mechanism leading to the phos-phorylation of LHC when PS II is over-excited (1,3). By means of this process and its underlying mechanism, photosynthetic oxygen evolving organisms can regulate the distribution of absorbed light energy between PS I and PS II. This helps maintain a maximum efficiency for non-cyclic electron transport. The mechanism by which LHC phosphorylation increases the quantal distribution of PSI at the expense of PS II is still unknown. Current thinking supported by experimental evidence, points to a lateral movement of part of the phosphorylated LHC associated with PS II in the appressed mem-brane regions (abundant in this region) towards the stroma exposed regions. There it comes in closer proximity with PS I pigment complexes.

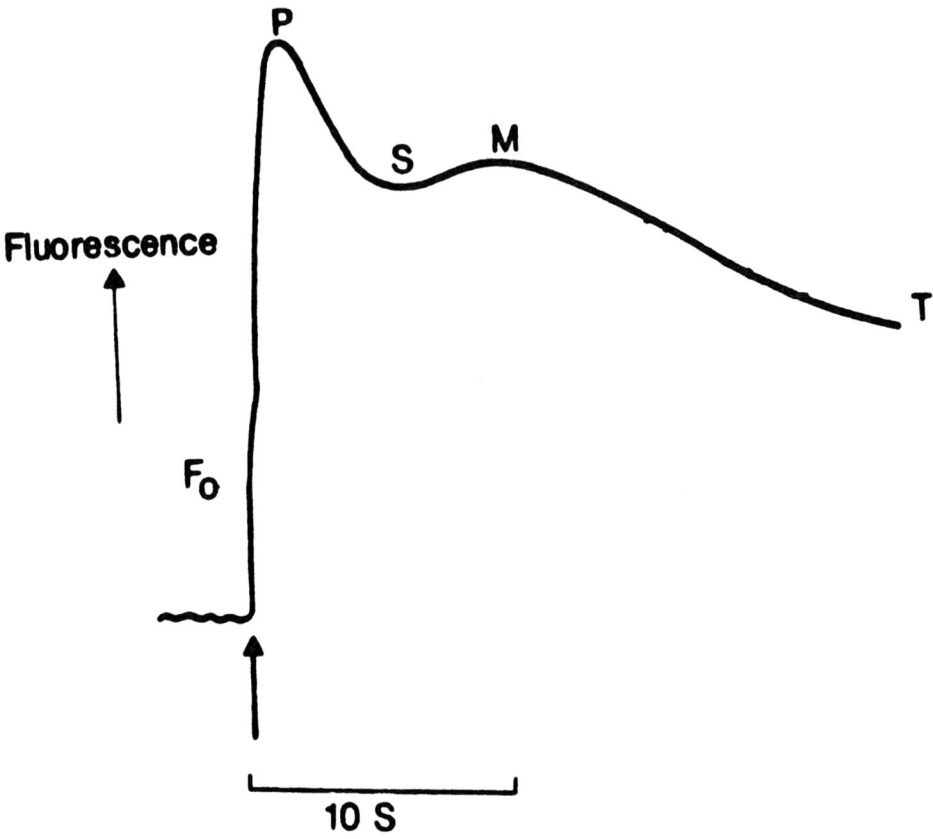

Figure 4. Fluorescence induction in a spinach leaf which has been dark adapted for 5 minutes and excited with blue light (filter combination F_1, as given in legend of Fig. 2) of high intensity. F_0, P, S, M, and T levels are indicated. Temperature 20°C.

The higher tendency of phosphorylated LHC to diffuse laterally is thought to be associated with the high negative surface charge density of the phosphorylated protein due to the addition of the charged P-groups on an exposed terminal of the protein (2,3).

   3.   Non-photochemical quenching due to the build-up of a pH gradient across the thylakoid membrane. The Δ pH-dependent or so-called energy quenching of fluorescence is assumed to be caused by increased thermal deactivation of excited states. This might be caused by structural membrane alterations due to H-Mg^{2+} exchange at the inner thylakoid surface (14).

   Ample evidence has been given (13) that in intact cells and leaves the PS-fluorescence decline for the major part is caused by Q- and energy (Δ pH)-quenching, which are not related to increased energy transfer to PS I (spill-over). The increase in spill-over, (state 1 - state 2 transition) as reflected by a lowering of the fluorescence at room temperature (increase in k_T), is masked by the larger effect of Q-oxidation and build-up of a pH gradient on the fluorescence yield (decrease in β due to increase in k_P and k_D, respectively). The control of quenching by the redox state of Q and by the Δ pH across the thylakoid membrane has been recently confirmed by experiments with intact leaves in which rapid changes in light intensity or CO_2-availability are induced (6,35). These changes lead to an altered photosynthetic rate measured by O_2 evolution and fluorescence emission. This rate decreases in an oscillating pattern with several oscillations during 5 to 10 minutes. There is a clear anti-parallel but phaseshifted relationship between Chl a fluorescence and photosynthetic CO_2 assimilation (O_2-evolution). The cause of these oscillations has been suggested to reside in imbalances introduced into the regulatory controls that govern CO_2 assimilation in the reductive pentose phosphate pathway (Calvin cycle). These lead to pulsating changes in the NADP level and consequently in the redox state Q. This, in turn, leads to changes in Q-quenching, followed by changes in ADP/ATP ratio and Δ pH quenching. For instance, an increase in CO_2 concentration causes an increase in NADH consumption (increase in O_2 evolution) and more oxidized Q (Q-quenching, i.e. decrease in fluorescence). Increase in CO_2 concentration also leads to an increase in PGA and, as a consequence, to an increase of ribulose-5-P which acts as an ATP sink. As a consequence, ADP concentration also increases with an associated decrease in Δ pH across the thylakoid membrane. This causes less Δ pH-quenching, i.e. an increase in fluorescence yield. Increase in ADP will cause a decrease in NADPH comsumption and O_2 evolution due to decrease in the PGA kinase reaction. Readjustment of the ATP/ADP ratio by cyclic or pseudocyclic electron transport will then reverse the entire sequence.

Evidence that the fluorescence decline during the P-S phase is mainly due to Q_A^--oxidation (via the PS I-driven oxidation of PQH_2) can be obtained from Figure 5. If illumination is interrupted between P andS one sees that at S the initial fluorescence upon re-illumination is much lower than at P. This indicates a rapid oxidation of Q_A^- during the P-S phase. Apparently the photochemical rate of PS I during the P-S phase is higher than that of PS II. A balance is likely to be reached during the S-M phase. We suggest that the fluorescence decline during the M-T phase is mainly due to the Δ pH-quenching.

PS II pool size and rate of photosynthesis

In a simple approach the maximal rate of CO_2 fixation (J_{CO2}) (i.e. the rate of photosynthesis) may be written as proportional to the concentration of an enzymic unit (U) involved in the rate-limiting step. These are related by a proportionality constant equal to the reciprocal of its turnover time (σ).

$$J_{CO2} = U/\sigma$$

122

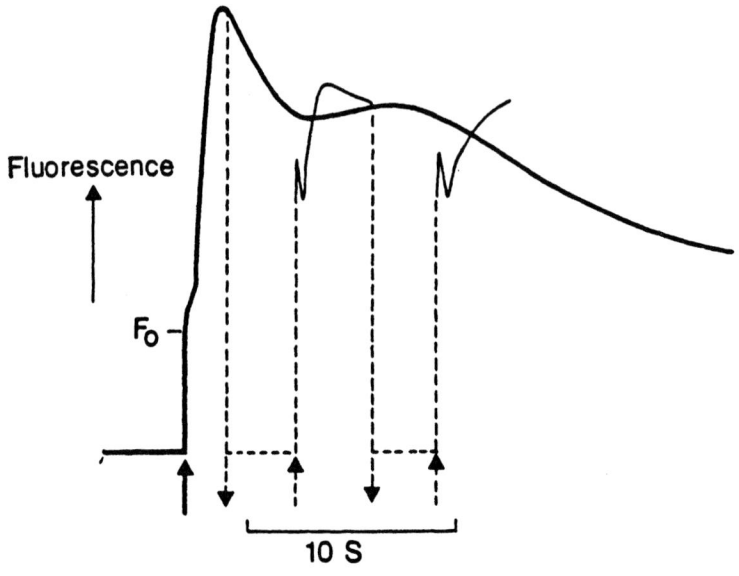

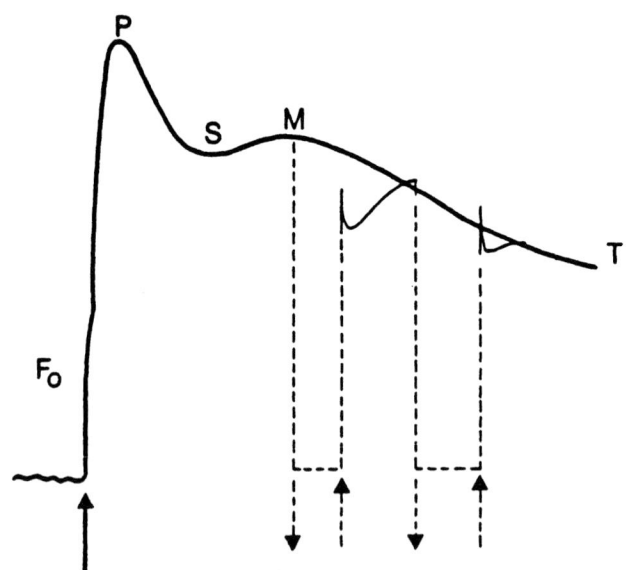

Figure 5. As Figure 4, but with interruptions of illumination indicated by broken lines and arrows at P (upper curve) and M (lower curve) levels.

For example, for leaves of Pertityle emoryi (Asteraceae), it was found (18) that $J = 1.5$ nmol CO_2 cm^{-2} s^{-1}, which on the basis of an estimated chlorophy concentration of 38 nmol cm^{-2}, corresponds to 0.039 nmol CO_2 nmol Chl^{-1} s^{-1}.

From the fluorescence induction area measured in the presence
of DCMU it was estimated, with proper corrections for geometrical
factors and attenuation of incident light and fluorescence emission
in the leaf, that $Q_o/Chl = 1.43.10^{-3}$ (i.e. a unit size of 700).
Thus, $\sigma = Q/J = 1.43.10^{-3}/39.10^{-3} = 26$ ms. This value corresponds
fairly well with the turnover time of approximately 20 ms of the
overall electron-transport chain at room temperature. For another
species with a smaller unit size of 215 (Chl/Q_o) JCO_2 was found to
be correspondingly higher $(JCO_2 = 8.6$ nmol CO_2 cm^{-2} $s^{-1})$ with $\sigma =$
33 ms. This illustrates that an estimation of PS II unit size by
means of the fluorescence induction method is a means to get information about the photosynthetic capacity of the intact leaf. It
should be emphasized that the proportionality between PS II unit
size and rate of photosynthesis ony holds when the rate of electron
transport through PS I under steady state light conditions is not
rate limiting, irrespective of its unit size (Chl/P700). This possibility has to be substantiated for each individual plant species.
    Photosynthetic unit sizes of PS II have been estimated for a
large variety of plant species using the fluorescence induction
method. Variation in sizes between 200 (Chenopodium) and 930 (Vida
flabella) have been reported (19). A pronounced difference has
been found between sun and shade plants. In different species of
sun plants, the PS II unit size varied between 220 and 480, in
shade plants unit sizes were between 630 and 940. In view of what
has been discussed before, sun plants apparently have adapted to
high light intensities and are able to perform a high photosynthetic
rate. However, there are reports which are at variance with the
aformentioned results. Melis and Harvey (21) have reported a low
PS II unit size of 160 to 220 in three diferent shade plants with
a relatively high Q/P700 ratio ranging from 2.4 to 3.9. The relatively high concentration of PS II units was shown to be accompanied
by a high density of grana, typical for shade plants. In these
plants, PS I reaction centers are turning over electrons faster than
PS II centers and the increased PS II/PS I ratio is suggested to
serve the offset of the energetic imbalance and to maintain an improved photosynthetic rate. Nevertheless, the conclusion may hold
that shade plants have less PS I complex and as a consequence have
a much lower rate of maximal photosynthesis.

P515 absorbance changes, kinetic patterns

    The competence of the chloroplast (thylakoid) membrane to store
electrochemical energy generated by light is of crucial importance
for the photosynthetic performance of the leaf. Generation of this
energy can be monitored by measuring the absorbance changes of the
electrical field-sensitive endogenous pigment moiety denoted as
P515, with a maximal absorbance change at 515 and 520 nm and a
characteristic absorbance difference spectrum. Short saturating
light flashes (half time approximately 10 μs) inducing a single
charge separation in all reaction centers cause changes in

transmembrane and inner membrane electrical fields. The kinetic parameters of generation and decay of these fields can be read from the rate constants of the multiphasic reaction kinetics of the absorbance change at 515 nm (P515 response). Examples will be illustrated with special emphasis on the kinetic profile of the P515 response. The P515 response is a relatively simple but accurate parameter for quantifying the potential of the photosynthetic apparatus to perform its energy transducing and accumulating function in the intact leaf.

The identity of the pigment(s) or process(es) responsible for the light-induced absorbance change in the 450 to 550 nm wavelength region, first discovered by Duysens (7) in Chlorella cells, has been a matter of discussion for a long time. The light minus dark difference spectrum of this so-called 515 nm absorbance change shows a minimum at about 490 nm and a maximum, depending on the species, at a wavelength between 515 and 520 nm. The pigment or pigment complex involved in the reaction leading to the 515 nm absorbance change has been called P515, or in some older literature, P518 or P520.

There is now ample evidence (12) that the P515 absorbance change in chloroplasts, intact cells and leaves is due to an electrochromic band shift, i.e., an electrical effect on the energy levels of pigments with a permanent or induced dipole which results in a red shift of the absorption band of the pigments. Thus the P515 response has been shown to be a competent sensitizer of trans- and inner membrane electric fields generated by the primary charge separation in PS I and PS II an by subsequent electron and proton transfer through the oriented loop (Fig. 1) in the membrane, respectively. Here we will focus on P515 measurements done in intact leaves with a single beam absorption apparatus. This is illustrated in Figure 2. We will only deal with its characteristics during and after separate short flash excitations which each cause a single turnover in the reaction centers of the photochemical systems. Under these conditions absorbance changes associated with changes in light-scattering properties which have spectral characteristics different from the spectrum of P515, are relatively small and can be accounted for.

Kinetics of P515 in single turnover light flashes

In intact cells, as well as in intact (class A) and carefully prepared fresh broken chloroplasts, the rise and decay kinetics of P515 have been found to be multiphasic (11,28,33). The rise kinetics often, although not regularly, not only show a fast, but also a slow rise. The latter taking one to several tens of milliseconds. This slow rise has been called phase b. The fast initial rise phase a (11). Analysis of the decay kinetics of the P515 response after single flash excitation has revealed that it consists of at least two components with different first order rate constants (27,28).

The P515 decay kinetics after flash excitation are distinctly different from the single exponential kinetics of the transmembrane potential measured in single chloroplasts of Peperomia metallica with microcapillary electrodes. This has been taken as an indication that the P515 absorbance change has a component which reflects a sensitivity of the P515 pigments to inner membrane processes. Owing to the fact that the spectrum of this component is identical to the spectrum characteristic for the electrochromic band shift, it has been concluded that these processes are also associated with or caused by changes in membrane electric fields.

A third slow decay contributing about 10% of the initial response at the start of the flash has been attributed to a scattering change, with a spectrum different from P515 (32). After a series of 2 to 5 subsequent flashes separated by 100 to 150 ms, a response is measured, called reaction 1 with a fast rise (< 0.5 ms, and probably in ns) and a single exponential decay with a rate constant of the order of 10 s^{-1}. This component is similar to what is measured with a microcapillary electrode, and has been attributed to the generation and decay of the transmembrane electric field. Apparently the reaction responsible for the slow component, called reaction 2, becomes saturated after a few flashes, at least in intact chloroplasts and in a leaf. The kinetics of the component responsible for the slow rise in the overall response have been resolved in chloroplasts by subtracting those of the fast reversible component, (reaction 1) seen in repetitive flashes with a high repitition rate (27,28) or in the presence of 2,5-dibromo-3-methyl-6-isoprophyl-p-benzoquinone (DBMIB) (23), from the overall response. An example for an intact leaf is illustrated in Figure 6 (reproduced from 15). These kinetics show that the slow rise of reaction 2 after a saturating flash occurs with a relaxation time of several tens of milliseconds, similar to isolated intact chloroplasts (27). The dark reversion takes several hundreds of milliseconds. This slow decay has been discussed to be too slow to be governed by passive charge dissipation of the electrogenetically charged membrane (28,33). Moreover, the decay of reaction 2 in chloroplasts appeared to be relatively insensitive to membrane modifying agents like ionophores. However, its magnitude was found to be altered, i.e. largely suppressed by valinomycin (28) and increased by nigericin (5,29).

These observations were taken as evidence that the slow component is a reflection of localized inner membrane conformational changes presumably including the ATPase. These changes are suggested to be accompanied by changes in strength and orientation of local inner membrane electric fields. The reversion of these conformational changes apparently are slow. An effect of H^+-binding to and H^+-release from specific membrane sites has been suggested to contribute to the process that is responsible for the slow component of the P515 response (16,29,34). It might be that one of these sites is near to, or identical with, the FeS-cytb-cytf complex at the inner thylakoid membrane surface. It has been suggested (34) that protons released at this site are screened by a rate-limiting diffuse layer from the bulk aqueous phase due to the existence of

surface charge dependent energy barriers in the interphase between
membrane surface and aqueous phase. The release of protons restricted in mobility to specific lateral domains in neutralizing permanent ions isolated from the aqueous phase might be the cause of
localized conformational changes in the membrane and viewd by P515
complexes in the vicinity (16,34).

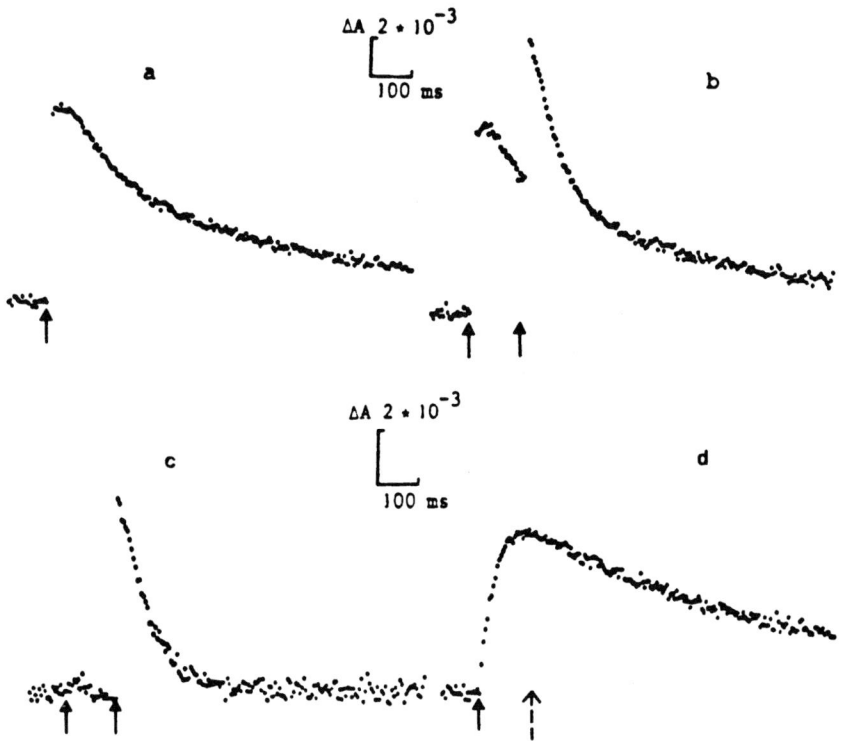

Figure 6. Absorbance changes at 518 nm in a dark-adapted spinach
leaf disc (diam ≈ 1 cm), induced by a single (a) or a double (time
difference 125 ms) saturating single turnover flash. The signals
are an average of 50 repititions done at a rate of 0.1 Hz and
carried in an apparatus described in ref. 15. Trace c is the difference between traces b and a, giving the net response upon the
second flash in trace b (reaction 1). Trace d is the difference
between the response of a and of c, indicating the slow component
(reaction 2) of the overall response (reaction 1 + reaction 2) in
the first flash. Apparently reaction 2 is absent in the second
flash (curve b) as indicated in curve d by the broken arrow.

P515 response in relation to growth conditions and preillumination

It has been found that during the initial growth of spinach
leaves under a normal light-dark regime, the lipid composiiton of

the thylakoid membrane changes. The change is most pronounced in the ratio between monogalactosyldiacyl glycerol (MGDG) and digalac-tosyldiacyl glycerol (DGDG)(9). A low MGDG/DGDG ratio was also found in chloroplasts isolated from leaves (harvested 7 weeks after germination) grown at a light intensity of 6 Wm^{-2} (26). Plants grown at an intensity of 60 Wm^{-2}, which at the time of harvesting had more fully expanded leaves had a ratio which was approximately 40% higher. This difference is manifest in the abundance of reac-tion 2 in the overall P515 response upon flash illumination as shown in Figure 7. Whereas no significant difference could be detected with respect to the kinetics of reaction 1, the low-light grown plants with a MGDG/DGDG ratio of about 1.0 had a small reaction 2 component with a magnitude which was less than 25% of the component in the high light grown plants.

As discussed elsewhere (26), these results illustrate the pos-sible function of MGDG in the organizational pattern and assemblage of proteins in the thylakoid membrane. This is of relevance in respect to its bioenergetic competence. In particular, a relation between the light-induced activation of the ATPase and the presence and function of H^+-binding and -stabilizing domains in or at the membrane-lumen interface is under consideration. It appears that information can be obtained in intact leaves by the non-evasive technique of measuring the kinetics of the P515 response upon single flash activation.

The effect of preillumination on a dark adapted leaf is shown in Figure 8. It appears that after a preillumination of 10 sec the reaction 2 component in the P515 response is completely lost. Its reappearance (Fig. 9) takes 10 to 20 minutes. Similar effects with different time relaxations are found in intact chloroplast and in broken chloroplasts in the presence of ATP and thiol agents. The latter are necessary in the absence of endogenous dithiol compounds like thioredoxin. This shows that the reaction 2 component of reaction 2 is a sensitive probe for the hydrolyzing activity of the light activates ATPase (25). It has to be established whether or not this preillumination effect on reaction 2 is concomitant with the induction and dark reversion of the Δ pH quenching of the fluorescence during the PSMT phase (see before).

In conclusion, we feel that owing to increased knowledge about the fundamental aspects of photosynthetic energy conversion in chloroplasts, and to availability of sophisticated optical and elec-tronic instrumentation, one is able to characterize and qualify the photosynthetic competence of photosynthetic organisms. Fluorescence and absorbance measurements appear to be extremely useful tools to study the photosynthetic performance of individual plant leaves under a variety of conditions during growth and maintenance.

128

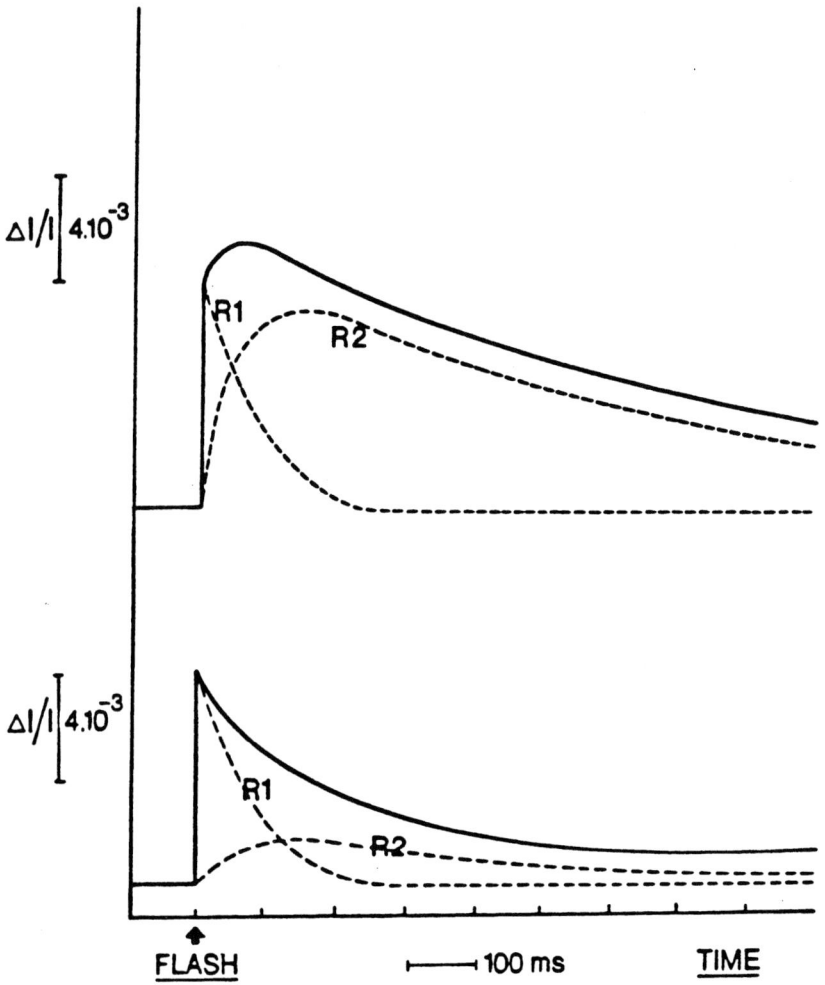

Figure 7. Absorbance changes at 518 nm in dark adapted spinach chloroplasts induced by a single turnover light flash (solid curves). Dashed curves are of reaction 1 and reaction 2, estimated by a procedure described in ref. 27 (see also Fig. 6). The response response is an average of 8 flashes fired at rate of 0.1 Hz.

Upper curve: chloroplasts isolated from plants grown at a light intensity of 60 W.m^{-2}.

Lower curve: chloroplasts isolated from plants grown at a light intensity of 6 W.m^{-2}.

Note the similarity in reaction 1(R1) but the large difference in reaction 2 (R2).

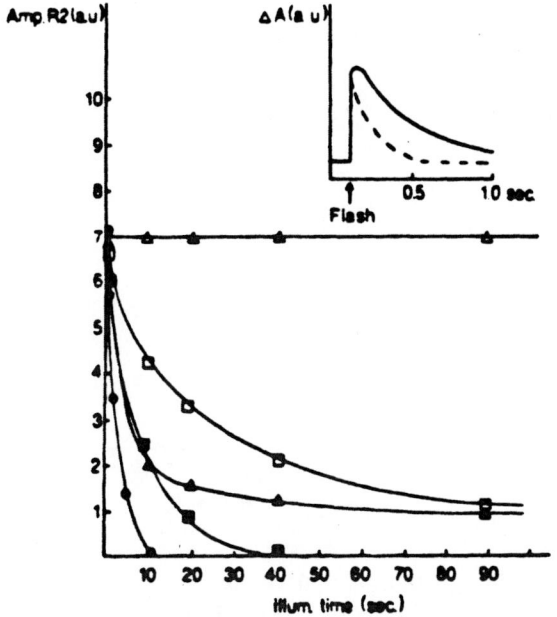

Figure 8. The effect of preillumination on the amplitude of reaction 2 of the flash-induced P515 response in intact leaves (●), intact chloroplasts (■), and broken chloroplasts in the absence (Δ) and presence of 300 μmol/liter ATP (□). The effect of preillumination in broken chloroplasts in the presence of 300 μmol/liter ATP and 1 mmol/liter DTE is also shown (▲). The inset shows the P515 response induced by a single light flash in intact chloroplasts before (solid curve) and after (dashed curve) a preillumination period of 40 sec. 1 a.u. corresponds to approximately 8×10^{-4} ($\Delta I/I$).

130

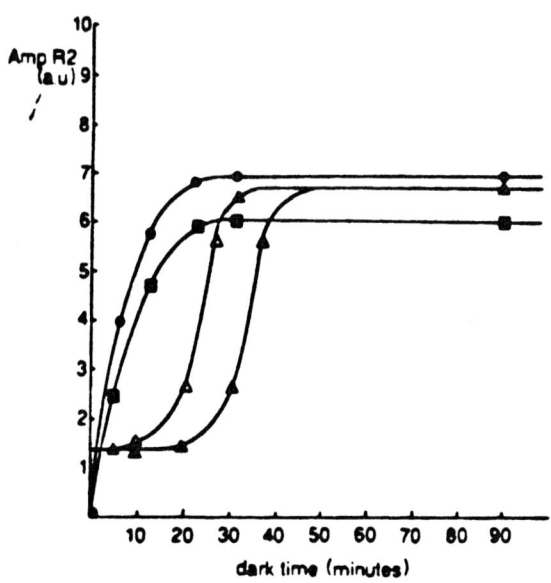

Figure 9. The effect of dark adaptation after a 60 sec preillumination on the recovery of the amplitude of reaction 2 in intact
leaves (●), intact chloroplasts (■), and broken chloroplasts in
the presence of 1 mmol/liter DTE, with 100 μmol/liter ATP (Δ) or
with 300 μmol/liter ATP (▲), respectively. 1 a.u. corresponds to
approximately 8×10^{-4} ($\Delta I/I$).

REFERENCES

1. Allen, J.F., Bennett, J., Steinback, K.E. and Arntzen, C.J.
    Nature 291:25-29, 1981.

2. Barber, J. Biochim. Biophys. Acta 594:253-308, 1980.

3. Bennett, J., Steinback, K.E. and Arntzen, C.J. Proc. Natl.
    Acad. Sci. USA 77:5253-5257, 1980.

4. Bonaventura, C. and Myers, J. Biochim. Biophys. Acta 189:336-
    383, 1969.

5. Crowther, D. and Hind, G. Arch. Biochem. Biophys. 205:568-577,
    1980.

6. Delieu, T.J. and Walter, D.A. Pl. Physiol. 73:534-541, 1983.

7. Duysens, L.N.M. Science 120:353-354, 1954.

8. Duysens, L.N.M. and Sweers, H.E. In: Studies on Microalgae and Photosynthetic Bacteria (H. Tamyia, ed.), Univ. of Tokio Press, p. 353-365, 1963.

9. Frey, R. and Tevini, M. In: Advances in the Biochemistry and Physiology of Plant Lipids (L.A. Appelqvist and C. Liljenberg, eds.) Elsevier Biomedical Press, Amsterdam, pp. 225-229, 1979.

10. Joliot, P. and Joliot, A. C.R. Acad. Sci. 258:4622-26, 1964.

11. Joliot, P. and Delosme, R. Biochim. Biophys. Acta 357:267-284, 1974.

12. Junge, W. Annu. Rev. Pl. Physiol. 28:503-536, 1977.

13. Krause, G.H., Vernotte, C. and Briantais, J.M. Biochim. Biophys. Acta 679:116-124, 1982.

14. Krause, G.H., Briantais, J.M. and Vernotte, C. Biochim. Biophys. Acta 712:169-175, 1983.

15. Van Kooten, O., Gloudemans, A.G.M. and Vredenberg, W.J. Photobiochem. and Photobiophys. 6:9-14, 1983.

16. Van Kooten, O., Leermakers, F.A.M., Peters, R.L.A. and Vredenberg, W.J. In: Advances in Photosynthesis Research (C. Sybesma, ed.) Vol. II, Martinua Nijhoff/Dr. W. Junk, Publ, The Hague, pp. 265-268, 1984.

17. Lavorel, J. and Etienne, A.L. In: Primary Processes in Photosynthesis (J. Barber, ed.) Vol. II, Elsevier Publ Company, Amsterdam, pp. 203-268, 1977.

18. Malkin, S., Armond, P.A., Mooney, H.a. and Fork. D.C. Pl. Physiol. 67:570-579, 1981.

19. Malkin, S. and Fork, D.C. Pl. Physiol. 67:580-583, 1981.

20. Melis, A. and Brown, J.S. Proc. Natl. Acad. Sci. USA 77:4712-4716, 1980.

21. Melis, A. and Harvey, G.W. Biochim. Biophys. Acta 637:138-145, 1980.

22. Melis, A. and Homann, P.H. Arch. Biochem. Biophys. 190:523-530, 1978.

23. Olsen, L.F., Telfer, A. and Barber, J. FEBS Letters 118:11-17, 1980.

24. Papageorgiou, G. In: Bioenergetics of Photosynthesis (E.A. Govindjee, ed.) Acad. Press, New York, pp. 319-371, 1975.

25. Peters, R.L.A., Bossen, M., van Kooten, O. and Vredenberg, W.J. J. Bioenerg. & Biomembr. 15:335-346, 1983.

26. Peters, R.L.A., van Kooten, O. and Vredenberg, W.J. Bioenerg. & Biomembr. 16:283-294, 1984.

27. Schapendonk, A.H.C.M., Vredenberg, W.J. and Tonk, W.J.M. FEBS Letters 100:325-330, 1979.

28. Schapendonk, A.H.C.M. Ph.D. Thesis Agriculture University, Wageningen, 1980.

29. Schuurmans, J.J., Peters, A.L.J., Leeuwerik, F.J. and Kraayenhof, R. In: Vectorial Reaction in Electron and Ion Transport in Mitochondria and Bacteria (F. Palmiere, E. Quagliariello, N. Siliprandi and E.C. Slater, eds.) Elsevier Publ Company, Amsterdam, 1981.

30. Strasser, R.J. In: Photosynthesis (G. Akoyunoglou, ed.) Vol. III, Balaban Int. Sci. Serv., Philadelphia, pp. 727-737, 1981.

31. Vermaas, W.F.J. and Govindjee Biochim. Biophys. Acta 680:202-209, 1982.

32. Vredenberg, W.J., van Kooten, O. and Peters, R.L.A. In: Advances in Photosynthesis Research (C. Sybesma, ed.) Vol. II, Martinua Nijhoff/Dr. W. Junk, Publ. The Hague, pp. 241-246, 1984.

33. Vredenberg, W.J. Physiol. Plant. 53:598-602, 1981.

34. Westerhof, H.V., Helgerson, S.L., Theg, S.M., van Kooten, O., Wikstrom, M.K.F., Skulachev, V.P. and Dancshazy, Z. Acta Biochim. Biophys. Acad. Sci. Hung. 18:125-150, 1984.

35. Walker, D.A., Sivak, M.N., Prinsley, R.T. and Cheesbrough, J.K. Plant Physiol. 73:542-549, 1983.

ENVIRONMENTAL INSTRUMENTATION

HUMICAP® THIN FILM HUMIDITY SENSOR

Eero Salasmaa
Pekka Kostamo
Vaisala Oy, Helsinki, Finland

1. INTRODUCTION

The humidity sensor described in this paper is a thin film capacitor. Its capacitance is dependent on the water absorption in the sensor's dielectrical material. The sensor is produced commercially under trade mark HUMICAP®. (Suntola 1974)

Most of the humidity sensors which display a change in an electrical parameter are resistive, i.e. the resistance of the sensor is a function of the humidity to which the sensor is exposed. The Humicap sensor was developed especially for Finnish radiosondes having capacitive transducers for all parameters (pressure, temperature and humidity). Therefore, it is quite natural that change of capacitance was adopted as a working principle of the new sensor from the beginning of the development work.

Various other capacitive humidity sensors are described in the literature (1,3) and some of these are known to be commercially available. For various reasons, such as high cost, fragility, unsuitable measuring or capacitance range, etc., none of these seemed to be suitable for the applications in question.

2. CONSTRUCTION AND TECHNOLOGY

The Humicap sensor is a small capacitor with organic polymer dielectric. In principle, a sensor like this can be made by providing a thin polymer sheet with electrically contacting but water vapor permeable electrodes on both sides of it. A sensor with this construction, however, suffers from instability and hysteresis which is probably due to mechanical weakness. It is fragile, and in any case the polymer sheet cannot be made thin enough to achieve very short response time.

In the Humicap sensor these difficulties are avoided by using a solid glass substrate to support the electrodes and the thin polymer film. The construction of a practical Humicap sensor is shown in Figure 1. The sensor is fabricated using thin film technology similar to that generally used in microelectronics. The lower electrodes (b) are first etched onto a metallized glass plate (a). The surface is then coated with an active polymer layer (c) using dipping techniques. This results in a uniform amorphous polymer layer. The polymer is about 1 μ thick. Great care and cleanliness is necessary to avoid pin holes. The upper electrode (d) is vacuum evaporated onto the polymer surface through a mechanical mask to form the active capacitor area. The upper electrode must be permeable to water vapor and is, therefore, very thin (approx. 10^{-2} μ). The thickness of the upper electrode must be chosen as a compromise between the response time of the sensor and the ohmic loss in the electrode. The latter tend to decrease the Q value of the capacitor especially because high frequency is used for measurements.

The electrode material can be of some "noble" metal and probably also carbon. The lifetime of the sensor depends on the material used. In this respect best results have been obtained with palladium, but gold is good enough for radiosonde sensors, and more easily evaporated.

A great number of sensors can be coated simultaneously. On a 50 x 50 mm glass plate, generaly used, there are about 100 sensors which after coating are separated by cutting the plate in 4 x 6 mm pieces. After the cutting process, electrical leads (3) are soldered to the lower metal electrodes of the sensor.

In this construction (Fig.1), there are two moisture sensitive capacitors in series in each sensor, so that lead contacting to the very thin upper electrode of the sensor is avoided.

3. THEORY OF OPERATION

At an early stage of the development of Humicap, sensor elements were made with dielectric of various different polymers. It is quite natural that the characteristics of the sensor very strongly depend on the nature of the polymer material and treatment of the sensor during fabrication. In general, independently of the exact type of polymer used, the dielectric film must be amorphous and non porous; these properties can be influenced by a controlled dipping procedure. The basic absorption process in different polymers is briefly described in a paper by Suntola and Antson (1973).

In the absorption process, water molecules form bonds with polymer molecules. These bonds have binding energies which depend on the type of the bond.

137

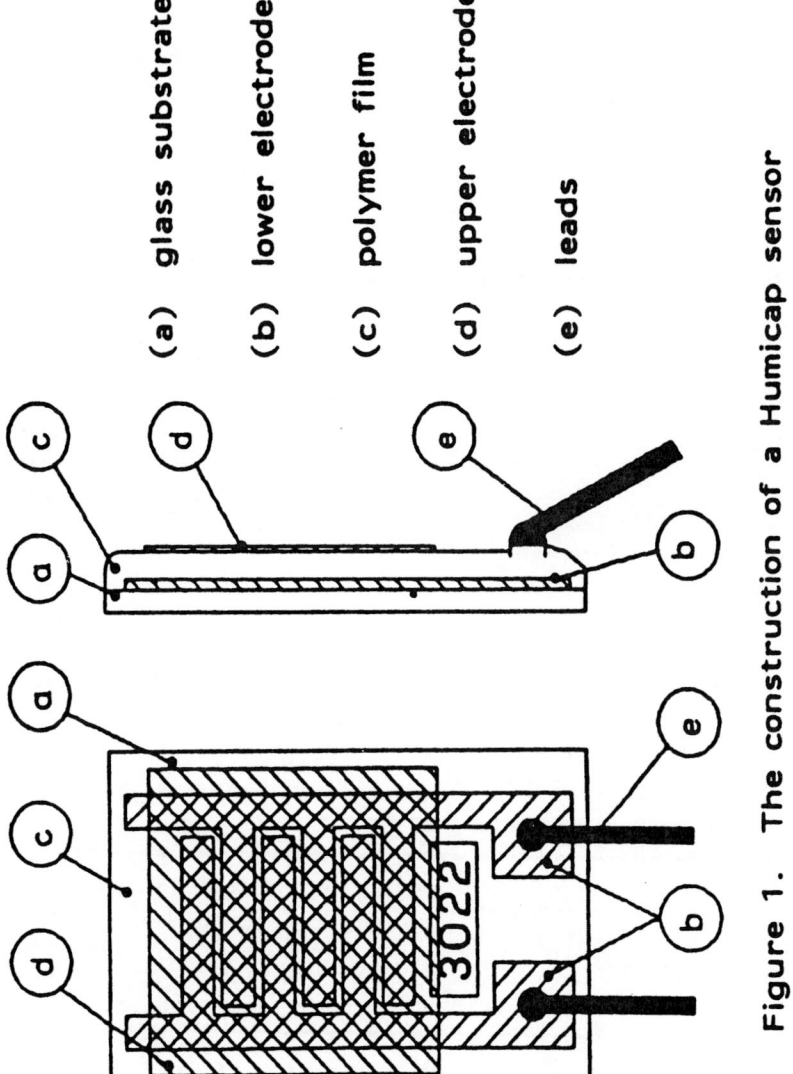

(a) glass substrate

(b) lower electrode

(c) polymer film

(d) upper electrode

(e) leads

Figure 1. The construction of a Humicap sensor

According to Suntola (after some simplifying assumptions) the relative number θ of occupied water molecule bonding sites in polymer is

$$\theta = \frac{\dfrac{R\,a\,P_0}{b'(2\pi mkT)^{1/2}}\,e^{(q(\theta)-q_0)/kT}}{1+\dfrac{R\,a\,P_0}{b'(2\pi mkT)^{1/2}}\,e^{(q(\theta)-q_0)/kT}} \qquad (1)$$

where
R = ambient relative humidity (0<R<1)
a = effective area f a bonding site
P_0 = partial pressure of saturated water vapor
$q(\theta)$ = binding energy of water molecule in the polymer
q_0 = binding energy of water molecule in water
b' = probability of emission of water molecules
k = boltzmann constant
T = absolute temperature

From equation (1) it can be seen that:

1. If $\theta \ll 1$, i.e. only a small fraction of possible bonding sites is occupied or the number of possible sites increases with absorption, when 0<R<1

2. If $q(\theta)$ is constant when θ<R<1 θ is a linear function of R

3. If still $q(\theta) = q_0$, i.e. the binding energy of water molecules in the polymer equals the bonding energy in water, the exponential thermal dependence of absorption disappears.

By selection of polymer so that the conditions 1,2 and 3 are nearly obtained, the linearity within 3% and thermal coefficient of 0.05% RH/°C is achieved in the Humicap sensor.

In practice, the situation is more complicated, e.g. the water mass absorption of the polymer is less linear, as measured by a vibrating quartz crystal coated with the polymer (Fig.2). The same non-linear curve is obtained when the capacitance of the humicap sensor is measured with low frequency but the curve is getting more and more linear when the measuring frequency is increased. The difference in these measurements can be explained by assuming other water molecule bonds which do not have the same influence on the dielectric properties. When the capacitance is measured with a low frequency instead of radio frequencies, the resulting capacitance curve approaches mass absorption curve form. The excess absorption thus forms dipoles with essentially longer relaxation time. This phenomenon is of high importance in practice. We can restrict the measured absorption to one bonding type only which in turn is a necessary condition to good linearity and to temperature independent absorption rate.

139

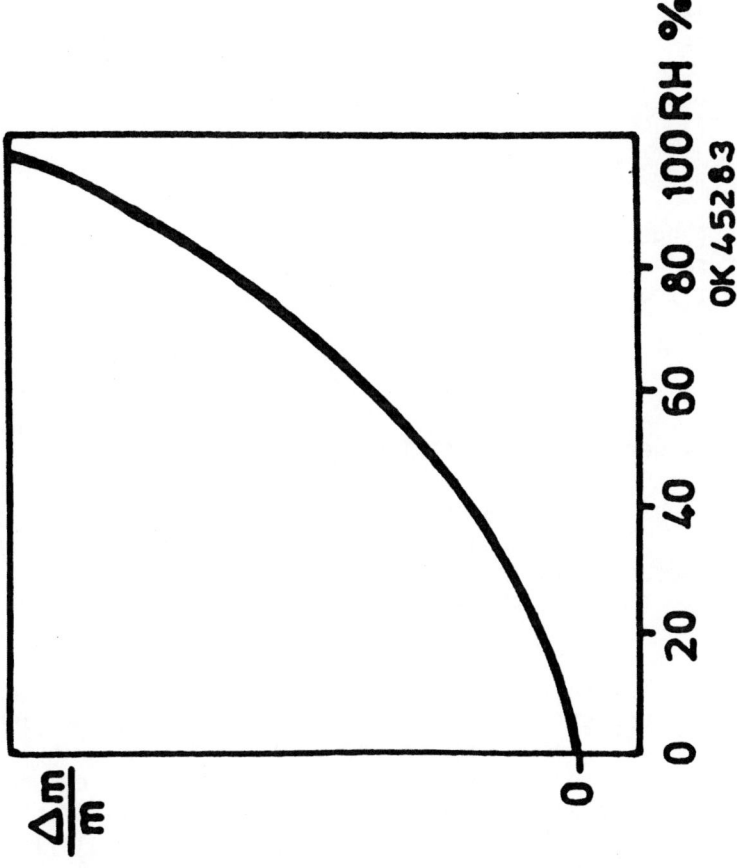

Figure 2. Mass absorpbtion curve of a polymer

Another deviation from the simplified theory is that a long
exposure of Humicap to high ambient humidity slightly and reversibly
alters the dynamical behavior of the sensor, i.e. the slope of the
C = f(R) curve. This phenomenon is discussed in some detail in the
following chapters.

4. CHARACTERISTICS OF HUMICAP SENSOR

Two types of Humicap sensors are now in commercial production:
a) C(0% RH) = 5.25 ± 0.55 pF
b) C(0% RH) = 45 ± 4 pH
Type a) is used in the Finnish (Vaisala) radiosondes and type b) is
for general purpose use, hygrometers, control systems, etc. The
general specifications of these sensors are:

- dynamic range : $\dfrac{\Delta}{C(0\% \text{ RH})}$ =22%
- humidity range: 0...100% RH
- temperature range: to 115°C
- response time: less than 1 sec. at +20°C to 90% of
                       final RH value
- accuracy: less than 2% for humidity excursion
                       20-80-20%; less than 3% for humidity
                       excursion 0-100-0%
- temperature dep.:   0.05% per 1°C for sensor

Typical step response curves are shown in Figure 3. The first step
of absorption occurs in less than one second. The time constant and
amplitude of the second step is highly dependent on the polymer material used. For a good sensor the second step is only a few percent
of the total response. It can be expected that the fast absorption
is due to bonding in the basic states when the polymer is held at a
constant humidity. The family of curves in Figure 4 indicates the
fast response characteristics after a sufficient stabilization time
(several minutes) at different humidities. The stabilization humidity point of each curve is marked with a small circle. The dashed
curve connecting these circles gives the C=f(RH) characteristics
measured with long waiting time (a few minutes at each humidity).
Apparently the increase in the number of bonding sites (which
increases the slope of the curves) is connected to swelling effects
in the material. Partial recovery during passage from one fast
response curve to another seems to be the basis of the hysteresis
effect which is generally less than 4% RH. The fast response
curves show no hysteresis.

The response time of the sensor is strongly dependent on temperature as indicated by Figure 5. Below 0°C the fast response step
cannot be separated from the total change any more. It should be
noted that the sensor response is proportional to the relative value
of water vapor pressure. The partial pressure above solid ice
gives therefore, a reduced reading.

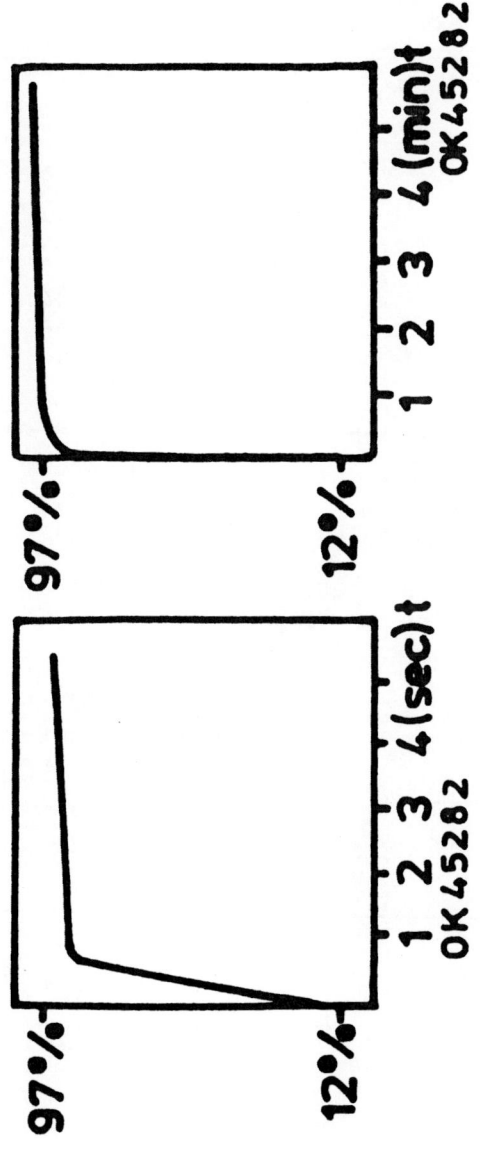

Figure 3. Step response curves of Humicap sensor

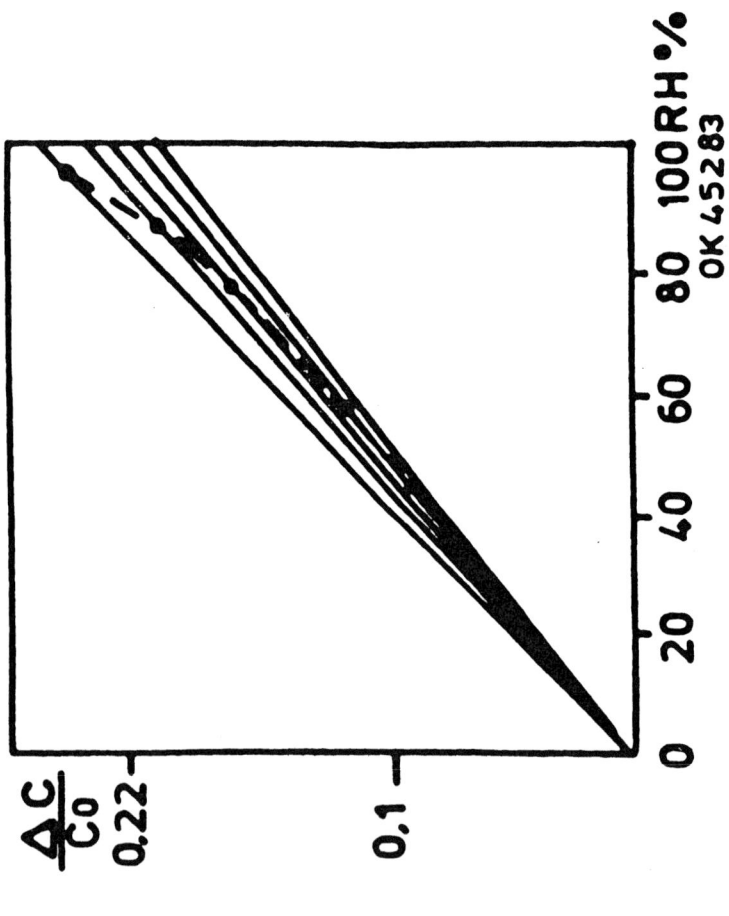

Figure 4. Fast response characteristics of a Humicap sensor

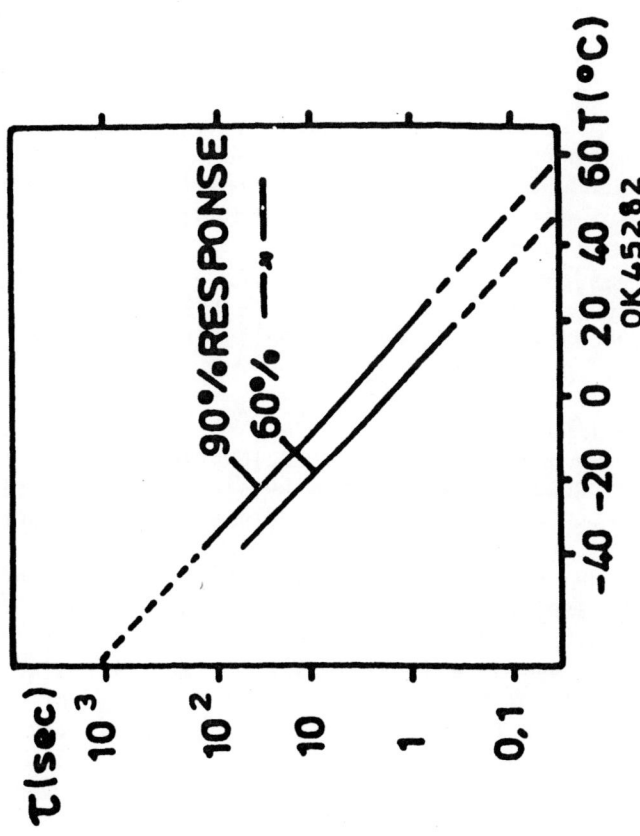

Figure 5.   Temperature dependence of the time constant

5. APPLICATION OF HUMICAP SENSOR

As mentioned earlier, the Humicap sensor was originally designed for the Finnish radiosonde. The small size of the Humicap
sensor (4 x 4 x 0.2 mm) is an extra advantage for radiosonde use,
since thermal mass is small and the sensor very closely and quickly
follows the ambient air temperature. This is obviously necessary
for obtaining true relative humidity values in the atmosphere. A
second example of the use of the Humicap sensor (4 x 6 x 0.2 mm) is
a humidity probe used in portable humidity meters, control systems,
etc.

6. OPERATIONAL FEATURES

In the radiosonde use, the sensor calibration can be verified
during the base line check prior to each observation, if considered
necessary. Therefore, there are no special requirements as regards
the long term calibration stability of the sensor. The shelf life
under normal storing conditions, however, must be long, up to 1-2
years. In most other applications, the long term stability for the
calibration is much more important, and the sensor's operational
life must be as long as possible.

The first Humicap sensors were produced in August 1972 and
some operational experience has been accumulated thereafter:

1. The shelf life of the sensor under normal conditions seems to
    be practically unlimited.
2. In use, i.e. measuring current on, the life of the sensor gets
    shorter, notably in humid environment with atmospheric pollu-
    tants. High concentration of certain pollutants may also
    shorten the life in storage.

The most serious usual pollutants are probably the small soot
particles with absorbed sulphur dioxide or sulphuric acid. These
particles, which are present in polluted city air, cause corrosion
to the thin upper electrode, if water is condensed to the sensor at
the same time.

Exposure of the Humicap to organic solvents is not recommended.
Some of them ruin the sensor immediately. Immersion to ethyl
alcohol irreversibly alters the calibration. Clean water does not
damage the Humicap sensor. It even resists temporary immersion to
boiling water. High ambient temperatures shorten the life of the
Humicap sensor, but in most cases it can safely be used in temperatures up to 115°C.

Upper air soundings have given valuable information on the
behavior of the Humicap sensor at low temperatures, where accurate
humidity calibration is difficult to carry out in the laboratory.
Figure 6 shows an example of a sounding. After the tropopause, the
measured humidity quickly drops very close to 0% RH when the radiosonde reaches the dry stratosphere.

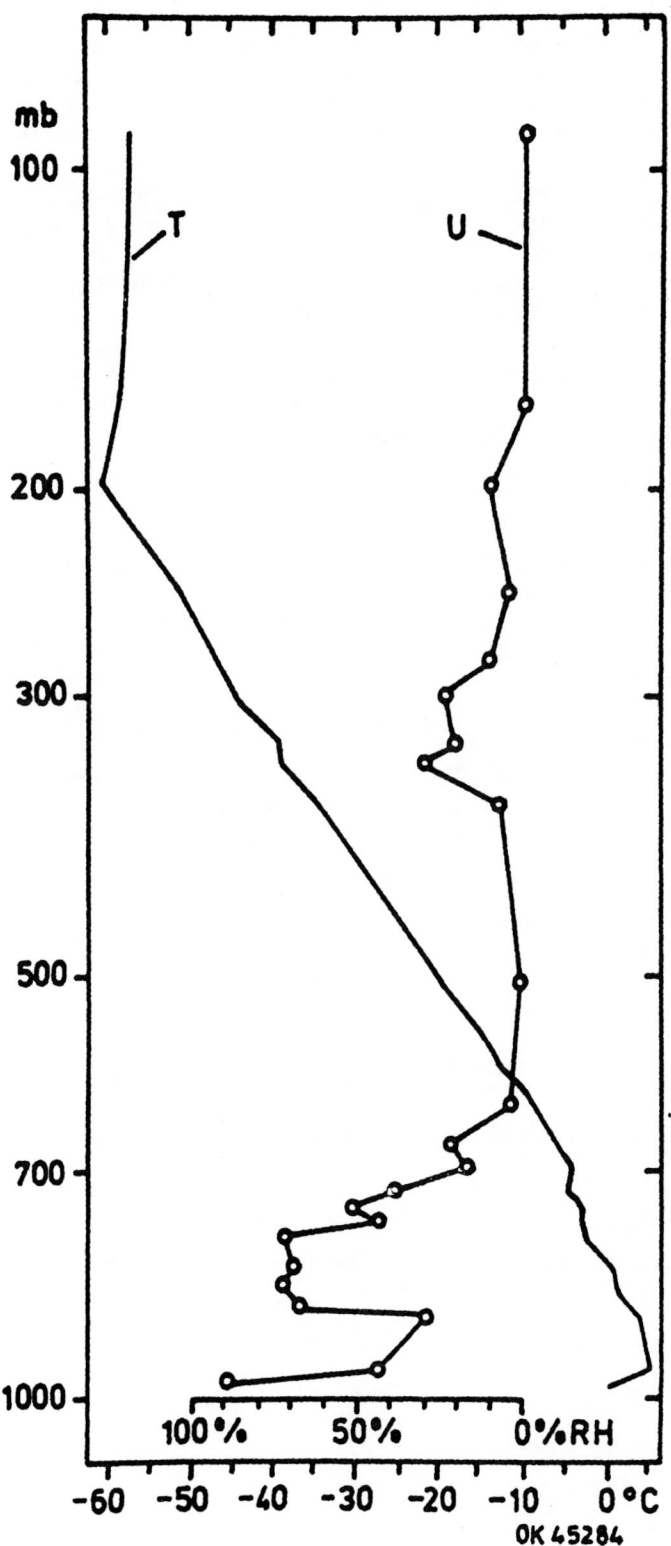

Figure 6

OK 45284

At lower levels the fine structure of the humidity profile is shown in much detail. When interpreting the results obtained in low temperatures, it must be noted that the sensor gives directly the relative humidity calculated from vapor pressure over water, not over ice.

The test carried out at outdoor city climates in the middle of Helsinki gives valuable information of the long term stability of the Humicap sensor. The test began in May 1979 and lasted two years. The reference in this test was a standard aspirated psychrometer to which 11 Humicaps were compared. Figure 7 shows the differences between the psychrometer and the mean of 11 humicaps. The single points are actual differences and the solid line is the mean of 10 consecutive differences. The dotted lines depict the standard deviation of the mentioned mean values.

7. CONCLUSION

The new Humicap sensor has many attractive features, fast response, good linearity, low hysteresis and small temperature coefficient. Due to its small size, it closely follows the temperature of ambient air and, in the measurements, it does not alter the system being examined. For the same reason, because of the fast response, the calibration is quickly and easily done, e.g. in small volume chambers with saturated salt solutions. The sensor can be produced in quantities at reasonable price. Further development of the sensor for various applications is going on.

REFERENCES

1. Misevich. Capacitive humidity transducers. IEE Trans. Indust. Elec. and Control Instrumen. 16:1, 1969.

2. Misevich, et al. U.S. patent No. 3,350,941, 1969.

3. Nelson. U.S. patent No. 3,168,829, 1969.

4. Suntola, T. A thin film humidity sensor. CIMO VI Scientific Discussions, 1973.

5. Suntola, T. Finish patent No. 48,299/1974; French patent No. 2,203,520/1974; United Kingdom patent No. 1418388/1976; USA patent No. 4164868/1979, 1973.

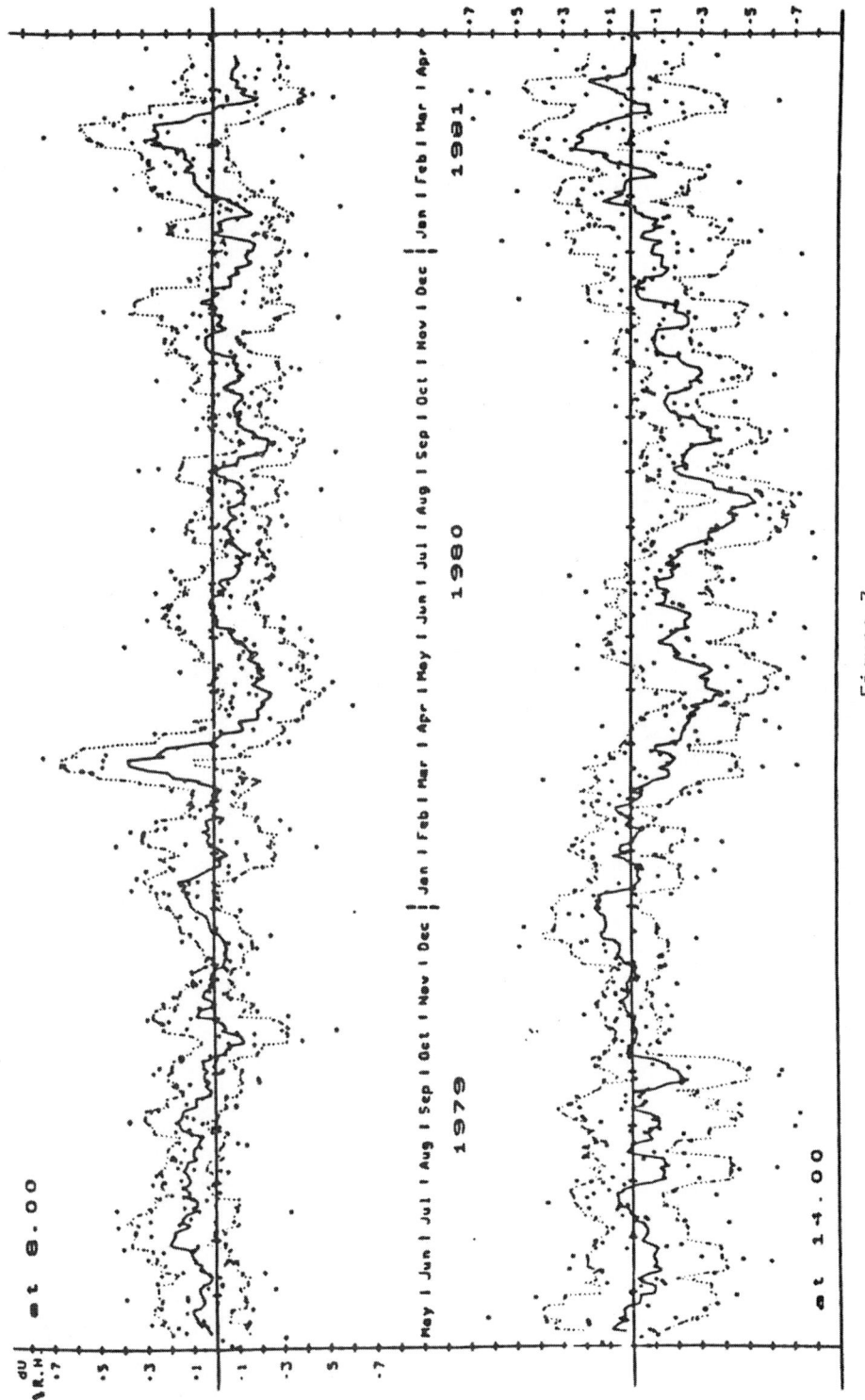

Figure 7

THE NEUTRON PROBE FOR SOIL MOISTURE MEASUREMENT

M. G. Hodnett

Institute of Hydrology
Wallingford, Oxon, UK

1. INTRODUCTION

Two types of neutron soil moisture meters are generally available: depth probes, which are lowered into lined access holes (access tubes) in the soil to make measurements at the required depths, and surface meters, which are placed on the surface of the soil and which measure the moisture content of the upper 0.10-0.15m of the soil profile. The latter are mainly used in civil engineering applications.

This chapter only deals with depth probes and covers theory, design, calibration, sources of error, data handling, use and applications with particular reference to the IH II neutron probe system. This was developed at the Institute of Hydrology and is produced by the Didcot Instrument Co., Ltd, Station Road, Abingdon, Oxon, UK.

The neutron soil moisture meter or neutron probe, which is widely used in agricultural and hydrological applications, currently provides the most accurate and convenient means of measuring the moisture content of soil profiles in the field. The major advantages of the method are that it is rapid and nondestructive: measurements are made in situ in access tubes installed in the soil. Once in place, these allow repeat measurements to be made at the same location at any time. The IH II probe is shown in use in the field in Figure 1.

The most useful general references covering the subject of neutron probes are the very detailed book Soil Moisture Assessment by the Neutron Method edited by E.L. Greacen (1), the report "Neutron Probe Practice" by J.P. Bell (2) and the International Atomic Energy Agency (IAEA) report "Neutron Moisture Gauges (3).

2. BACKGROUND

The principle of measuring soil moisture content by the neutron moderation effect was first proposed in the 1940's (4,5) and much of the pioneering work on the instruments was carried out in the U.S. in the early 1950's (6,7). Neutron probes became commercially available in the late 1950's and early 60's but these early probes were heavy, unreliable, electronically unstable, and complex to operate. In many cases two people were required to carry the equipment, which was generally not very weatherproof and therefore ill suited to field use. One observer is reported to have unkindly commented that some models were better at measuring soil temperature than soil moisture!

However, rapid advances in electronics have now resulted in vastly improved equipment which is light, rugged, reliable and easy to operate (see Fig. 1). In many cases, the equipment now incorporates a microprocessor and memory which allows data to be stored for direct transfer to a computer, or processed into moisture contents directly.

Figure 1. The IH II Neutron Probe System in use at a field site.

3. PRINCIPLE OF OPERATION

3.1 Outline

The neutron probe relies on the principle of neutron moderation and on the fact that hydrogen is by far the most effective neutron moderator found in soils. As most of the hydrogen in soil is in the form of water, the moderation is principally due to the soil moisture.

A small radioactive source in the probe emits high energy neutrons which are scattered and moderated (ie. lose energy) in collisions with the atomic nuclei of the soil, becoming "thermal" or "slow" neutrons. These diffuse through the soil and are absorbed, or captured, by the nuclei of the elements present. The processes of scattering and absorption form a cloud of thermal neutrons in dynamic equilibrium around the source and the density of this cloud, measured as a count rate by a thermal neutron detector close to the source, is a function of the volumetric hydrogen content of the soil.

The principles of the method are presented in more detail below but for a thorough treatment of neutron moisture meter theory, the reader is referred to the IAEA report entitled "Neutron Moisture Gauges" (3), to the work of Couchat (8,9) and Olgaard (10,11) or to a nuclear physics text.

3.2 Neutrons

Fast neutrons from the sources most commonly used have a spectrum of energies ranging from 0 to 11 MeV, with a mean energy of about 5 MeV. Neutrons can be classified, on the basis of their energy, into four groups (9): fast neutrons (>1 keV), intermediate neutrons (2 eV - 1 keV), epithermal neutrons (0.5 ev- 2 ev) and thermal neutrons (0.025 eV - 0.5 eV). Thermal neutrons are so-called because they have reached thermal equilibrium with the molecules of the surrounding medium.

Neutrons interact with the nuclei of the elements in a medium in one or two main ways, either scattering or capture. The type of interaction that occurs depends only on the energy of the incident neutron, and the isotope of the soil element involved.

3.3 Neutron Scattering

This is the process in which a neutron collides with, and transfers kinetic energy to, a nucleus without being absorbed by it. The neutron is therefore slowed down and its course of travel is also changed. Scattering may be elastic, where kinetic energy and momentum are conserved, or inelastic, where some of the energy of the incident neutron is absorbed by the nucleus involved, causing the emission of a lower energy neutron and a photon of gamma radiation. Over the range of neutron energy levels involved in neutron meter theory, elastic scattering is by far the dominant mode of interaction, and the effects of inelastic scattering can be ignored.

In elastic scattering, it can be shown that, since kinetic energy and momentum are conserved, the energy lost in a single "head-on" collision with a nucleus is given by:

$$E_1 - E_2 = \frac{4A}{(A + 1)^2} E_1 \tag{1}$$

Where E_1 and E_2 are the energy of the neutron before and after the collision and A is the mass number of the nucleus. The neutron mass is taken to be 1.

It can be seen from Eq. 1 that the energy loss is inversely proportional to the mass of the nucleus involved, and that the maximum possible energy loss (100%), will occur in a head-on collision with a hydrogen nucleus. For comparison, the maximum possible energy losses in collisions with oxygen, silicon, aluminium and iron are 22.1%, 13.8%, 13.3% and 6.8%, respectively.

However, collisions between neutrons and nuclei occur at all angles, and it can be shown that the following average number of collisions are required to moderate 2 MeV neutrons to thermal energies (0.025 eV):

Nucleus	No. of collisions	
H	18	
O	152	The commonest soil
Si	252	constituent
Al	279	elements.

Hydrogen is therefore by far the most effective moderator found in soils. Although the other, heavier elements do not contribute greatly to the slowing down power of the soil, they tend to keep the scattered neutrons in the vicinity of the source (3).

3.4 Neutron Absorption, or Capture

Two types of capture reaction are important in moisture gauge theory, radiative capture (in nuclear shorthand, the n, γ reaction) and capture with the expulsion of a charged particle, either an alpha particle or a proton (the n, α and n,p reactions, respectively).

Radiative capture occurs when a neutron combines with a nucleus, causing the emission of the excess energy as a photon of gamma radiation. This process only becomes significant at epithermal energy levels and below, and occurs in the soil, resulting in the removal of thermal neutrons from the system.

Capture with the emission of a charged particle occurs mainly at very high neutron energy levels, but in the case of the nuclei of ^{10}B, ^{6}Li and ^{3}He, this type of capture reaction has a very high probability of occurrence at epithermal energy levels and below. This type of reaction forms the basis of thermal neutron detectors.

3.5 Interaction or Reaction Cross-sections

The interactions which a neutron may undergo have different probabilities of occurrence, which depend on the energy of the neutron and on the nucleus involved. An interaction cross-section can be defined (IAEA 1970) as "a nuclear property of a given isotope that is proportional to the probability that its nuclei will undergo a given type of reaction." Cross-sections are expressed in barns, which are units of area. One barn is 10^{-28} m^2.

The cross-section of a single nucleus is known as its microscopic cross-section, denoted by the symbol σ . Cross-sections are quoted for given reactions and neutron energies, or ranges of neutron energies. For example, the neutron absorption cross-section, denoted by σ_a might be quoted for 0.025 eV neutrons or the elastic scattering cross-section, denoted by σ_s might be quoted for 2 MeV neutrons.

The total cross-section of unit volume (1 m^3) of a given isotope is known as its macroscopic cross-section, Σ:

$$\Sigma = N\sigma \times 10^{-28} \quad m^{-1} \tag{2}$$

where σ is the microscopic cross-section for the reaction at a given energy level, and N is the number of nuclei in one m^3 of the isotope (3). If N is calculated from the density of the material, ρ, Avogadro's number, N_0, and the atomic weight of the isotope, M, equation 2 becomes:

$$\Sigma = N_0 \, \sigma/M \quad m^{-1} \tag{3}$$

For a medium such as soil, which contains many elements, the macroscopic cross-section is given by:

$$\Sigma = N_0 \sum_1^n w_i \sigma_i/M_i \quad m^{-1} \tag{4}$$

where there are n elements present and w_i, σ_i and M_i are the percentage by weight, the microscopic cross-section and the atomic weight of the i th element respectively (3). It can be seen from Eq. 4 that the microscopic cross section of the soil for a given neutron interaction depends on its bulk density and on which elements are present and in what proportions.

For fast neutrons, the microscopic elastic scattering cross-sections, σ_s of most elements are mainly within the range 1-5 barns and increase gradually with decreasing neutron energy. Below about 1 eV, σ_s for some elements increases fairly suddenly, in the case of hydrogen, from 20 to about 80 barns.

The neutron capture cross-sections, σ_a for most elements only become significant at epithermal energy levels and below and show a very wide range of variation. For most elements, the capture cross-sections for thermal neutrons generally vary as the inverse of the square of the neutron energy, although some show pronounced resonance absorption peaks at epithermal energy levels and above. Most of the common soil elements have a σ_a of 0.1 - 1.0 barns for 0.025 eV neutrons, but some of the less common elements may contribute significantly to the macroscopic cross-section of a soil as their σ_a (shown in parentheses) is particularly large. Among these are

Gadolinium (46000), Cadmium (2450), Boron (755), Chlorine (33), Manganese (13.2), Titanium (5.8) and Iron (2.6). Although σ_a of iron is not very great, some tropical soils may be very iron rich. The presence of significant amounts of strong neutron absorbers reduces the thermal neutron density around the source, lowering the count rate.

4. NEUTRON SOURCES AND DETECTORS

4.1 Fast Neutron Sources

Sources of fast neutrons contain a compacted mixture of an alpha particle emitter, usually Americium-241 (as oxide), and Beryllium metal powder. The fast neutrons are produced by the reaction between the alpha particles and the Beryllium target material:

$$\,^4_2He + \,^9_4Be \;\;\text{---}\!\!> \;\,^{12}_6C + \,^1_0n + 5.5 \text{ MeV}$$

^{226}Ra-Be sources were widely used in the past but have been superseded by Am-Be sources, as the gamma output from ^{241}Am is much lower than that from ^{226}Ra for an equivalent neutron emission. ^{241}Am has a half-life of about 460 years, ensuring an almost constant neutron output. Activities of Americium sources used nowadays typically range from 0.37 - 3.7 GBq[1] (10 - 100 mCi) although 7.4 GBq (200 mCi) (12) sources have been used in the past. 1.85 GBq (50 mCi) is probably the most commonly used activity at present.

The source materials are particularly hazardous and are therefore double encapsulated in stainless steel. Source capsules have to meet stringent international design specifications (ISO and IAEA) and are usually designated as "Special Form" Radioactive Material. The sources used in neutron probes are either cylindrical or annular in form.

Other sources have recently been evaluated (13), the most promising of which, from the aspect of safety, are Curium-244 Beryllium and Californium-252. Both source types have very low gamma radiation emissions. The half-life of ^{244}Cm is 17.4 years. ^{252}Cf is notable because the neutrons (mean energy 2 MeV) are produced by spontaneous emission and a very low activity (1 MBq or 27 μCi) is required to produce a neutron output equivalent to a 1.85 GBq (50 mCi) Am-Be source. However, ^{252}Cf has a very short half-life (2.6 years) which means that sources would require replacement approximately every 3 years. This would be inconvenient and also expensive.

---

[1] 1GBq = 10^9 disintegrations per second.

4.2 Thermal Neutron Detectors

Neutrons, having no charge, are detected by means of the ion-
ization caused by the charged particles produced by their interaction with isotopes such as ^{10}B, ^{3}He and ^{6}Li which, as mentioned
previously, have very large neutron capture cross-sections at
epithermal and thermal energy levels. Detectors utilizing these
isotopes are insensitive to fast neutrons.

In the case of ^{10}B, the capture reaction (n,α) is:

$$^{10}_{5}B + ^{1}_{0}n ---> ^{7}_{3}Li + ^{4}_{2}He$$

The most commonly used detectors are Boron trifluoride (BF_3)
and ^{3}He gas filled proportional counter tubes. These are tubular
in form but their dimensions, and the gas pressures within them,
vary. ^{6}Li enriched glass scintillation counters, which are also
efficient detectors of gamma (γ) radiation, are sometimes used in
dual purpose instruments which measure moisture content by neutron
moderation, and bulk density by the γ-backscatter method. All three
types of detector system require a very high and stable voltage
supply to operate (0.8 - 2.3 kV). The various detectors all detect
γ radiation and efficient pulse discrimination circuits are required
to ensure that only the pulses arising from the detection of thermal
neutrons are counted.

5. INSTRUMENT DESIGN

The complete neutron probe equipment consists of 3 main components:
1. The depth probe
2. Counter unit
3. Transport shield
A schematic diagram of the equipment is shown in Figure 2.

5.1 The Depth Probe

The probe consists of a waterproof cylindrical stainless steel,
aluminium or brass casing which contains the source and detector and
an amplifier. Some probes also contain the pulse discrimination
circuitry and the high voltage generator for the detector tube. The
IH II probe is of the latter type. Different designs of probe vary
in length from 0.30 - 0.75 m, and the most common diameters are 38
and 47 mm, designed for use in 44.5 mm (1 3/4") and 51 mm (2") OD
access tubes respectively. Aluminium is arguably the best material
for probe casings because it is a very poor neutron absorber compared to iron and copper, both of which will reduce thermal neutron
count rates by a few percent. The IH II probe casing is 38 mm in
diameter and 0.75 m long but is of stainless steel because of its
considerably superior strength and resistance to corrosion.

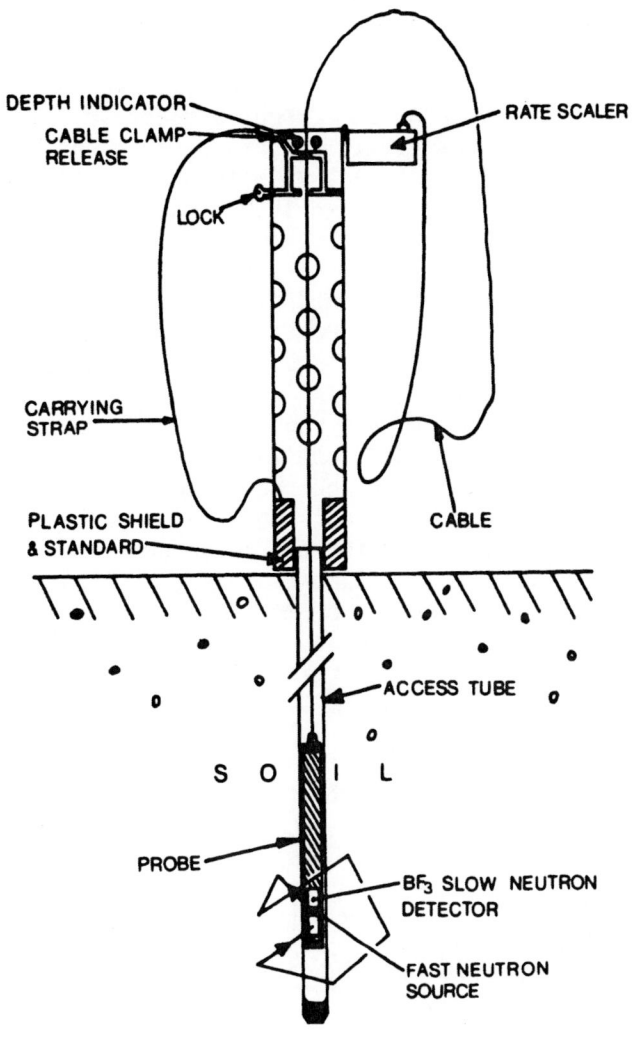

Figure 2. Schematic diagram of the neutron probe system in use.

The source and detector are located at the lower end of the probe casing and various source-detector geometries are used in different probes.

When BF_3 and ^3He detectors are used, the optimum geometry is achieved by locating the source at the center of the sensitive length of the detector. This results in a linear calibration relationship between thermal neutron count rate and volumetric moisture content. Annular sources, which fit like a short sleeve around the center of the detector tube, are generally used in this geometry. This is the arrangement used in the IH II probe.

Cylindrical sources are sometimes used, located beside the detector, but side placed sources result in a slightly assymetrical neutron distribution which may not be desirable in nonhomogenous soils.

In some designs, the source is placed below the detector. This arrangement gives rise to non linear calibration curves, although this may not be a serious problem if the detector is short (12). However, the center of sensitivity of this arrangement is not clearly defined when there is a gradient of moisture content because the detector is not symmetrically placed in the thermal neutron field. Cylindrical sources, which are less expensive, are used.

For probes with scintillation detectors, the optimum source-detector distance for sensitivity and linearity of the calibration is between 0.05 and 0.10 m (11). Part of the space between the two is occupied by a lead spacer to shield the detector from direct γ radiation from the source.

The probe is connected to the counting unit by a cable which is often also used to lower the probe into the access tube. The cable carries power to the probe and the pulses from it. The IH II probe cable carries only a 12v supply for the circuitry housed in the probe. If the high voltage for the detector is carried to the probe in the cable, the latter is more prone to faults. These usually arise due to the penetration of moisture into the connections or because flexing of the cable gradually causes the insulation to fracture. The cable is usually terminated by a demountable waterproof connector at each end, but in some designs the cable emerges directly from the top of the probe. With the latter arrangement, cables are not easily changed, if for example a cable fails or a longer length is required for a different application.

5.2 The Counter Unit

The counter unit is usually a ratescaler, although rate meters and scaler-timers are also used. A ratescaler counts the pulses from the probe for a preset time interval and then displays the mean count rate in counts per second. The data are sometimes displayed standardized to a particular counting period (eg 16 seconds). Various preset counting times can be selected, providing a choice of counting precision. Typical count times offered might be 8, 16, 32 and 64 seconds. The IH II ratescaler offers 16 and 64 second and 16 and 64 minutes counting times. The latter two are provided for very high precision counting, for standard counts for example.

Scaler timers, in contrast to ratescalers, operate by timing the interval required to count a preset number of counts. Rate meters produce a voltage analogue of the count rate which is displayed on a moving coil meter and can be output to a chart recorder. Rate scalers and scaler timers are preferred to ratemeters as the precision of the readings is considerably better (0.1% or better compared to 1-2%). There is also no subjective bias in making the readings and dead time corrections are unnecessary. Ratemeters are now little used.

The counter unit usually also contains rechargeable cells to power the system, but removeable rechargeable packs are sometimes used. Fully charged cells should run the equipment continuously for at least 10 hours. The IH II system has the ability to operate from both rechargeable and dry cells. The latter may be used regularly or carried for emergency use. The counter unit should be waterproof and able to operate reliably over a temperature range from -15 to at least +50°C.

The most recent development in ratescaler technology is the use of microprocessors. These use low power consumption CMOS circuitry and the level of sophistication varies; some units can process the count rate data into moisture content directly given the correct calibration data, others can store the data, and some can do both. When the data has been stored in the ratescaler memory, it can be output via an RS-232 standard interface to a printer, or to a computer for storage and further processing. The data may also be output to a modem and transmitted over telephone lines to a distant computer. The facility to store the data and output it directly to a computer considerably reduces the scope for the introduction of errors into the data by eliminating manual field recording and manual transfer of the data.

In the case of the IH Memory Ratescaler, shown in Figure 3, the microprocessor can also log the depth of the probe directly from an electronic depth counter unit. The software for the microprocessor has been written to allow a "directory" of access tube numbers and their associated reading depths to be entered and stored. After each reading has been made, the next reading depth is displayed to prompt the operator to lower the probe.

Figure 3. The IH Memory Ratescaler.

The depth of the probe is monitored and shown on the ratescaler display, and counting can only proceed when the probe has been lowered to the correct depth. If required, observations may also be made without using the directory facility; in this case, the operator lowers the probe to the desired depths, which are logged immediately before the count rate readings are made.

The unit has 8k of RAM and can store an amount of data equivalent to that from 65 access tubes with 10 readings in each. The date, time, tube number and other details are stored with the count rate data from each tube.

5.3 The Transport Shield

The transport shield is used to house the probe for carrying purposes. It is fitted with radiation shielding at the lower end and a locking mechanism, or provision for locking, to prevent unauthorized removal of the probe from the shield. The transport shield may, as in the IH II system, also house a depth measuring mechanism and cable clamp. The radiation shielding usually consists of a sphere or block of hydrogen-rich material, usually paraffin wax or polypropylene, with a tube through the center to house the end of the probe containing the source. The base of the transport shield is fitted with a socket which locates on the access tube.

In use, the transport shield, with the probe in place, is located on an access tube and the probe is then lowered from the shield into the tube, suspended by the cable. Occasionally, a lightweight chain, or a thin stranded steel cable with stops at fixed intervals (eg 0.10 m) is used to suspend the probe.

Depth measurement and clamping methods vary. The IH II system uses a mechanical depth counter which is driven by a friction wheel held against the cable by a pinch wheel. The counter is calibrated in centimeters, and the latest models incorporate an electronic shaft encoder which interfaces with the Memory Ratescaler system. The probe is held at the required reading depth by a built-in clamp released by a lever. Other methods include the use of depth marked cables with a removable spring loaded clip to hold the probe in position, and the use of cables with stops which can be fixed firmly at the required intervals. With both of the latter two methods, cables must be checked for possible stretching in use.

The ratescaler is usually attached to the top of the transport shield. In the IH design it is quickly detachable by the removal of a threaded hinge so that the ratescaler and the probe in its transport shield may be kept and used separately, for example to charge the batteries or unload data from the memory ratescaler. This facility can reduce the time that the operator and others have to spend near the probe itself (see Section 12, Safety).

The IH II probe transport shield is longer than most other designs but it has significant advantages. In use, (see Fig.1) the controls are conveniently and ergonomically placed at waist level and the operator need not bend or squat down (which may disturb a crop) to make observations. The equipment has been designed to be carried by a shoulder strap so that the shielded source may be kept at least 0.40 m from any part of the users body or limbs. The shield is also provided with lugs around which the cable can be wound for storage. Figure 4 shows the probe being carried in the field.

At a field site, the probe is carried in its transport shield, but for storage, carriage in a vehicle, or freighting, the complete instrument is packed in a carrying case. This must comply with various safety standards and provides some additional shielding, usually by virtue of its size.

Figure 4. The IH II neutron probe system being carried in the field.

6. RESOLUTION AND MEASUREMENTS NEAR THE SURFACE

6.1 The Sphere of Importance

A knowledge of the size and shape of the volume of soil which affects the count rate reading is important. In a uniform soil, the interactions which influence the thermal neutron count rate occur in a virtually spherical zone around the source within which the soil nearest the source has the greatest influence on the count rate. Olgaard (10) proposed the term "sphere of importance" for this spherical zone and defined it as "the sphere around the source, which, if all soil and water outside the sphere were removed, would yield a (thermal) neutron flux at the source that is 95% of the flux obtained if the medium is infinite."

The radius of the sphere of importance decreases with increasing moisture content but also depends to a lesser extent on the soil bulk density and chemistry, the neutron energy spectrum (ie source type) and on the length of the detector used. The radius decreases with increasing bulk density and amount of thermal neutron absorbers and increases with neutron energy. Increasing the detector length tends to elongate the sphere in the vertical sense.

For an Am-Be source, the radius of the sphere of importance has been calculated to range from about 0.22 m at a moisture content of 0.30 by volume, to 0.47 - 0.53 m at a moisture content of 0.05 for a range of Danish soils. In pure water, the radius is about 0.10m (10).

6.2 Measurements Near Interfaces in the Soil

If the sphere of importance intersects an interface betwen soil at different moisture contents, the sphere will become distorted and the count rate will reflect the influence of both layers of soil.

This is illustrated in Figure 5 which shows the apparent moisture content indicated by a Rothamsted neutron probe (12) when measurements were made through an interface between a wet clay soil and the same soil at various lower moisture contents.

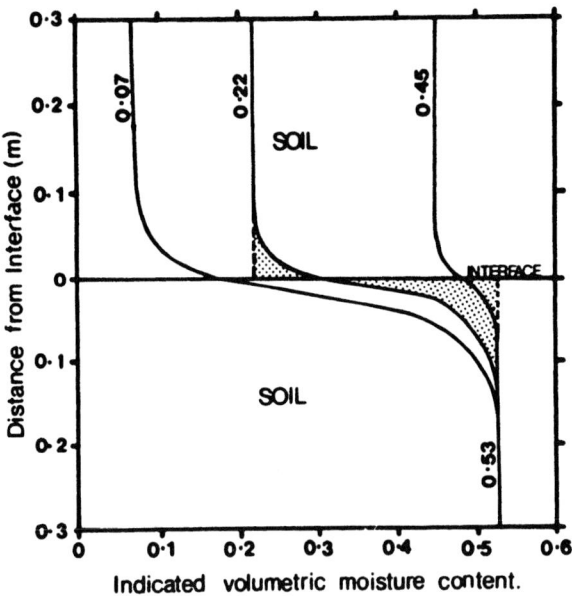

Figure 5. Interface effect between clay soil at various moisture contents and at field capacity. (Redrawn from Long and French, 1967)

Different probes will produce slightly different curves depending on their source-detector geometry, but the general form of the curves is common to all probes. It can be seen that as the moisture content difference between the layers increases, the distance from

the interface at which the count rate is affected increases. The
moisture content is overestimated in the drier layer and underestimated in the wetter layer. However, the overestimation in the drier
layer is always less than the underestimation in the wetter layer
(represented by the shaded areas in Fig.5) and the discrepancy
increases as the difference between the moisture contents of the
layers increases.

These data infer that the total moisture content of a soil profile will always be underestimated if there are gradients of moisture content within it. Symmetrical source-detector geometries, ie
those with centrally placed sources, are preferable from this point
of view. It is also apparent that the neutron probe cannot be used
to measure the exact moisture content profile in a soil. If an
exact profile is required, other methods of moisture measurement
must be sought.

In field soils, moisture gradients are generally not as abrupt
as those illustrated, and the underestimation is usually of minor
importance and tends to be ignored. Since the depth resolution of
the neutron probe is limited, there is generally no advantage to be
gained in making measurements at depth intervals of less than 0.1 m.

6.3 Measurements Near the Surface

When the sphere of importance intersects the soil surface, some
neutrons will escape and the count rate will be reduced so that the
calibration relationship for the bulk soil cannot then be used; this
is illustrated in Figure 6.

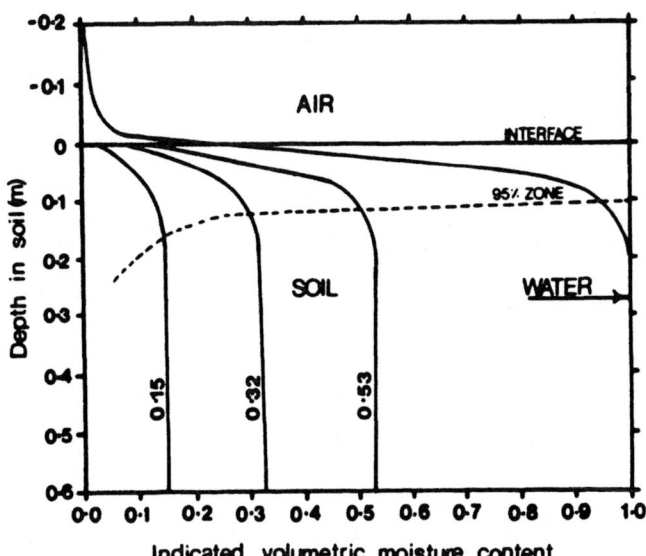

Figure 6. Air soil interface effect in Rothamsted clay soil at
various moisture contents. (Redrawn-Long & French,1967)

Various methods have been used to take account of this problem and these include the use of surface extension trays and plastic neutron reflectors, the derivation of individual calibrations for the surface layers and the use of empirical corrections.

A surface extension tray(2) is a means of temporarily extending the soil surface upwards using a tray of topsoil at approximately the same moisture content as that of the surrounding soil so that when readings are made at a depth of 0.10 m, for example, the sphere of importance remains within soil. However, they are heavy to handle and in many cases cannot be used because of the damage they would cause to any crop around the tube. They now appear to be little used.

Plastic neutron reflectors (3) are designed to function in a similar way, using a hemispherical layer of plastic, eg. polyethylene, to contain the thermal neutron field. Generally reflectors will only satisfactorily compensate the count rate over a limited range of moisture contents and other correction procedures may still be required. In addition, they suffer from the same disadvantages as surface extension trays and do not appear to be very popular.

The use of individual calibrations derived for the surface layers seems to be one of the most favored methods and has been used with considerable success, for example by Greacen et al. (14), Kristensen (15) and Luebs et al. (16). Methods for establishing the calibration relationship are given in Section 7.

The empirical surface correction methods of Harris (17) and Grant (18) and the use of surface extension trays have been tested by Parkes and Siam (19) who favor Grant's method. Both empirical methods are based on field measurements and assume uniformity of moisture content in the upper part of the profile although Grant describes an iterative procedure for correcting for non-uniformity. Grant's method involves the measurement of count rates at, for example, 0.1 m intervals from the surface to a depth of about 0.6 m, followed by the removal of the upper .10 m of the soil around the tube over an area at least 1.0 m in radius. Count rates are measured again at the same depth intervals. Correction factors are derived from the ratios of the count rates before and after the removal of the upper layer of soil by a process of cumulative upward correction from the 0.6m depth (which is assumed to be unaffected by the surface). The procedure is repeated at a range of moisture contents to derive the relationship between correction factors and moisture content for that soil.

7. CALIBRATION

7.1 The Calibration Relationship

The relationship between thermal neutron count rate and volumetric moisture content depends on the soil and on the probe design. For most modern probes, with the source placed at, or close to the center of the detector, the relationship between thermal neutron count rate (R) and volumetric moisture content (θ) for non-swelling soils is linear and is usually expressed in the form:

$$\theta = a.R/R_s + b \qquad (6)$$

where a is the slope, R_s is the count rate in a standard, and b is the intercept. The use of a standard is described in Section 8.

θ is generally used to denote "free" water, which is that driven off when the soil is dried at 105°C. Different soils also contain varying amounts of equivalent or "bound" water (for example hydrogen in organic matter and water of hydration) which is not driven off at 105°C. This is denoted by θe (14). The measured thermal neutron count rate is actually related to the total water content of the soil θt where $\theta t = \theta + \theta e$, but calibrations need not be expressed in terms of θe as for a given soil θe does not change and is incorporated in the intercept term. However, Greacen et al. (14) demonstrate that scatter in calibration data is reduced if the calibration is expressed in terms of θt.

The main soil factors influencing the calibration are dry bulk density and the amounts of strong neutron absorbers present. The bulk density affects both the macroscopic scattering and absorption cross sections of the soil (Eq. 5). The effect of bulk density on the calibration is shown in Fig 7, after Jensen and Somer (20). Density effects have been discussed by many authors, among them Holmes (21), Olgaard and Haahr (22), and Greacen and Schrale (23). Soil texture affects the calibration indirectly through variations in bulk density, chemistry and the amount of bound water.

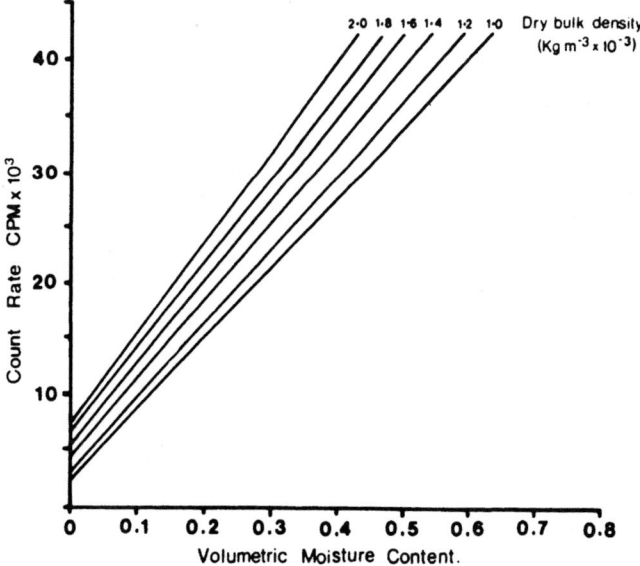

Figure 7.   Calibration curves computed by the Olgaard method for the NEA-BASC neutron probe in soils with different dry bulk densities. (After Jensen and Somer, 1967)

The most important absorbers commonly found in soils are B, Cl, Fe, Mn and K. The Chlorine content of saline soils for example may significantly affect the calibration. The rare earth elements, particularly Gadolinium, are extremely strong absorbers and will cause marked effects even if very small variations in concentration occur. Nicolls et al. (24) have shown that 5 ppm of Gadolinium makes the same contribution to the macroscopic absorption cross section of the soil as 21 ppm of Boron, 67 ppm of Cadmium, 0.15% of chlorine or 3.2% of iron. They found Gadolinium concentrations ranging from 0 - 36 ppm in some Tasmanian soils.

The calibration is also influenced by the diameter, wall thickness and material of the access tubes used; the calibration slope decreases with increasing tube diameter and wall thickness.

7.2 Derivation of the Calibration

There are 3 main methods of deriving the calibration relationship:
1. Theoretical methods
2. Drum calibrations carried out in the laboratory
3. Field calibrations

Theoretical methods using neutron capture models such as those of Olgaard (10) and Couchat (8) require accurate values of the macroscopic neutron interaction cross sections of the soil for which the calibration relationship is required. These may either be determined from a detailed chemical analysis of the soil, or by direct measurement.

Olgaard's method requires the volumetric concentration of at least 20 different elements in the soil to be accurately measured (to a few ppm in the case of some elements) so that the interaction cross sections can be calculated from published microscopic cross sections using Eq. 5. The calibration is then derived using a mathematical simulation which includes the source and detector characteristics and geometry of the particular probe. Olgaard's method has been extensively tested by Greacen and Schrale (24) and has been found to be satisfactory for soils with low macroscopic absorption cross sections (<0.0004 m^2/kgm) but unsatisfactory for soils with large absorption cross sections. Problems may arise because some key elements may be inadvertently missed out of the chemical analyses, which are time consuming and expensive.

Theoretical methods using direct measurements of the neutron interaction cross sections of the soil appear more promising as chemical analyses are not involved and the effects of all the elements present are automatically taken into account. This approach has been used by Couchat et al. (25), McCulloch and Wall (26) and Williamson and Turner (27), using different experimental methods to measure the neutron interaction cross sections. Vachaud et al. (28) compared calibrations derived gravimetrically with those derived by a neutron capture model using direct measurements and found that the methods gave comparable results, but that the latter approach was particularly suited to "difficult" soil types in which the gravimetric method cannot be applied with confidence.

Drum calibrations involve the careful and representative collection of large amounts of soil (approximately 1 tonne, but up to 4 or 5 tonnes in the case of sands). This is then uniformly packed into a large drum which must be of sufficient size to contain the sphere of importance over the required range of moisture contents. Figure 8 shows the minimum drum radii required to give an accuracy of 0.005 by volume moisture for weakly and strongly neutron absorbing soils both with a density of about 1400 kg/m^3.

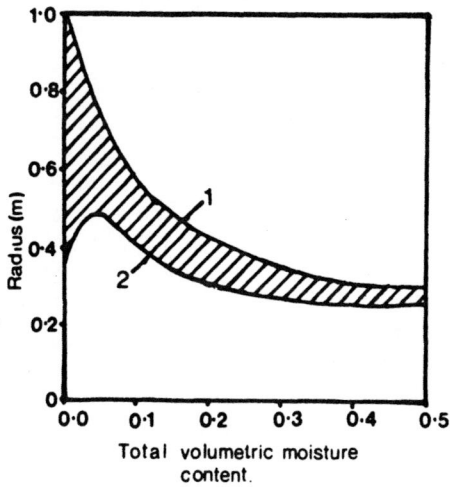

Figure 8.   Radius of a sample giving an accuracy not worse than 0.005 by volume moisture for a soil with a dry bulk density of about 1400kg m^{-3}. Curves are shown for: 1. a mixture of pure SiO_2, water and air, and 2. a soil with a high concentration of strong neutron absorbers. (Redrawn from International Atomic Energy Agency, Neutron Moisture Gauges, Technical Report Series No. 112, IAEA Vienna, 1970, p. 63)

The soil must be packed at a known moisture content and density. An access tube is then installed in the center of the drum using the field installation technique and the neutron probe count rate is measured precisely. The process is repeated over a range of water contents. The drum need not be packed to the field density as Greacen and Schrale (23) have shown that the count rate, n, for each packing of the drum (density ρ) can be satisfactorily corrected to the count rate, n_f, at the mean field density, ρ_f using the square root correction equation:

$$n_f = n\ (\ \rho/\ \rho_f)^{1/2} \qquad\qquad (7)$$

The calibration relationship can then be derived from the pairs of count rate ratio and volumetric moisture content data by linear regression. Greacen et al. (14) indicate that the calibration will contain less bias if the count rate ratio is regressed on the moisture content and the expression is then inverted.

Field calibration requires the installation of a temporary access tube at the location of interest which ideally should be as close as practicable to the field sites for which the calibration is required. Precise neutron counts are taken at the required depths, and 5 or 6 large undisturbed core samples of known volume are then taken, centered on those depths, from as close to the tube as possible. The volumetric moisture content of the core samples is determined gravimetrically. The process must be repeated over as wide a range of moisture contents as possible. Details of field calibration are given in Bell (2) and Greacen et al.(14). Vachaud et al.(28) and Haverkamp et al.(29) used a gamma soil density probe to measure the bulk density, so that core samples of known volume were not required. This simplifies the sampling procedure but may introduce additional errors arising from the calibration of the density probe. The square root correction (23) can be used to correct the individual count rates to the mean soil density, and this has been shown to reduce scatter in the data. The regression is carried out in the same way as for the drum calibration data. Greacen and Hignett (30) have reported a method for the correction of moisture content data for the effects of cracking in swelling clay soils.

Calibrations carried out for different probes in a range of soils are also reported in Lal (31,32), Luebs (15), Shirazi and Isobe (33) and Nakayama and Reginato (34). The latter authors describe a technique for intercalibrating different probes using a range of plastic standards.

Field calibrations are the most widely used although carefully carried out drum calibrations generally give the most precise results (23), although they are not necessarily the least biased. Bias may arise, for example, if all of the soil for the calibration is taken from a single pit so that the small scale areal variability of the calibration relationship is not taken into account. Field calibration is generally simpler. The calibration equation may be the source of considerable bias in neutron probe data and great care and attention to detail is required when carrying out calibration work, in order to minimize this bias.

The effort and expense put into the derivation of the calibration ultimately depends on the precision required. Manufacturers calibrations should generally be treated with caution as some are derived using artificial soils containing mixtures of neutron absorbers. These are unsatisfactory. For the IH probe, using 44.5 mm OD, 1.5 mm wall access tubing and a water drum as a standard, some examples of field derived calibrations are given below:

Sandy, silty or gravelly soils	$\theta = 0.790 \ R/R_s - 0.024$
Loams	$\theta = 0.867 \ R/R_s - 0.016$
Clay	$\theta = 0.958 \ R/R_s - 0.012$

8. SOURCES OF ERROR

The neutron probe is generally used to measure changes of moisture content with time, rather than absolute moisture contents. In most applications, the mean moisture content change within a particular area (eg a plot, field or catchment) is required and the aim should be to minimize the error in this value. Very small changes of moisture content may be resolved as the precision of measurement can be very high. If measurements are made in a small area of uniform soil, the accuracy of the measurements is mainly limited by the quality of the calibration, but over larger areas, estimates of the mean change in moisture content are influenced by spatial variability, which is independent of the soil moisture measurement technique adopted.

Sources of error and their statistical treatment have been covered in detail by Hewlett et al. (35) and more recently by Sinclair and Williams (36), Williams and Sinclair (37), Vachaud et al. (38) and Haverkamp et al. (29). Neutron probe soil moisture data are subject to both random and systematic errors.

8.1 Random Errors

The main components of the random error arising from the measurement of the areal mean soil moisture content or moisture content change using the neutron probe are:

1. The instrument component, due to random counting and depth placement errors.
2. The location component, due to the areal variability of soil moisture and moisture change.
3. The calibration component, due to variance in the slope of the calibration relationship.

Random counting error arises from the random nature of radioactive decay. Neutron count rates conform to the Poisson distribution, and the random counting error σ_R of a given count rate R (counts s^{-1}), is given by:

$$\sigma_R = (R/t)^{1/2} \tag{7}$$

where t is the preset count time in seconds. From this, it can be seen that the counting error can be reduced by increasing t and, for a given value of t, the error increases with count rate. When the error is expressed in terms of moisture content, θ , the error in the standard count (which is generally very small), should be included:

$$\sigma_\theta = a.R/R_s \ (1/R.t + 1/R_s.t_s)^{1/2} \tag{8}$$

where a is the gradient of the calibration, R_s is the standard count, and t and t_s are the count times in the soil and the standard, respectively. If the following typical values are taken (for an IH neutron probe, sandy soil and water standard): $a = 0.79$, $R = 300$, $t = 16$, $R_s = 900$ and $t_s = 640$, the moisture content error, σ_θ is 0.004 (68% probability) at a moisture content (from Eq. 6) of 0.239. From this example and Eq. 8, it can be seen that, for measurements made at a single location using the correct calibration, the precision of measurement is high and limited only by the counting times t and t_s and ultimately by the stability of the instrument. If for example, t and t_s are both increased to 16 minutes, σ_θ is reduced to 0.0006.

Depth placement errors occur when the probe is not exactly relocated at the correct depth when repeat measurements are made. They are greatest when there are steep gradients of moisture content in the soil profile, for example at interfaces and particularly near the surface (see Figs 5 and 6). The need for accurate and reproducible depth placement is very important when changes are being measured. Given precise depth placement, moisture content changes as small as 0.002 can be resolved fairly readily.

The location component is dependent on the areal variability of soil moisture content and moisture content change over the area of study. Under grass and crops with narrow row spacings, moisture abstraction and input are likely to be fairly uniform, but in taller and more widely spaced row crops, orchards and tree plantations, moisture abstraction will be very variable due to uneven root distribution. Moisture inputs will also be areally variable due to variations in throughfall and stemflow.

The locational component can be reduced by increasing the number of sampling sites (ie access tubes). If the areal variability of moisture content and moisture content change is known, the number of tubes required for a given level of precision can be calculated using the relationship below, given by Peterson and Calvin (39):

$$N = t^2 S^2 / D^2 \tag{9}$$

where N is the number of sampling sites (tubes), t is the value of Student's t at the probability level required, S^2 is the variance of the moisture content (or change) and D is the precision specified, for example, ± 5 mm.

In general, when a program of soil moisture monitoring is being planned for an experimental plot, and the time available for observations is limited, a balance has to be struck between increasing the counting time to reduce the random counting error, and increasing the number of access tubes in order to reduce the locational error. Locational errors are usually very much greater than random counting errors (36), and it is therefore better to install a large number of tubes and read them using short counting times. Additionally, the benefits of establishing high precision calibration equation(s) should be considered.

8.2 Bias

Measurements made with the neutron probe are subject to sampling and measurement bias. Sampling bias occurs when the access tube sites selected are not representative of the area under study, or if, during installation work, the sites are made unrepresentative through disturbance to the soil or vegetation. This may be due to poor installation technique or to carelessness.

Measurement bias largely occurs through the use of an incorrect calibration equation and through the effects of instrumental drift. Calibration bias may arise if a calibration is assumed, if the calibration procedure is badly carried out, or if observations are made in a different type of access tube to that used to derive the calibration. In general, the range of calibration slopes for most probes is not great: for the IH II probe, slopes rarely fall outside the range from 0.65 - 1.00 (using 44.5 mm Al access tubes and water standard).

Instrumental drift is usually caused by ageing of the detector and other components, but may be due to component changes due to repairs or to slight shifts in the discriminator or EHT settings due to shock. Bias due to instrumental drift is removed by standardization of soil count rates using the count rate in a standard absorber (ie use of the count rate ratio, R/R_s).

8.3 Standard Counts

The preferred standard is a drum of water, at least 0.5 m diameter and 0.6 m deep with an access tube fitted vertically in the center. A water standard provides an effectively infinite absorber in which the count rate is largely independent of temperature, and also provides a common reference standard, allowing probes of the same type to be used interchangeably, without the need for individual calibration.

A long, precise count should be taken in the drum at regular intervals (ideally on each day that observations are made). These data should be recorded in a book kept for the purpose so that probe performance can be monitored. Standard counts should be highly stable and any sudden changes in the standard count should be investigated (2).

Many manufacturers recommend the use of the hydrogen rich shielding of the transport shield as a standard but this is unsatisfactory for several reasons:

a)   Transport shields are small and therefore not "infinite." Count rates within them are affected by the proximity of surrounding (particularly hydrogenous) material so that the counts must be taken with the probe in as reproducible position as possible.

b)   The probe must also be locked in an exactly reproducible position in the shield.

c)   Count rates in the shield are temperature dependent.

The high precision of the standard counts is important as all of a day's readings are standardized using this count. If this varies from day to day as a result of either poor counting precision or bias introduced by careless use of the shield as a moderator, an additional source of error will be introduced into the data.

9. ACCESS TUBES

9.1 Access Tube Materials

The dimensions of, and materials used for, access tubes vary considerably. The most important requirements for the material are transparency to neutrons, mechanical strength and resistance to corrosion. Dimensionally, the radial clearance between the probe and the tube wall should be as small as possible, but should be large enough to avoid the probe sticking in a slightly distorted tube. The wall thickness of the tube has to be a compromise between an increase in strength and a decrease in count rate as the thickness increases.

The most commonly used material is probably aluminium (or Al alloy), but stainless steel, brass, galvanized iron, and polyethylene are also used. PVC may be used but is not recommended as count rates are reduced because of its high chlorine content. Aluminium is preferred as it is virtually transparent to neutrons. The choice of access tube material and thickness depends on several factors, including soil type, soil chemistry (corrosiveness), depth and method of installation and availability.

9.2 Installation of Access Tubes

The installation of the access tube is one of the most important aspects of the field work. A badly installed tube will give rise to permanently biased readings. During installation work, the points below must be adhered to in order to avoid rendering the soil and surroundings unrepresentative of the area under study.

a) Disturbance of the soil surface around the installation point, for example, through compaction or puddling in wet conditions, must be kept to a minimum.

b) If installation is carried out in a vegetated area, the vegetation must not be damaged. Work may have to be carried out from a raised platform to minimize disturbance.

c) Disturbance of the soil (eg compaction) in the profile around the tube must be avoided. Any density changes caused will affect the calibration and may influence the processes of moisture movement around the tube, in the zone which has the greatest affect on the count rate.

d) The access tube must be a tight fit in the hole. There must be no cavities or annular voids around the tube. The effect of poor installation has been examined by McGowan and Willams (40), who found that most of the damage was restricted to the upper 0.3 m

of the profile. Under unsaturated conditions, small voids around
the tube will remain empty of water and will have only a minor
effect on the count rate, but if saturated conditions occur, these
voids will fill and have a pronounced and erroneous effect on the
count rate.

Most of the recommended installation methods are variations on
the following technique, which has been described by Bell (2) and
which is shown in use in Figure 9.

Figure 9. Installation of an access tube using an auger and guide
           tube (reamer).

An undersized "pilot" hole is drilled using an auger working
through a steel guide tube (or reamer) whose outer diameter is the
same as, or very slightly smaller (eg 0.1 mm) than that of the
access tube to be installed. The pilot hole is drilled up to 0.15 m
ahead of the guide tube, which is then driven down with a slide
hammer to ream out the hole to the correct diameter. The guide tube
is sharpened on the inside so that the cuttings move inwards as the
hole is reamed, avoiding compaction of the soil around the tube.
The guide tube is 1.15 m long, and once a depth of about 1.0 m has

been reached, the tube is withdrawn and replaced by a tube 1.0 m longer than the first. The handle of the auger is extended and the process continues using progressively longer guide tubes until the desired depth is reached. In favorable conditions, a depth of up to 6 m can be reached using this method. Greater depths can be reached using sectional guide tubes. The guide tube is then removed (a jack is usually required) and an access tube is inserted. The top of the access tube is cut off to leave the required length extending above the surface. Tubes should be sealed at the bottom if saturated conditions are likely to occur. A cap is also required to keep out rainwater.

If the method described above is used to install a deep tube, particularly in a heavy, sticky soil, the repeated withdrawal of guide tubes may enlarge the top of the hole, and in this instance, it may be advisable to use the access tube itself as a guide tube. A disposable steel cutting edge may be required for aluminium tubes or steel access tubing may be used. After installation, the tube must be cleaned thoroughly inside and sealed at the bottom. Prebble et al.(41) describe several sophisticated techniques, for very deep installations (5 to 10 m or more) and for rocky or stony soils.

When access tubes are located in cultivated soils, they may be designed so that the top section, from about 0.25 m below the soil surface, can be removed so that cultivation can be carried out unimpeded. When the top is to be removed, the lower part of the tube is plugged and its position is recorded by triangulation so that it can be relocated. Alternatively, a metal detector may be used. Once the tube has been relocated after cultivations, a hole is carefully excavated to allow the top section to be reattached and the soil is then replaced.

The method of installation adopted will depend on the circumstances prevailing, as no single method is suitable for all conditions. Methods involving driving tubes directly into the soil or into augered holes should be avoided. When readings are being made, great care must be taken to minimize any disturbance to the soil and vegetation.

10. DATA PROCESSING

10.1 Manual Data Recording

Data should be recorded onto carefully designed, printed field data sheets which should have clearly labelled boxes in which to enter the data. In addition to the reading depths and readings, the following data should also be recorded: access tube number, date and time, serial numbers of the probe and ratescaler used, observer, crop, counting times used (in soil and in the standard), standard count and number of depths read. A space in which comments can be entered is also useful. If the data are to be processed by a computer, the observer can sum the count rate readings for all of the depths and record this on the field data sheet. When the data

are keyed into the computer, the machine can sum the count rate
readings entered and compare this sum with that calculated manually
by the observer. If the two values agree, the data have probably
been entered correctly.

10.2 Data Logging

When data are logged by the ratescaler, they must be stored in
such a way that few ambiguities can arise when the data are transmitted to a computer or are shown on the built-in display. The scope
for introducing human error into the data must also be limited. For
this reason, the IH Memory Ratescaler software has been designed so
that each set of readings from an access tube is identified by a 4
digit tube number, and the date and time, and each count rate reading is stored with the depth at which it was made. Using the directory facility, readings cannot be made at the wrong depths or missed
out inadvertently. If the latter occurs, and the count rate data
alone are stored, it can be a very tedious and time consuming task
to locate where the error occurred and subsequently edit the data,
particularly after it has been transferred for processing and storage. The problem is made even more difficult if large quantities of
data are being handled.
Some ratescalers provide the facility to convert the readings
into moisture contents as they are taken. This facility is useful
if the moisture content in a particular layer needs to be known
almost instantly, but in many cases, for example in irrigation
scheduling, the integrated profile water content is required and
this has to be calculated independently from the depth and moisture
content data. This may be readily done using a cheap programmable
calculator, but it is then no less complicated to calculate the
result directly from the count rate data. In most cases the data
will be processed and stored for further analysis using a computer,
so that there is often little advantage in field processing. Field
processing also requires the use of the correct calibration equation
and an up to date standard count. Data are best stored in the "raw"
form, as count rates, so that there can be no confusion as to which
calibration or standard count was used.

10.3 Data Processing

When the data are processed by a computer, or microcomputer,
the processed listing should ideally show:

Tube no., date, time, etc.
The depths
The readings (counts s^{-1}) at each depth
The volumetric moisture contents (θ) at each depth
The moisture storage in each layer*
The total moisture storage in the profile down to each depth

The change in layer moisture storage since the last reading date

The total storage change in the profile down to each depth

The random counting error for each layer moisture content is also useful. An example of data processed using a microcomputer is shown in Figure 10.

OBS. 42 TUBE 30188   DATE 300683   RDG NO. 1

TIME 1300 GMT   PROBE NO. 0169   COUNT TIME 64S   WATER COUNT 940

DEPTH (CM)	READING	MVF	LAYER (CM)	WATER IN LAYER(MM)	CUM (MM)	CHANGE SINCE 1/230683 LAYER (MM)	CUM 1208 (MM)
10	164	.164	0- 10	16.4	16.4	- 2.3	- 2.3
20	203	.147	10- 20	15.5	32.0	- 2.4	- 4.8
40	229	.168	20- 40	31.5	63.5	- 4.2	- 8.9
60	215	.157	40- 60	32.5	96.0	- 2.1	- 11.0
80	201	.145	60- 80	30.2	126.2	- .8	- 11.8
100	164	.114	80-100	25.9	152.0	- .5	- 12.2
120	236	.174	100-120	28.8	180.9	- 1.0	- 13.3
140	383	.298	120-140	47.2	228.1	- 1.8	- 15.1
160	428	.336	140-160	63.4	291.4	- 1.3	- 16.4
180	474	.374	160-180	71.0	362.4	- .8	- 17.3
200	397	.310	180-200	68.4	430.9	- .7	- 17.9
220	446	.351	200-220	66.0	496.9	- .3	- 18.3
240	427	.335	220-240	68.6	565.5	- .2	- 18.4
260	390	.304	240-260	63.9	629.3	+ .2	- 18.3
280	405	.316	260-280	62.0	691.3	+ .9	- 17.3
300	413	.323	280-300	63.9	755.3	+ .3	- 17.1
320	406	.317	300-320	64.0	819.3	- .7	- 17.8
340	376	.292	320-340	60.9	880.2	- .9	- 18.7
360	376	.292	340-360	58.4	938.6	- 1.1	- 19.8

Figure 10. An example of data processed using a microcomputer.

*The reading at each depth is usually taken to represent the moisture content of a layer extending halfway to the adjacent reading depths. Alternatively, layers may be regarded as extending from one reading depth to the next and their moisture content is then the mean of the moisture contents at those depths.

The layer storage is $\theta*d$, where θ is the moisture content and d is the thickness of the layer. The total profile storage is obtained by summing the individual layer storages. This in effect integrates a step function of the moisture content. Haverkamp et al. (29) indicate that the use of Simpson's approximation reduces the errors involved in the calculation of the profile moisture storage.

11. ALTERNATIVE METHODS OF SOIL MOISTURE MEASUREMENT

11.1 Gravimetric Determination

This is the method against which the neutron probe is often calibrated. The main advantage of the method is that it is simple and requires little capital outlay.

The disadvantages are that undisturbed and volumetrically accurate core sampling is difficult to achieve so that bias may arise.
Furthermore, the method is very destructive and time consuming,
particularly when large numbers of replicate samples are required to
gain adequate precision, and when frequent measurements are required.
Deep sampling is difficult and, for the measurement of moisture
changes, the precision of the method is less than that of the neutron probe. Hewlett and Douglass (42) found that between 10 and 20
pairs of samples were required to estimate volumetric moisture
content to a precision of ± 0.01.

11.2 Estimation from Measurements of Soil Matric Potential (Suction)

This method requires the moisture characteristic of the soil
(the relationship between matric potential and volumetric moisture
content) to be known very precisely. The moisture characteristic
varies with soil texture and structure and the relationship is hys-
teretic (ie dependent on whether the soil is wetting or drying). In
view of this, the method is only satisfactory when the soil is drying. Matric potential can be measured using tensiometers, resistance
blocks, heat dissipation blocks (43) or thermocouple psychrometers,
but in general, two types of instrument are required to cover the
full range of moisture content found in the field. The method has
been used by, among others, LaRue et al.(44) and Nielsen et al.(45).

11.3 Dual Tube Gamma Method

This method uses the principle of gamma attenuation and requires
paired access tubes to be installed exactly parallel to each other
20-50 cm apart. A gamma source is placed in one tube and a detector
is placed in the other and the two are lowered in unison to the required depths. The moisture content is measured very precisely in
a thin (5 - 10 mm) layer between the tubes. The equipment is complex
and not readily portable and is more suited to determinations of
moisture in soil columns in the laboratory than for regular field
applications. Bridge and Collis-George (46) describe apparatus for
use with soil columns and Giesel et al. (47) have reported the successful use of the method in the field over a 3 year period. The
latter authors found that, when calibrated, the error in the absolute measurement of volumetric moisture content was ± 0.01 and the
precision ± 0.0015.

11.4 TDR and Capacitance Methods

Both of these methods are fairly recent developments and are
based on the fact that the dielectric constant of a soil, Ka is
strongly dependent on its moisture content. Topp et al. (48) have
shown that Ka varies from about 3 in dry soil to about 40 at θ =
0.55 and is only weakly dependent on soil type, bulk density and
temperature over a wide range of frequencies.

The TDR, or Time Domain Reflectometry (48), technique measures the apparent dielectric constant of the soil by transmitting an electromagnetic pulse into a closely spaced pair of parallel rods (which form a parallel transmission line) in the soil and measuring the time taken for the pulse to reach the end of the line. This is dependent on the length of the line and the dielectric constant. The average volumetric moisture content over the length of the line can then be obtained from the empirically derived "universal" K - θ relationship derived by Topp et al. (48). A series of separate transmission lines of various lengths (which may be installed vertically or horizontally) are required to measure moisture contents in individual layers. The technique may also be used to determine the depth of wetting fronts in the profile (49). TDR equipment has recently become commercially available and provides a direct moisture content readout. Errors of absolute moisture content of ± 0.02 and a precision of readings of 0.0025 are claimed.

The capacitance method generally utilizes an oscillator circuit in which a pair of electrodes form the plates of a capacitor with soil as the dielectric. Variations of moisture content alter the capacitance and therefore the frequency of the oscillator, which is measured. Soil calibrations relating frequency to volumetric moisture content are required. The method has been described by Kuraz et al. (50) and is described in detail by Schurer elsewhere in this book. Both this and the TDR method show considerable promise, but further development and evaluation appear to be required.

12. SAFETY

The safety aspects of the use of the neutron probe have been discussed by IAEA (3), Bell (2), Gee et al. (51), and Watson and Honeysett (52). Owners and users of neutron probes must be aware of, and comply with, the relevant regulations concerning the storage, transport and use of equipment containing a neutron source. These are generally based on IAEA guidelines, but do vary from country to country. The User's Handbook for the IH Neutron Probe System (53) states that "the regulations (in Britain) in essence have four objectives:

-to prevent the loss of the source
-to prevent breakage and leakage of the source
-to prevent persons not involved with the use of the probe
 from being exposed to radiation levels exceeding 7.5 μSv hr[-1]
-to prevent probe users from receiving more than a known,
 controlled and acceptable level of radiation

Modern neutron probes are generally designed so that when clearly defined precautions and operating procedures are adhered to, the radiation hazard to the user is small. Regular users who adhere to these procedures may receive total yearly radiation doses which are not much higher than the limits set for the general public.

This is one tenth of the limit set for occupational exposure (whole body dose 50 mSv/yr). However, poor operating procedures and carelessness (often due to unawareness of the hazards), can result in much higher radiation doses being received. Operator training is therefore very important.

The main radiation hazard from Am-Be probes is due to the fast neutrons (51). There are also minor contributions from the slow neutrons and from the gamma output. Probes with Ra-Be sources have very much higher gamma outputs. Figure 11 shows the variation of fast neutron dose rate with distance from the source, measured around shielded and unshielded probes with two different Am-Be source activities.

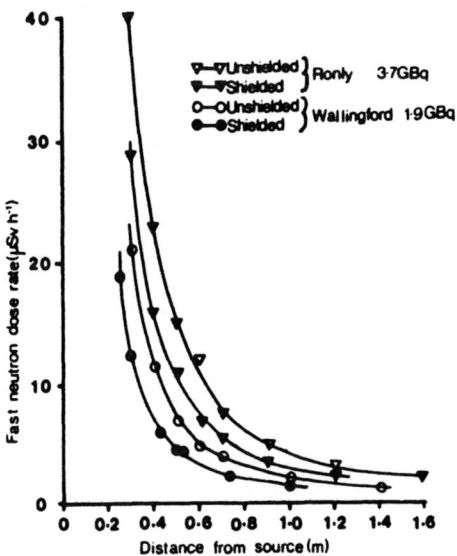

Figure 11.   Neutron dose rate as a function of distance from shielded and unshielded ^{241}Am-Be sources. (Redrawn from Watson and Honeysett, 1981)

These data, and those of Gee et al. (51) confirm that the dose rates largely obey the inverse-square law and also show that the shielding only reduces dose rates by about 35%. The amount of shielding provided is necessarily a compromise, because totally effective shielding would make the equipment extremely large and heavy and would severely prejudice its portability.

There are 3 possible methods of minimizing radiation doses from neutron probes, which are (52):
a) minimizing exposure time,
b) maximizing the distance from the source, and
c) the provision of extra shielding.
In practice, the aim should be to ensure that the doses received are kept to the absolute minimum. Extra shielding is impractical for

field use although it may be of benefit for storage of the equipment
or during transport. In general, maximizing the distance from the
source provides the best and most convenient method of minimizing
dose rates and the use of a trolley has been recommended for carrying 100mCi source probes in the field (50). As mentioned previously,
the IH II probe system has been designed so that the operator may
carry and operate the equipment while remaining at least 0.4 m from
the shielded source. Once the probe itself has been lowered more
than about 0.3 m into the soil, dose rates are almost negligible so
that for much of the time the probe is being used there is virtually
no radiation hazard. In a vehicle, the source end of the probe
should always be placed as far from the occupants as possible. In
the field, exposure time cannot be reduced below the minimum required
to carry the instrument to the site and make the readings, but unnecessary exposure should be avoided.

13. APPLICATIONS

The neutron probe provides the ability to measure, in situ, the
amount of moisture in a soil profile and its distribution with depth.
Subsequent observations in the same access tube(s) allow the calculation of the changes of moisture content and profile moisture
storage with time. The method finds applications in many fields of
agricultural, forestry and hydrological research, major examples of
which are:

- the measurement of profile moisture storage changes arising
  from different experimental treatments
- the measurement of soil moisture deficits
- the measurement of the available water capacity of soils
- water balance studies and the estimation of evapotranspiration and drainage
- studies of plant-water relations
- the investigation of the processes of soil water movement.

The probe is a particularly useful tool for the improvement of water
use efficiency in both irrigated and rainfed situations and is being
increasingly used in the commercial sector, mainly for irrigation
scheduling.

13.1 Changes of Moisture Storage

Measurements made with the neutron probe alone can be used to
compare the changes of moisture storage and moisture distribution
under plots subjected to different agronomic forestry or irrigation
practices, although in most cases, they cannot be used to quantify
the processes of moisture movement.
In many farming systems, particularly dryland farming, the
conservation of moisture in the soil during fallow periods is
essential to the success of the following crop and the neutron

probe has been used to monitor the effectiveness of different mulching treatments, for example by Unger and Jones (54), Smika (55) and Hundal and DeDatta (56). Figure 12 shows profiles of moisture content under 4 different treatments and a control at the beginning and end of a 6 week dry period during the fallow season. The data show the very considerable effectiveness of all the treatments compared with the control, but the differences between the treatments are also clearly demonstrated, not only by the amounts of water lost but also through the differences in the patterns of moisture loss.

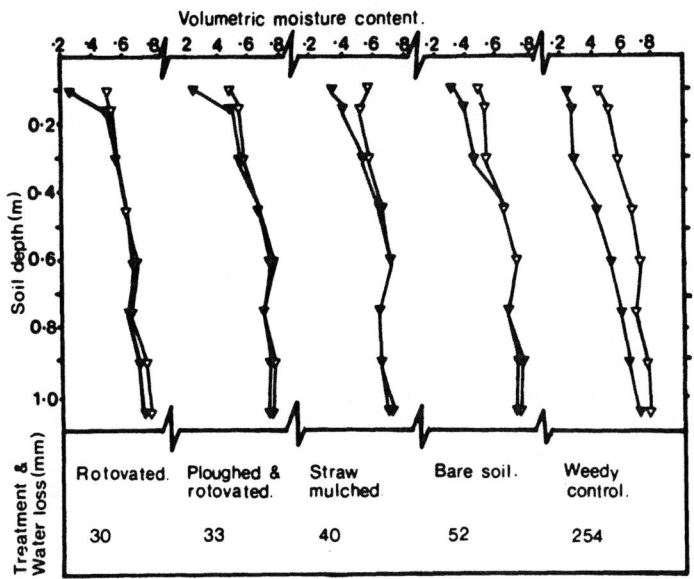

Figure 12. Soil water depletion patterns during a 6 week period under various dry season soil management systems. The moisture contents at 0.1 m were determined gravimetrically. (Redrawn from Hundal and De Datta, 1982)

Under irrigated conditions, the neutron probe may be used to investigate the distribution of water applied by different methods. Groot Obbink and Alexander (57) have studied distributions under an orchard subjected to flood and drip irrigation, and Akhtar Bhatti (58) has assessed the effects of different application rates on the uniformity of water distribution under furrow irrigation. The scheduling of irrigation using moisture measurements made with the probe has been described by Cull (59) and the method has recently been evaluated by Dragovic (60) for sugar beet and Rawitz et al. (61) for cotton.

Soil moisture depletion under cut and uncut areas of hardwood forest have been compared by Sartz (62) and Rogerson (63) and Power et al. (64) have investigated the effects of different thicknesses of top and sub soil on moisture content and crop production on reclaimed mine soils.

13.2 Measurement of Moisture Deficits

Crop response, yield and actual evapotranspiration rates are often related to moisture deficits, which are the amount by which the moisture storage in the profile has decreased below "field capacity." This is defined as the moisture storage in the profile below which drainage effectively ceases, although the concept is invalid for some soils as this situation is never attained. For large land areas, deficits are usually estimated from potential evaporation data and an appropriate moisture abstraction model, such as the MORECS system used in the UK, described by Field (65). Gardner and Field (66) report a study where deficits have been measured directly using neutron probes to assess the performance of the MORECS system and Black (67) has related evapotranspiration rates measured by the energy balance method to the amount of water stored in the soil profile. Field capacity values (when the concept is valid) and other soil moisture reservoir characteristics such as the abstraction limit profile and the available water capacity can be measured in situ under field conditions (2).

13.3 The Water Balance

Changes in the amount and the vertical distribution of moisture in a soil profile reflect the combined effects of inputs of rainfall and/or irrigation water, and losses due to evaporation from the soil surface, root abstraction and drainage. These are the components of the water balance of the soil profile, which can be expressed in the form:

$$R - Q = E + D_z + \Delta S_z \qquad (10)$$

where, over a given period, R is the rainfall or irrigation input, Q is the surface runoff (R - Q represents the net input of water to the profile), E is the actual evapotranspiration loss, D_z is the drainage (recharge to groundwater) through depth z in the profile and ΔS_z is the change in moisture storage above depth z. The neutron probe provides the best, and most commonly used, means of measuring ΔS_z. The balance is most often used to estimate E or D by measuring R and ΔS and selecting periods when the other components can be either quantified or assumed to be zero.

13.4 Estimation of Evapotranspiration and Drainage

The use of the water balance method to estimate actual evapotranspiration, using the neutron probe to measure soil moisture storage changes and assuming that drainage is negligible, has been evaluated by Rouse and Wilson (68), and has been compared with the lysimeter method by Wight (69) under semi-arid rangeland conditions. Rouse and Wilson found that under their conditions, moisture changes could be estimated to ± 1.6 mm so that with evapotranspiration rates of about 4 mm/day, the moisture changes had to be measured at inter-vals of 4 days or more in order to estimate the evapotranspiration to better than 10%. Results will be overestimated if the assumption that drainage is negligible is invalid. This method of estimating evapotranspiration from crops has also been used by Laurenroth and Sims (70) for grass subjected to different nitrogen and irrigation treatments, by McGowan and Williams (40) for cereals and by Hodnett and Bell (71) and Wallace et al. (72) for various crops in Central India. Figure 13 shows the actual evapotranspiration rate derived from soil moisture depletion data using the water balance. Potential evaporation data are shown for comparison.

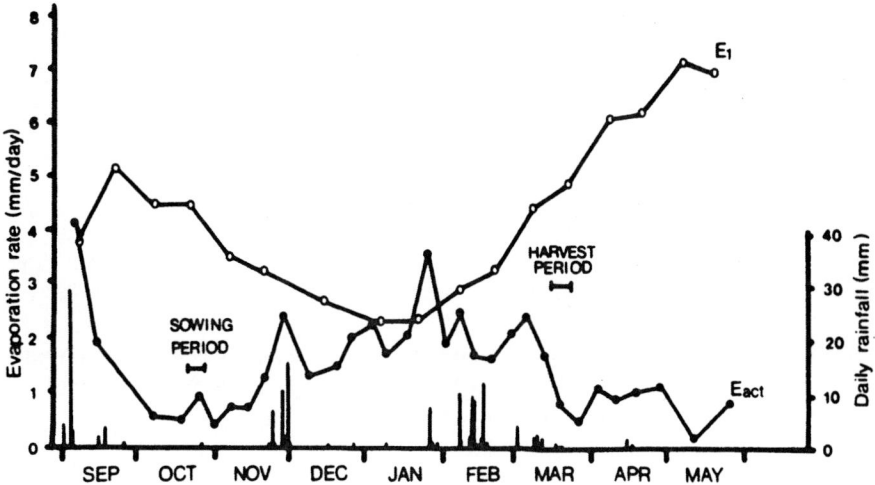

Figure 13. Actual evapotranspiration (Eact) from a black cotton soil derived by the water balance using the neutron probe to measure the soil moisture storage changes. Potential evapotranspiration (E_t) is shown for comparison. (After Hodnett and Bell, 1981).

The evapotranspiration from a spruce forest has been determined by Calder (73) from the water balance of a "natural lysimeter" by measuring the net rainfall and the drainage from the lysimeter and using the neutron probe to measure the changes of moisture storage within it.
When drainage cannot be assumed to be negligible and evapo-transpiration and drainage cannot be independently or reliably estimated, the "zero flux plane" (ZFP) method, used for example by DeBoodt et al. (74), Arya et al. (75a) and Wellings and Bell (76) or the method described by McGowan and Williams (40) may be used to separate these components. The latter is a graphical method of analysis using the moisture content data for each depth plotted in time series to locate the maximum depth of root abstraction, below which drainage can be assumed to be occurring.

The ZFP method requires measurements of total hydraulic potential (eg using tensiometers) at regular depth intervals down the
profile to allow the determination of the hydraulic gradients, which
cause the movement of water in the soil. Using these data, the
direction of water movement can be determined. During dry periods
for example, water movement in the upper part of the profile is
upward, towards the roots, while in the lower part it is downward.
The zones of upward and downward movement are separated by a plane
where the potential gradient is zero, known as the "zero flux plane."
When the location of the ZFP is known, the changes in moisture
storage in a soil profile can be separated into the components due
to drainage and evaporation. The drainage flux, Q_z at any depth,
z, below that of the ZFP, z_0, is given by (75):

$$Q_z = \int_{z}^{z_0} (\partial \theta / \partial t) \, dz \tag{11}$$

Figure 14 shows weekly mean drainage rates at a depth of 2.85 m
in a chalk soil profile in the UK determined by the water balance
method for the period from September to May and the ZFP method from
May onwards. The water balance method was used when there was no
ZFP in the profile and when the actual evapotranspiration could be
reliably assumed to equal the potential evapotranspiration (calculated from meteorological data). The gradual decline of the drainage
rate after the appearance of the ZFP is of note.

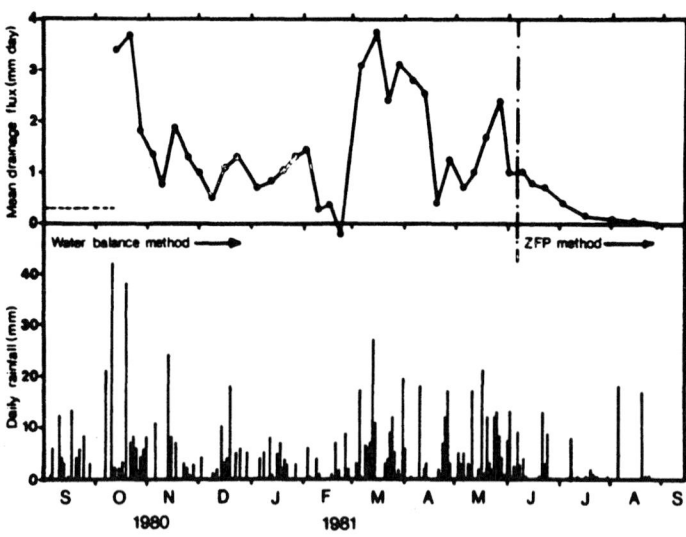

Figure 14. Mean drainage rates at 2.85 m depth in a chalk profile
under grass determined by the water balance method for
the period from September to May and by the ZFP methods
from May onwards (M.G. Hodnett, unpublished data).

The ZFP and McGowan and Williams' methods cannot be used to estimate the drainage from moisture content measurements during wet periods when drainage is occurring throughout the profile and there is no ZFP, but the procedure described by Scholl (77) may be used. This requires the moisture characteristic (the relationship between moisture content and matric potential) and the conductivity characteristic (the relationship between moisture content and hydraulic conductivity) of the soil to be known but measurements of hydraulic potential are not essential.

Measurements of drainage losses are required for many purposes. Wellings (78) has used the ZFP and water balance methods to measure the movement of nitrates, applied as organic and inorganic fertilizers, through the unsaturated zone towards a deep groundwater table. A knowledge of the losses of nitrate is essential in fertilizer uptake, and groundwater pollution studies. Leaching losses of nitrate from clay soils have been measured by White et al.(79) and the uptake of nitrogen by wheat has been studied by Strebel et al. (80). In irrigation studies, measurements of drainage may be used to estimate water use efficiency, to monitor the leaching fraction for salinity control and to prevent the excessive rise of the water table. The water balance and ZFP methods have been used by Cooper (81) to estimate the drainage (recharge to groundwater) beneath forest and grassland areas.

13.5 Measurement of the Hydraulic Properties of Soils

The combined use of field moisture content and hydraulic potential measurements allows the in situ derivation of various fundamental soil hydraulic properties, such as the moisture characteristic and the unsaturated conductivity characteristic. The advantages of determining these properties in situ are that unacceptable disturbances are avoided (for example, during the taking of cores for laboratory measurements) and the results are from a larger and therefore more representative volume of soil. A knowledge of the hydraulic properties of soils is essential in the understanding and modelling of the processes of water movement of soils.

The moisture characteristic may be derived for various depths in a soil profile by measuring the moisture content and the matric potential using the neutron probe and tensiometers, respectively. Measurements may be made under natural conditions over a long period, or the profile may be saturated and then allowed to dry out by drainage and evaporation.

The conductivity characteristic for different depths in the soil profile may be derived using the "instantaneous profile" method of Watson (82). In this method, the soil profile is saturated and then covered to prevent inputs and losses of moisture through the soil surface. Moisture content and hydraulic potential measurements are made at regular depth intervals, frequently at first, when drainage rates are high but then at gradually increasing time intervals. The moisture flux, Q_z, at any depth, z, can be calculated using Eq. 11, where z_0 is the soil surface. The potential gradient $\partial\psi/\partial z$, is calculated from the tensiometer data and the unsaturated conductivity, k, is then derived using Darcy's Law:

$$Q_z = k.\psi\partial/\partial z \tag{12}$$

The conductivity is then expressed as a function of moisture content $(k-\theta)$ or of matric potential $(k-\psi_m)$.

These methods of measuring the hydraulic properties of soils in situ have been widely used, for example by Van Bavel et al. (83a,b), Allmaras et al. (84a,b) and Arya et al. (75a,b,c) for studies of soil moisture movement and the uptake of water by roots, by Belmans et al. (85) as inputs into a model to simulate moisture regimes under a potato crop and by Hundal and DeDatta (86) to investigate the water loss by drainage from rice fields.

Arya et al.(75a) have used the ZFP method to measure the moisture fluxes used to calculate k, and Wall and Miller (87) have recently described a method of deriving the conductivity and moisture characteristics from measurements of soil moisture depletion made using a neutron probe, using a model of soil water drainage and a parameter optimization procedure.

ACKNOWLEDGEMENTS

The author would like to thank John Bell and Cate Gardner for their helpful comments and David Hutchings for preparing the diagrams.

REFERENCES

1. Greacen, E.L. (ed.) Soil Water Assessment by the Neutron Method. CSIRO Australia, 1981.

2. Bell, J.P. Neutron Probe Practice. Institute of Hydrology Report 19, Institute of Hydrology, Wallingford, Oxon, UK, 1976.

3. International Atomic Energy Agency. Neutron Moisture Gauges. Technical Reports Series No. 112, IAEA Vienna, 1970.

4. Brummer, E. and Mardock, E.S. A neutron method for measuring saturations in laboratory flow experiments. American Institute of Mining & Metall. Engineers, Los Angeles meeting, 1945.

5. Pieper, G.F. The Measurement of Soil Moisture by the Slowing Down of Neutrons. Ph.D. Thesis, Cornell University, Ithaca, New York, 1949.

6. Belcher, D.F., Cuykendall, T.R. and Sack, H.S. The Measurement of Soil Moisture by Neutron and Gamma Ray Scattering. Civil Aeronautics Adm. Tech. Dev. Report 127, 1950.

7. Gardner, W.R. and Kirkham, D. Determination of soil moisture by neutron scattering. Soil Science 73:391-401, 1952.

8. Couchat, P. Mesure Neutronique de l'Humidite des sols. Ph.D. Thesis, L'Universite de Paul Sabatier de Toulouse, 1974.

9. Couchat, P. Les Applications de la Methode Neutronique dans la Recherche Agronomique. Int. Symp. on Isotope & Radiation Tech. in Soil Physics and Irrigation Studies, Aix-en-Provence 1983 (IAEA Vienna).

10. Olgaard, P.L. On the Theory of the Neutronic Method for Measuring the Moisture Content of Soil, Report No. 97, Danish Atomic Research Institute, Riso 1965.

11. Olgaard, P.L. and Haahr, V. Comparative experimental and theoretical investigations of the DM neutron moisture probe. Nuclear Engineering and Design 5:311-324, 1967.

12. Long, I.F. and French, B.K. Measurement of soil moisture in the field by neutron moderation. J. Soil Science 18: 149-166, 1967.

13. Sicamois, D. Etude comparitive des sources ^{241}Am-Be, ^{244}Cm-Be et ^{252}pour equiper les humidimetres a neutrons. Bulletin Francais Humidimetres Neutronique 8:35, 1980.

14. Greacen, E.L., Correll, R.L., Cunningham, R.B., Johns, G.G. and Nicolls, K.D. Calibration, Chapter 6. In: Soil Water Asessment by the Neutron Method (E.L. Greacen, ed.) CSIRO Australia, 1981.

15. Kristensen, K.J. Depth intervals and topsoil moisture measurement with the neutron depth probe. Nordic Hydrology 4:77-85, 1973.

16. Luebs, R.E., Brown, M.J. and Laag, A.E. Determining water content of different soils by the neutron method. Soil Science 106:207-212, 1968.

17. Harris, W. A method for improving the accuracy of measuring soil moisture near the surface with a neutron meter. ARC Letcombe Lab Report for 1973, pp. 56-58.

18. Grant, D.R. Measurement of soil moisture near the surface using a neutron moisture meter. J. Soil Science 26: 124-129, 1975.

19. Parkes, M.E. and Siam, N. Error associated with the measurement of soil moisture change by the neutron probe. J. Agri. Eng. Res. 24:87-93, 1979.

20. Jensen, P.A. and Somer, E. Scintillation techniques in soil moisture and density measurements. Int. Symp. on Isotope & Radiation Tech. in Soil Physics and Irrigation Studies, Istanbul 1967 (IAEA Vienna).

21. Holmes, J.W. Influence of bulk density of the soil on neutron moisture meter calibration. Soil Science 102: 355-360, 1966.

22. Olgaard, P.L. and Haahr, V. On the sensitivity of neutron moisture gauges to variations in bulk density. Soil Science 105:62-64, 1968.

23. Greacen, E.L. and Schrale, G. The effect of bulk density on neutron moisture meter calibrations. Aust. J. Soil Res. . 14:159-169, 1976.

24. Nicolls, K.D., Hutton, J.T. and Honeysett, J.L. Gadolinium in soils and its effect on neutron moisture meter calibration. Aust. J. Soil Res. 15:287-291, 1977.

25. Couchat, P., Carre, C., Marcesse, J. and Le Ho, J. The measurement of thermal neutron constants of the soil. Proc. Conf. on Nuclear Cross Sections Technology, October 1975, Schrack and Bowman, eds.

26. McCulloch, D.B. and Wall, T. A method of measuring the neutron absorption cross sections of soil samples for the calibration of the neutron moisture meter. Nuclear Instruments and Methods 137:577-581, 1976.

27. Williamson, R.J. and Turner, A.K. Calibration of a neutron moisture meter for catchment hydrology. Aust. J. Soil Res. 18:1-11, 1980.

28. Vachaud, G., Royer, J.M. and Cooper, J.D. Comparison of methods of calibration of a neutron probe by gravimetry or neutron capture model. Journal of Hydrology 34:343-356, 1977.

29. Haverkamp, R., Vauclin, M. and Vachaud, G. Error analysis in estimating soil water content from neutron probe measurements: 1. Local standpoint. Soil Science 137:78-90, 1984.

30. Greacen, E.L. and Hignett, C.T. Sources of bias in the field calibration of a neutron meter. Aust. J. Soil Res. 17:405-415, 1979.

31. Lal, R. The effect of soil texture and density on the neutron and density probe calibration for some tropical soils. Soil Science 117:183-190, 1974.

32. Lal, R. Concentration and size of gravel in relation to neutron and density probe calibration. Soil Science 127: 41-50, 1979.

33. Shirazi, G.A. and Isobe, M. Calibration of neutron probe in some selected Hawaiian soils. Soil Science 122:165-170, 1976.

34. Nakayama, F.S. and Reginato, R.J. Simplifyng neutron moisture meter calibration. Soil Science 133: 48-52, 1982.

35. Hewlett, J.D., Douglas, J.E. and Clutter, J.L. Instrument and soil moisture variance using the neutron scattering method. Soil Science 97:19-24, 1964.

36. Sinclair, D.F. and Williams, J. Components of variance involved in estimating soil water content and water content change using a neutron moisture meter. Aust. J. Soil Res. 17:237-247, 1979.

37. Williams, J. and Sinclair, D.F. Accuracy bias and precision. In: Soil Moisture Assessment by the Neutron Method (E.L. Greacen, ed.), Chapter 5, CSIRO Australia, 1981.

38. Vauclin, M., Haverkamp, R. and Vachaud, G. Analyse des erreurs liees a L'utilisation de L'humidimetre neutronique. Int. Symp. on Isotope & Radiation Tech. in Soil Physics and Irrigation Studies, Aix en Provence (IAEA Vienna 1983).

39. Peterson, R.G. and Calivn, L.D. Sampling. In: Methods of Soil Analysis (C.A. Black, ed.) American Society of Agronomy, Madison, Wisconsin, USA, 1965.

40. McGowan, M. and Williams, J.B. The water balance of an agricultural catchment: 1. Estimation of evaporation from soil water records. Journal of Soil Science 31:217-230, 1980.

41. Prebble, J.A., Forrest, J.A., Honeysett, J.L., Hughes, M.W. McIntyre, D.S. and Schrale, G. Field installation and maintenance. In: Soil Moisture Assessment by the Neutron Method (E.L. Greacen, ed.) CSIRO Australia, 1981.

42. Hewlett, J.D. and Douglas, J.E. A method of calculating error of soil moisture volumes in gravimetric sampling. Forest Science 7:265-272, 1961.

43. Phene, C.J., Hoffman, G.J. and Rawlins, S.L. Measuring soil matric potential in situ by sensing heat dissipation in a porous body. Proc. Soil Sci. Soc. Amer. 35: 27-33, 1971.

44. LaRue, M.E., Nielsen, D.R. and Hagan, R.M. Soil water flux below a ryegrass root zone. Agronomy Journal 60:625-629, 1968.

45. Nielsen, D.R., Biggar, J.W. and Erh, K.T. Spatial variability of field-measured soil water properties. Hilgardia 42:215-260, 1973.

46. Bridge, B.J. and Collis-George, N. A dual source gamma ray traversing mechanism suitable for the non-destructive simultaneous measurement of bulk density and water content in columns of swelling soil. Aust. J. Soil Res. 11:83-92, 1973.

47. Giesel, W., Lorch, S. and Renger, M. Water flow calculations by means gamma absorption and tensiometer field measurements in the unsaturated soil profile. In: Isotope Hydrology 1970, Proceedings of a Symposium on the Use of Isotopes in Hydrology, IAEA Vienna, 1970.

48. Topp, G.C., Davies, J.L. and Annan, A.P. Electromagnetic determination of soil water content: measurements in coaxial transmission lines. Water Resources Research 16:574-582, 1980.

49. Topp, G.C., Davies, J.L. and Chinnick, J.H. Using TDR water content measurements for infiltration studies. Land Resources Research Institute Contribution No. 83-45, Agriculture Canada, Ottawa, Canada 1983.

50. Kuraz, V., Kutilek, M. and Kaspar, I. Resonance capacitance soil moisture meter. Soil Science 110:278-279, 1970.

51. Gee, G.W., Stiver, J.F. and Borchert, H.R. Radiation hazard from Americium-Beryllium neutron probes. Soil Sci. Soc. Amer. J. 40:492-494, 1976.

52. Watson, C.L. and Honeysett, J.L. Safety aspects. In: Soil Moisture Assessment by the Neutron Method (E.L. Greacen, ed.) CSIRO, Australia, 1981.

53. Anon. User's Handbook for the Institute of Hydrology Neutron Probe System. NERC Institute of Hydrology Report No. 79. Institute of Hydrology, Wallingford, Oxon, UK, 1981.

54. Unger, P.W. and Jones, O.R. Effect of soil water content and a growing season straw mulch on grain sorghum. Soil Sci. Soc. Amer. J. 45:129-134, 1981.

55. Smika, D.E. Soil water change as related to position of wheat straw mulch on the soil surface. Soil Sci. Soc. Amer. J. 47:988-991, 1983.

56. Hundal, S.S. and DeDatta, S.K. Effect of dry season soil management on water conservation for a succeeding rice crop. Soil Sci. Soc. Amer. J. 46:1081-1086, 1982.

57. Groot Obbink, J. and Alexander, D. McE. Observations of soil water and salt movement under drip and flood irrigation in an apple orchard. Agricultural Water Management 1:179-190, 1977.

58. Akhtar Bhatti, M. Use of neutron soil moisture probe to determine water distribution uniformity in furrow irrigation. Int. Symp. on Isotope & Radiation Tech. in Soil Physics and Irrigation Studies, Aix en Provence, IAEA Vienna, 1983.

59. Cull, P.O. Irrigation scheduling with a neutron probe. Irrigation Farmer 7:2-3, 1980.

60. Dragovic, S. Sugar beet irrigation scheduling and the possibilities of using a neutron probe to measure soil moisture in northeastern Yugoslavia. Int. Symp. on Isotope & Radiation Tech. in Soil Physics and Irrigation Studies, Aix en Provence, IAEA Vienna, 1983.

61. Rawitz, A., Marani, A., Mahrer, Y. and Berkovich, D. Evaluation of different methods of measuring evapotranspiration as a scheduling guide for drip irrigated cotton. Int. Symp. on Isotope & Radiation Tech. in Soil Physics and Irrigation Studies, Aix en Provence, IAEA Vienna, 1983.

62. Sartz, R.S. Soil water depletion by a hardwood forest in southwestern Wisconsin. Proc. Soil Sci. Soc. Amer. 36:961-964, 1972.

63. Rogerson, T.L. Soil water deficits under forrested and clearcut areas in northern Arkansas. Soil Sci. Soc. Amer. J. 40: 802-804, 1976.

64. Power, J.F., Sandoval, F.M., Ries, R.E. and Merrill, S.D. Effects of topsoil and subsoil thickness on soil water content and crop production on a disturbed soil. Soil Sci. Soc. Amer. J. 45:124-129, 1981.

65. Field, M. The meteorological office rainfall and evaporation calculation system - MORECS. Agricultural Water Management 6:297-306, 1983.

66. Gardner, C.M.K. and Field, M. An evaluation of the success of MORECS, a meteorological model, in estimating soil moisture deficits. Agricultural Meteorology 29:269-284, 1983.

67. Black, T.A. Evapotranspiration frcm Douglas fir stands exposed to soil water deficits. Water Resources Res. 15: 164-170, 1979.

68. Rouse, W.R. and Wilson, R.G. A test of the potential accuracy of the water budget approach to estimating evapotranspiration. Agricultural Meteorology 9:421-446, 1972.

69. Wight, J.R. Comparison of lysimeter and neutron scatter techniques for measuring evapotranspiration from semi-arid rangelands. Journal of Range Management 24:390-393, 1971.

70. Laurenroth, W.K. and Sims, P.L. Evapotranspiration from a shortgrass prairie subjected to water and nitrogen treatments. Water Resources Res. 12:437-442, 1976.

71. Hodnett, M.G. and Bell, J.P. Soil physical processes of groundwater recharge through Indian black cotton soils. Institute of Hydrology Report No. 77, NERC Institute of Hydrology, Wallingford, Oxon, UK, 1981.

72. Wallace, J.S., Batchelor, C.H. and Hodnett, M.G. Crop evaporation and surface conductance calculated using soil moisture data from central India. Agricultural Meteorology 25:83-96, 1981.

73. Calder, I.R. The measurement of water losses from a forested area using a "natural" lysimeter. Journal of Hydrology 30: 311-325, 1976.

74. DeBoodt, M., Hartmann, R. and DeMeester, P. Determination of soil moisture characteristics for irrigation by neutron moisture meter and air purged tensiometers. Int. Symp. on Isotope & Radiation Tech. in Soil Physics and Irrigation Studies, Istanbul, IAEA Vienna, 1970.

75a. Arya, L.M., Farrell, D.A. and Blake, G.R. A field study of soil water depletion patterns in presence of growing soybean roots: I. Determination of hydraulic properties of the soil. Proc. Soil Sci. Soc. Amer. 39:424-430, 1975.

75b. Arya, L.M., Farrell, D.A. and Blake, G. R. A field study of soil water depletion patterns in presence of growing soybean roots: II. Effect of plant growth on soil water pressure and water loss patterns. Proc. Soil Sci. Soc. Amer. 39: 430-436, 1975.

75c. Arya, L.M., Farrell, D.A. and Blake, G. R. A field study of soil water depletion patterns in presence of growing soybean roots: III. Rooting characteristics and root extraction of soil water. Proc. Soil Sci. Soc. Amer. 39:437-444, 1975.

76. Wellings, S.R. and Bell, J.P. Movement of water and nitrate in the unsaturated zone of upper chalk near Winchester, Hants, England. Journal of Hydrology 48:119-136, 1980.

77. Scholl, D.G. Soil moisture flux and evapotranspiration determined from soil hydraulic properties in a Chaparral stand. Soil Sci. Soc. Amer. J. 40:14-18, 1976.

78a. Wellings, S.R. Recharge of the upper chalk aquifer at a site in Hampshire, England: I. Water balance and unsaturated flow. Journal of Hydrology 69:259-273, 1984.

78b. Wellings, S.R. Recharge of the upper chalk aquifer at a site in Hampshire, England: II. Solute movement. Journal of Hydrology 69:275-285, 1984.

79. White, R.E., Wellings, S.R. and Bell, J.P. Seasonal variations in nitrate leaching in structured clay soils under mixed land use. Agricultural Water Management 7:391-410, 1983.

80. Strebel, O., Grimme, H., Renger, M. and Fleige, H. A field study with nitrogen-15 of soil and fertilizer nitrate uptake and of water withdrawal by spring wheat. Soil Science 130: 205-210, 1980.

81. Cooper, J.D. Measurement of moisture fluxes in unsaturated soil in Thetford Forest. Institute of Hydrology Report No. 66, NERC Institute of Hydrology, Wallingford, Oxon, UK, 1980.

82. Watson, K.K. An instantaneous profile method for determining the hydraulic conductivity of unsaturated porous materials. Water Resources Research 2:709-715, 1966.

83a. Van Bavel, C.H.M., Stirk, G.B. and Brust, K.J. Hydraulic properties of a clay loam soil and the field measurement of water uptake by roots: I.Interpretation of water content and pressure profiles. Proc. Soil Sci. Soc. Amer. 32:310-317, 1968.

83b. Van Bavel, C.H.M., Stirk, G.B. and Brust, K.J. Hydraulic properties of a clay loam soil and the field measurement of water uptake by roots: II. The water balance of the root zone. Proc. Soil Sci. Soc. Amer. 32:317-321, 1968.

84a. Allmaras, R.R., Nelson, W.W. and Voorhees, W.B. Soybean and corn rooting in southwestern Minnesota: I. Water uptake sink. Proc. Soil Sci. Soc. Amer. 39:764-771, 1975.

84b. Allmaras, R.R., Nelson, W.W. & Voorhees, W.B. Soybean and corn rooting in southwestern Minnesota: II.Root distributions and related water inflow. Proc. Soil Sci. Soc. Amer. 39:771-777, 1975.

85. Belmans, C., Dekker, L.W. and Bouma, J. Obtaining soil physical field data for simulating soil moisture regimes and associated potato growth. Agricultural Water Management 5: 319-333, 1982.

86. Hundal, S.S. and DeDatta, S.K. In situ water transmission characteristics of a tropical soil under rice based cropping systems. Agricultural Water Management 8:387-396, 1984.

87. Wall, B.H. and Miller, A.J. Optimization of parameters in a model of soil water drainage. Water Resources Research 19: 1565-1572, 1983.

THERMOCOUPLE PSYCHROMETERS FOR WATER POTENTIAL MEASUREMENTS

Ralph Briscoe

Wescor Inc
Logan, Utah

1. INTRODUCTION

Water potential is defined as the measure of the free energy
of water in a system compared with the free energy of pure water at
the same temperature and pressure. Water potential is expressed in
terms of the energy per unit mass (joules kg^{-1}) or pressure (pas-
cals) and is negative if the free energy is lower than that of pure
water at the same temperature and pressure (1). Total water potential, Ψ, is the sum of a number of components including matric
potential, Ψ_π, osmotic potential, Ψ_τ, and pressure potential, Ψ_p.

$$\Psi = \Psi_\pi + \Psi_\tau + \Psi_p$$

Water potential is an accepted term for quantifying water status
in terms of free energy. Water moves from regions of algebraically
higher potential to regions of algebraically lower potential.
Plants respond to water according to the energy required to assimi-
late the water into the plant.
In 1951, D.C. Spanner published his classical work on the
measurement of water potential using the Peltier effect (2). Since
that time, a great deal of research has been accomplished utilizing
this method of water potential measurement. Since thermocouple
psychometers/hygrometers allow the measurement of total water potential and not just some components of water potential, they are
extremely valuable in determining water status in relation to plant
growth.

THE RELATIONSHIP BETWEEN WATER POTENTIAL AND RELATIVE HUMIDITY

The theoretical relationship between water potential and relative humidity is given by the Kelvin equation:

$$\Psi = \frac{RT}{V_w} \ln \frac{e}{e_0}$$

Where:

Ψ is the water potential (Pa)
R is the universal gas constant (8.3143 J mol^{-1}K^{-1})
e/e_0 is the relative humidity expressed as a fraction
T is the absolute Temperature (°K)
V_w is the molar volume of water (1.8 x 10^{-5} m^3 mol^{-1})

The equilibrium relative humidity of soil air is near 1 for water potentials which will support plant growth (greater than -1.5 MPa). Various relative humidity values corresponding to various values of water potential are shown in Table 1.

Table 1. Relative Humidity Values Corresponding to Various Water Potential Values

RH	(e/e_0)	0.999926	0.999926	0.9926	0.9296	0.48
Ψ	(MPa)	-0.01	-0.1	-1.0	-10.0	-100.0
Ψ	(bars)	-0.1	-1.0	-10.0	-100.0	-1000.0

The equivalent values for water potential units found in the literature are shown in Table 2.

Table 2. Equivalent Water Potential Units

UNIT	EQUIVALENT VALUE
bar	-1.0
N m^{-2} (pascal)	1 x 10^5
KPa	-100
MPa	-0.1
erg cm^{-3}	-1 x 10^6
J kg^{-1}	-1 x 10^2
cm H$_2$0	1020
mm Hg	750
atmosphere	-0.987

Using a Taylor series expansion it can be shown that the ln (e/e_0) is approximately equal to $1-e/e_0$ for relative humidities near 1. Thus, for water potentials in the plant growth range ψ (in MPa)= 135 $(1-e/e_0)$. A change in relative humidity of 0.01 is equivalent to a water potential change of about 1.4 MPa or 14 bars. Plants and associated microorganisms often respond to changes in water potential of less than 0.1 MPa (1 bar). This corresponds to relative humidity changes of 0.001 or less. Standard techniques for measuring relative humidity in air are not capable of achieving this accuracy. However, thermocouple psychrometers/hygrometers under isothermal conditions are capable of achieving an accuracy of 0.005 MPa (0.05 bars).

2. METHODS OF OPERATION

Two important principles underlie the usefulness of thermo-couple psychrometers/hygrometers. The Seebeck Effect is the phenom-enon that permits a thermocouple to be used for temperature measure-ment. A thermocouple is formed when two different metals are joined together. If both ends of the wire are joined to form a closed loop, electrical current will flow through the wires whenever the junc-tions are at different temperatures. The magnitude of the voltage produced is dependent upon the temperature difference between the junctions.

The Peltier effect is the phenomenon which allows a thermo-couple junction to be cooled by passing an electrical current through the junction. When current flows across the junction of two dissimilar metals, heat will be either absorbed or liberated at the junction. If the current flows in the same direction as the current produced by the Seebeck Effect at the hot junction, heat is absorbed. If the current flows in the opposite direction, heat is liberated.

Two methods have been developed for the measurement of equil-ibrium relative humidity from which the water potential can be determined. These are the psychrometric (wet bulb) and the hygro-metric (dew point) methods. The same sensors are used for either method but the electronic control and measuring apparatus operate differently.

3. IN SITU SOIL PSYCHROMETERS/HYGROMETERS

Figure 1 illustrates the construction of a typical soil psy-chrometer/hygrometer. The junction of the chromel and constantan wires is the thermocouple junction used in the measurement of the equilibrium relative humidity. The points of contact at the gold pins form the reference junction for this measurement. Very fine chromel and constantan wires (25 μm dia.) are used so that cooling of this junction will not significantly change the temperature of the larger reference junction. Another thermocouple is formed by the junction of the copper and constantan leads.

196

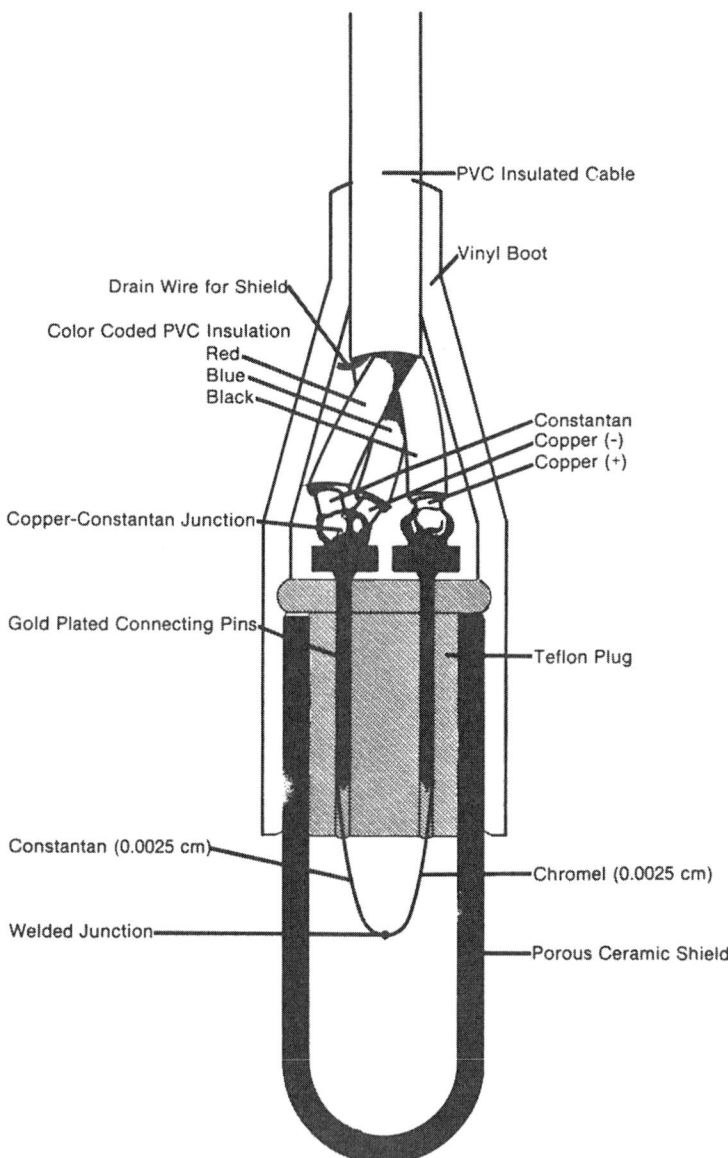

Figure 1   Peltier Psychrometer / Hygrometer with porous ceramic thermocouple shield.

This junction is used to determine the temperature of the sensor. An electronic reference junction, simulating a copper-constantan junction in an ice bath, is located in the readout device enabling the instrument to determine the sensor temperature accurately regardless of changes in the ambient temperature.

3.1 Psychrometric Mode

The recorded output for the psychrometric (wet bulb) mode of operation is shown in Figure 2. The difference between the two curves in this figure is the result of the readout and control devices used. The device used to produce curve A is capable of cooling and monitoring the temperature of the thermocouple simultaneously. Curve B was produced by an instrument which is not able to measure the thermocouple output while cooling is taking place.

The sequence of events that takes place in a psychrometric water potential measurement is as follows:

1. The sensing thermocouple is allowed to come to temperature equilibrium with the air space surrounding it. The air space must also be in temperature and vapor equilibrium with the soil or sample to be measured.

2. Current is passed through the thermocouple causing cooling of the junction by the Peltier Effect. The output of the thermocouple during this time is illustrated by the portion of the curve between a and b in part A of Figure 2. The magnitude and duration of the cooling current must be sufficient to cool the junction below the dewpoint temperature of the equilibrated air. When the temperature of the junction is below the dewpoint, water will condense upon the junction from the air.

3. The Peltier current is discontinued and the thermocouple output is monitored as the temperature of the thermocouple returns to ambient. The temperature changes rapidly toward the ambient temperature until it reaches the wet bulb depression temperature corresponding to point c of the curves. At this point evaporation of the water from the junction produces a cooling effect upon the junction that offsets the heat absorbed from the ambient surroundings. This continues until point d is reached. At this point the water is depleted and the thermocouple temperature returns to the ambient temperature at point e.

The output of the thermocouple at wet bulb temperature is approximately 45 $\mu V/kPa^{-1}$ at 25°C.

3.2 Dewpoint Mode

The dewpoint method is somewhat more complex. This method was first reported by Neumann and Thurtell (3). Campbell and Barlow (4) later reported on the development and evaluation of a system to determine the dewpoint temperature in a single operation. If a thermocouple were thermally isolated so that the latent heat from the water were the only mechanism for transferring heat to or from the junction, the wet junction would automatically converge to the dewpoint temperature.

198

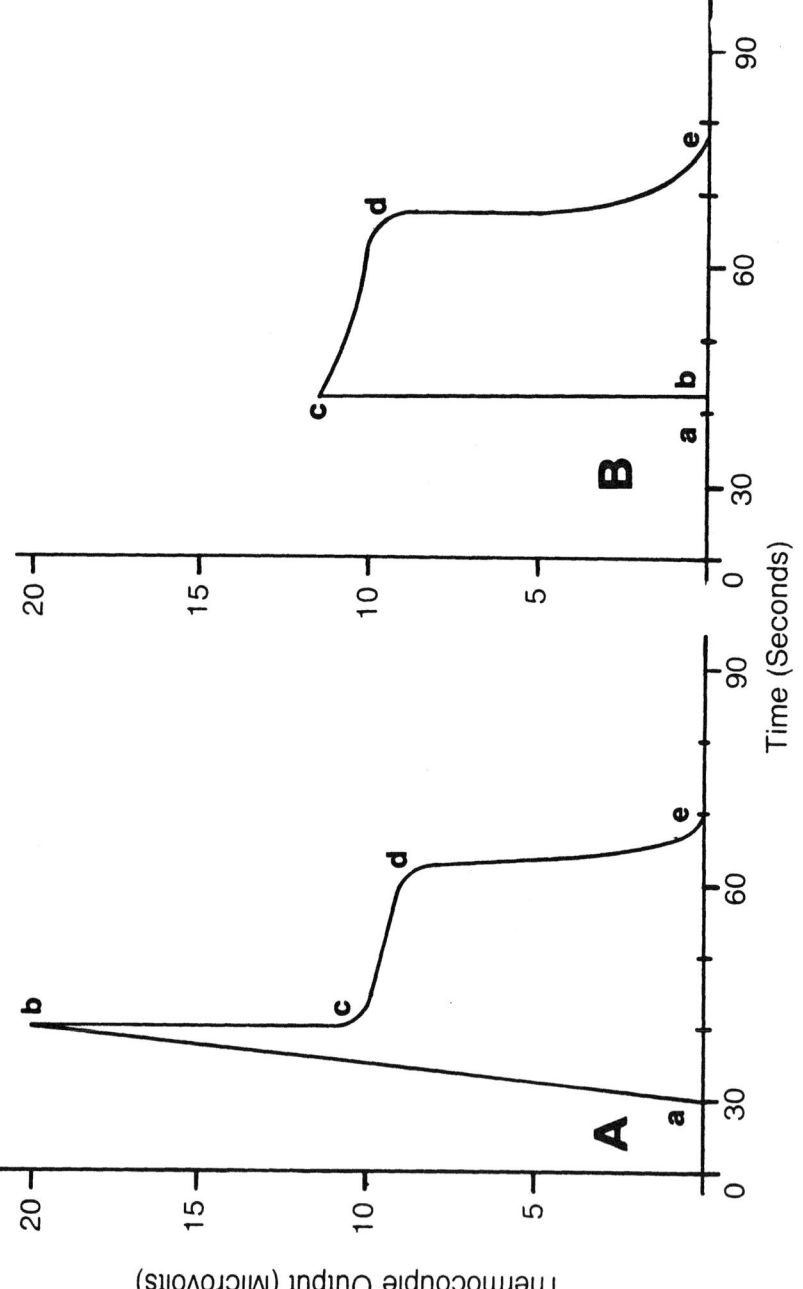

Figure 2   Recorder Output of Thermocouple Psychrometer/Hygrometer (Psychrometric Mode) with Different Readout Devices

If the junction temperature were below dewpoint, the latent heat of condensation would warm it. If the junction were above the dewpoint temperature, the latent heat of evaporation would cause it to cool to the dewpoint. In actual practice, it is not possible to isolate the thermocouple from other mechanisms of heat transfer. It is, however, possible to determine the amount of heat gained from other sources and offset that heat with Peltier cooling.

The dewpoint microvoltmeter, commercially available from Wescor, Inc., has a time sharing circuit which allows the electronics to monitor the temperature depression of the thermocouple and to provide cooling proportional to the temperature. The cooling current is automatically adjusted to provide the amount of cooling needed to offset any heat transferred from the environment. The latent heat exchange causes the thermocouple to converge to the dewpoint temperature. The output of the thermocouple quantifies the dewpoint temperature (and hence the relative humidity) and is proportional to the water potential. The proportionality constant is approximately 75 $\mu V/kPa^{-1}$.

The amount of water on the thermocouple while operating in the dewpoint mode does not change if the cooling coefficient has been properly set. Therefore, it is possible to obtain a wet bulb measurement after the dewpoint measurement is complete by switching to amplifier output and observing the thermocouple output. A typical hygrometric output curve and a curve which illustrates the combination of hygrometric and psychrometric output is shown in Figure 3.

One advantage of the dewpoint mode is that it provides a continuous output rather than a falling plateau. Since the time of reading is not critical, it is easier to obtain accurate measurements. The hygrometer will not converge to dewpoint unless the cooling coefficient is accurately set to the proper value. The cooling coefficient changes by 0.7 microvolts per degree Celsius.

4. SAMPLE CHAMBERS

Sample chambers are used in the measurement of the water potential of samples of soil, plant or other tissues, or the osmolality of solutions.

Some of the sample chambers are constructed so that in normal laboratory or field environments no special care is needed to control the temperature of the chamber. This is accomplished by providing a large thermally conductive heat sink surrounding the sample. This heat sink is thermally attached to the reference junction which improves the performance by assuring that the reference junction remain at a constant temperature during cooling of the measurement junction. A cross-sectional view and three dimensional representation of Wescor's C-52 sample chamber is shown in Figure 4 and Figure A1 in the Appendix.

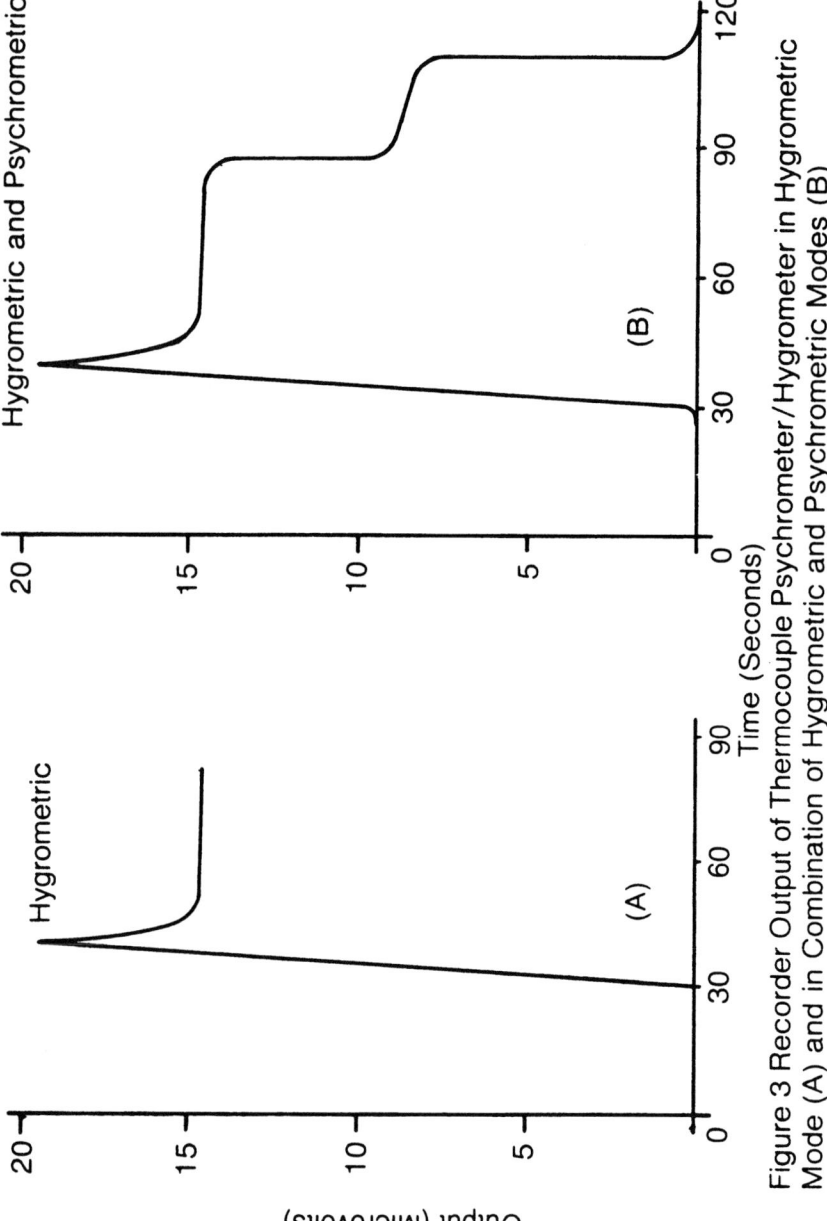

Figure 3 Recorder Output of Thermocouple Psychrometer/Hygrometer in Hygrometric Mode (A) and in Combination of Hygrometric and Psychrometric Modes (B)

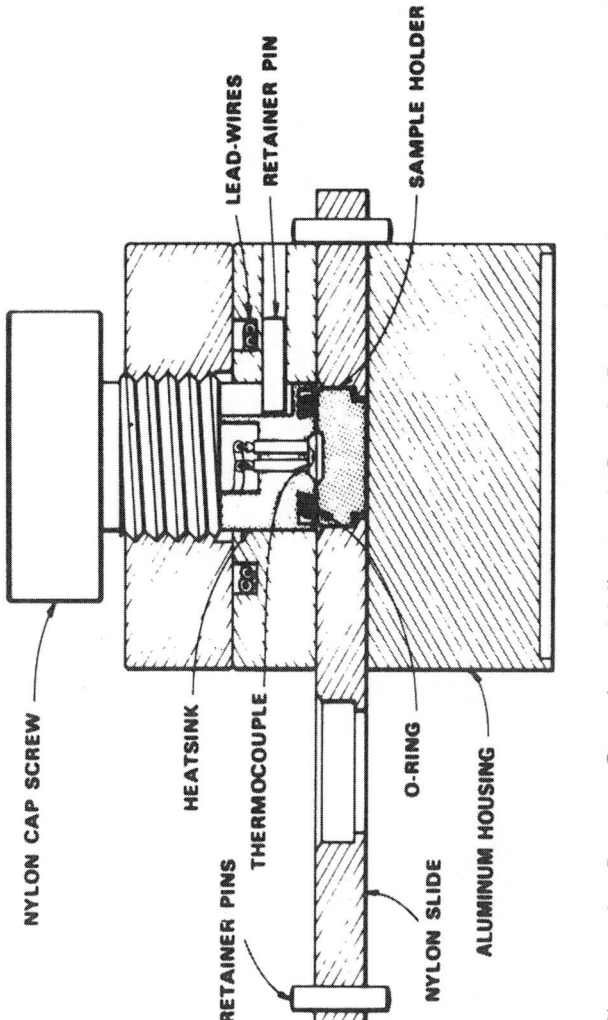

NYLON CAP SCREW

LEAD-WIRES

RETAINER PIN

SAMPLE HOLDER

RETAINER PINS

THERMOCOUPLE

HEATSINK

NYLON SLIDE

O-RING

ALUMINUM HOUSING

Figure 4   Cross Sectional View of C-52 Sample Chamber

Other sample chambers generally utilize a soil psychrometer sealed in a cavity which has space provided for a sample. A cross-sectional view of such a chamber reported by Brown and Collins (5) is shown in Figure 5.

An in situ leaf psychrometer/hygrometer was developed by Wescor, Inc. and evaluated by Campbell and Campbell (8) in 1973. This sensor allows the measurement of water potential in a growing leaf with a minimum of disturbance to the natural environment of the leaf.

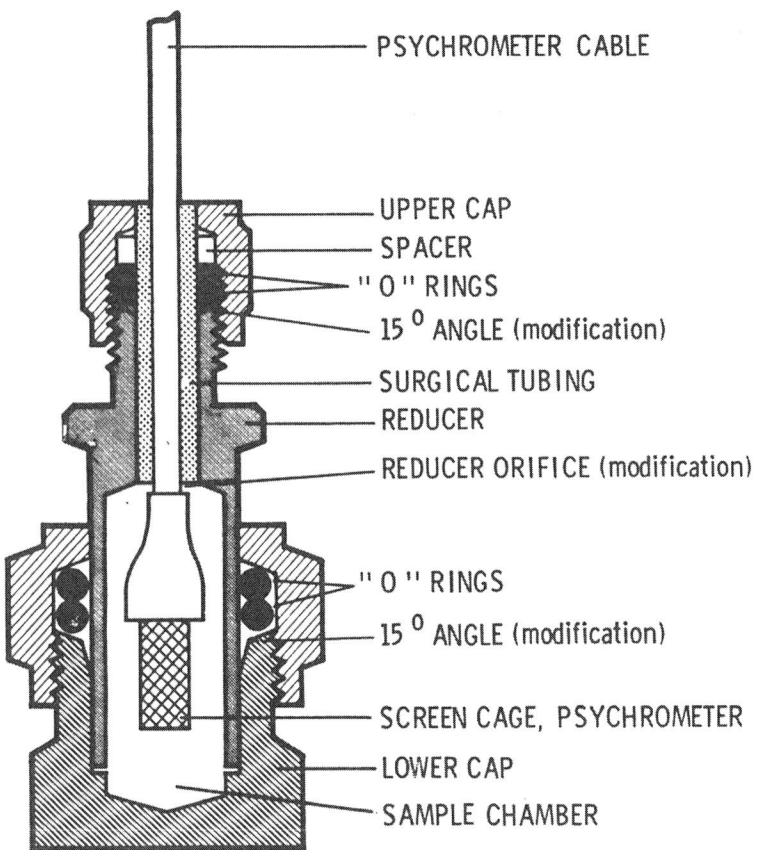

Figure 5 Stainless steel calibration chamber with sealed psychrometer in place (after Brown and Collins 1980).

5. CALIBRATION

The calibration of a thermocouple psychrometer/hygrometer system consists of performing measurements of samples of known water potential. Since the magnitude of the water potential is only one of several factors which affect the output of the sensor, it is important to duplicate the conditions which will exist in the field measurements as closely as possible while calibrating.

Solutions of NaCl or KCl are generally used in the calibration. Tables are available which give the relationship between water potential and molality for solutions of these salts.

For sample chambers which are thermally compensated such as the C-52, calibration consists simply of placing filter paper discs with solutions of a known concentration within the sample chamber and measuring the voltage output. This is repeated for a range of concentrations and a plot of voltage output vs water potential is obtained. The resulting plot should be a straight line. The time required for vapor and thermal equilibration is usually less than 2 or 3 minutes after the sample is loaded.

It is important that the temperature of the chamber remains constant during such measurements. Slight variations in air temperature will not affect the calibration appreciably, but care should be taken to avoid touching the metal parts of the chamber since heat transferred from a person's hand may have a considerable effect.

The amount of water condensed on the thermocouple will have some effect upon the voltage output. Therefore, cooling current and cooling time should be standardized.

The proximity of the sample to the thermocouple can also change the output. If a deep sample holder is used during the field measurement, it should also be used during calibration. If the sample fills the chamber, several layers of saturated filter paper can be used to approximate the volume of the sample.

The temperature of the sample changes the voltage output considerably even if the entire system is isothermal. The output dependence upon temperature is well established theoretically and empirically. The temperature correction must be applied to calibration measurement as well as field measurement.

Soil psychrometers/hygrometers are more difficult to calibrate because no large heat sink is attached to the thermocouple to stabilize the temperature. In a typical laboratory or greenhouse, the variations in temperature are sufficient to make it impossible to obtain accurate measurements unless steps are taken to stabilize the temperature of the sensor. This can be accomplished by placing the sensors in a well-insulated container with a seal against air drafts. At least 0.5 m lead should be within the container to prevent heat flux from traveling along the lead wire to the measuring thermocouple.

Many experimenters use a water bath to control the temperature of the sensors during calibration. Precise temperature is not important but thermal gradients must be minimized. Water can sustain considerable thermal gradients for a short period of time and in some cases the heat source for the water bath is turned off for several minutes prior to the calibration measurement, this gradient is generally reduced or eliminated.

The calibration configuration for soil psychrometers/hygrometers which will most nearly approximate field conditions is to immerse the sensors in the calibration solution. The pore size of the ceramic shield or the stainless steel shield surrounding the thermocouple is sufficiently small to prevent liquid water from entering the cavity. However, if temperature gradients exist within the unit, the vapor can condense. If the calibration is done by immersion, the psychrometer must be thoroughly cleaned following calibration to eliminate any salt residue.

In situ leaf psychrometers/hygrometers have a thermal attachment between the thermocouple, the cylinder, and the mounting block. It is possible to achieve a seal between the cylinder and an aluminum envelope containing saturated filter paper (6). Calibration can also be accomplished by suspending the cylinder into a container of solution. An air bubble formed around the thermocouple provides a cavity. The calibration should be done in a room where the temperature is stable, but it is not necessary to provide an insulated container because the cylinder has sufficient thermal mass to stabilize the temperature.

6. FIELD OR LABORATORY USE

Outdoor operation of thermocouple psychrometers/hygrometers exacerbate the thermal problems. With a sample chamber it is necessary to provide shade from the sun and protection from the wind. The soil provides fairly good insulation to keep the temperature of soil psychrometers/hygrometers uniform, particularly at large depths. The soil temperature at a depth of one meter or more remains nearly constant. However, the soil surface can change temperature over a wide range during a 24-hour period. If the soil is in direct sunlight, the surface temperature can change as much as 20 to 30°C from early morning to mid-afternoon. Therefore, there will be a heat flux out of the soil during the night and into the soil during the day. Accurate soil water potential measurement near the surface will be difficult, if not impossible, during the periods of maximum flux. Because the heat flux changes directions twice a day, there will be two time periods when the gradients are near zero at any particular depth. Thermal gradient problems will be alleviated if the measurements are made during those periods.

Soil psychrometers/hygrometers should, generally, be installed so that the axis of the sensor is parallel to the surface of the soil and perpendicular to the direction of radiant heat flux. The total thermal gradient across the sensor will be less because of the smaller dimension of the cavity in this direction.

The same type of problem can exist in a greenhouse environment. Often additional problems are encountered because of the venting or cooling systems used. Typically the air is allowed to cycle several degrees between maximum and minimum temperatures. Another problem in the greenhouse is that the plants are often grown in small containers. As the air at the sides and top of the container changes temperature, gradients not only can affect the measurement of water potential but can also change the water potential itself. If vermiculite or another insulating material is used between the potted plants, the temperature gradients from the sides of the pots will be reduced.

In situ leaf psychrometers/hygrometers have a special problem when used out of doors, particularly if there is a breeze. The leaf is inserted through a slot in the mounting block. The slot provides a pathway for heat flux that is not protected by the heat sink of the block. Protection to eliminate air currents and shade to eliminate direct sunlight from striking the sensor is necessary if satisfactory measurements are to be obtained. Savage, et al. (6), report that insulating the entire sensor with a few layers of foam and then covering with aluminized mylar tape reduced thermal problems significantly. They also covered a length of the lead wire emerging from the cylinder top with aluminized mylar tape.

Two additional factors must be dealt with in the use of leaf psychrometers/hygrometers. The leaf must be sealed to the base of the cylinder or vapor equilibration will not take place. The sealing can be accomplished by placing a ring of petroleum jelly or a beeswax-lanolin mixture around the rim of the cylinder where it will touch the leaf. The second factor is that the substomatal vapor must diffuse through the leaf cuticle due to presumed stomatal closure under dark chamber conditions. In another publication, Savage, et al. (7) report a number of methods tested to reduce cuticular resistance by abrasion of the cuticle or other means and compare the results of these methods. Typically, two hours or more are required for vapor equilibration to occur after installation of the in situ hygrometer/psychrometer on the leaf. Several recent publications give evaluation of the leaf psychrometer/hygrometer and additional information on operational techniques.

7. CONTAMINATION

The accurate measurement of water potential with a thermocouple psychrometer requires a uniform evaporation rate of the water from the thermocouple junction and a uniform vapor concentration throughout the cavity. If contaminants are present on the junction, the rate of evaporation is changed, the output is often reduced, and satisfactory precision cannot be obtained.

The difference in the psychrometric output from a clean psychrometer and the output from a contaminated psychrometer is shown in Figure 6. The difference between the hygrometric (dewpoint) output for a clean and for a contaminated sensor is more subtle. The output in that case is usually lower and less stable but appears much like the output from a clean hygrometer.

Contamination in sample chambers can be avoided by careful loading of samples. If the chamber does become contaminated, cleaning is a simple matter. Cleaning instructions are usually given in the manual for commercial chambers.

In situ leaf psychrometers/hygrometers are subject to contamination by the sealing compound which is not water soluable.

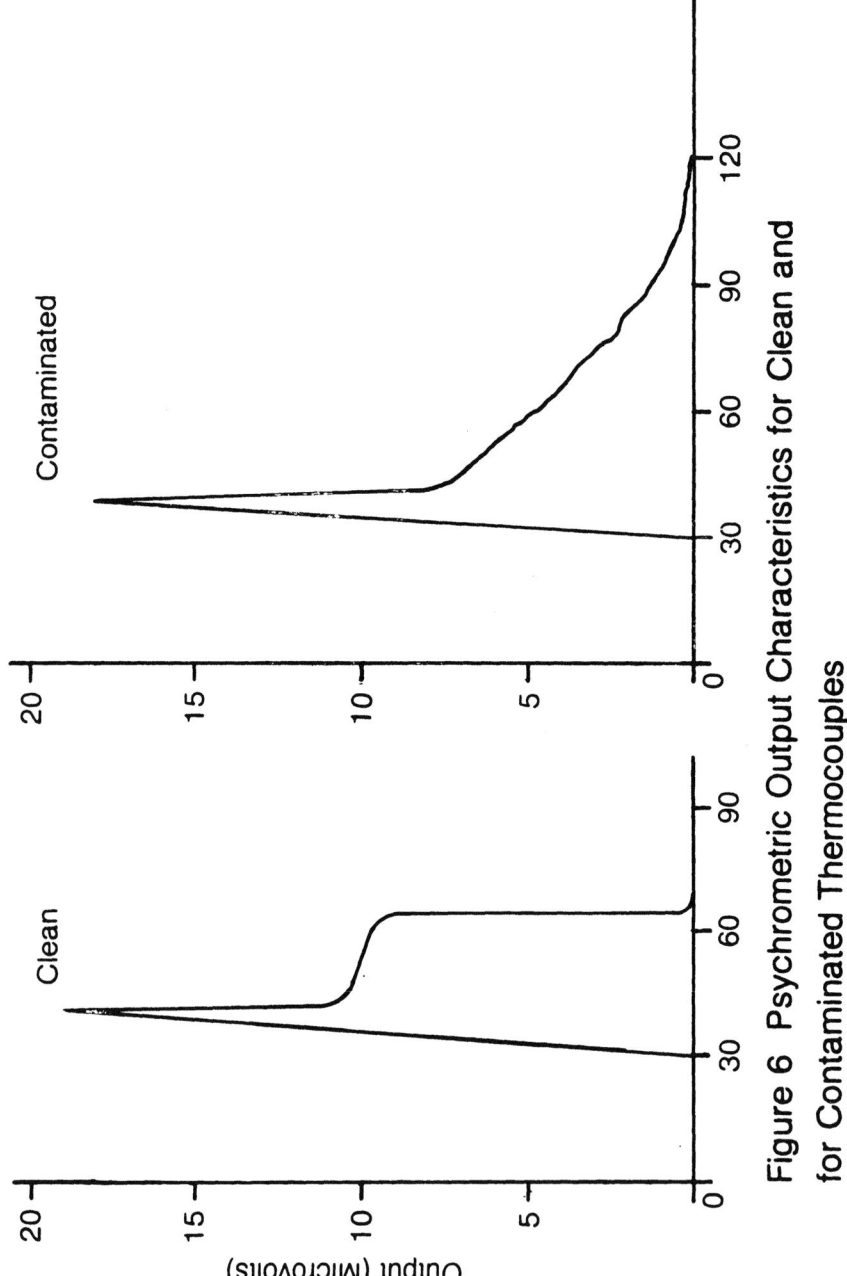

Figure 6 Psychrometric Output Characteristics for Clean and
for Contaminated Thermocouples

Solvents such as acetone or ammonia solutions help in removing such contaminants. Solvents which are acidic may attack the thermocouple and should be avoided.

Soil psychrometers with ceramic shields have a pore size approximately 3 microns in diameter. Those with stainless steel screen shields have a pore size from 20 to 30 microns in diameter. Most contaminating substances, like soil particles, are not able to pass through the ceramic or stainless steel screen shields. Contamination, however, is a common problem with the soil psychrometers/hygrometers. Dissolved particles of salt can pass through to the inner surface of the shield. The salt and the presence of high humidity often cause corrosion of the thermocouple. This can result in extremely poor precision. Sometimes the corrosion is severe enough to destroy the wires. Some psychrometers/hygrometers are built to facilitate the removal of the shield so that the thermocouple can be easily examined and cleaned.

8. CONCLUSION

Peltier-cooled thermocouple psychrometers/hygrometers provide a valuable tool in determining the water potential of solutions, tissues, and soils. Each experimenter should become familiar with the principles and techniques needed for successful operation under the conditions in which he will be required to take measurements. When proper procedures and techniques are used, the measurement is simple and accurate.

REFERENCES

1. Brown, R.W. and Van Haveren, B.P. (eds.) Psychrometry in Water Relations Research. Utah Agric Res Stat and Utah State Univ, p. 328.

2. Spanner, D.C. The Peltier effect and its use in the measurement of suction pressure. J. Exp. Bot. 2:145-168, 1951.

3. Neuman, H.H. and Thurtell, G.W. A Peltier cooled thermocouple dewpoint hygrometer for in situ measurement of water potential. In: Psychrometry in Water Relations Research (R.W. Brown and B.P. van Haveren, eds.) Utah Agric Exp Stat and Utah State Univ., pp. 103-112.

4. Campbell, E.C., Campbell, G.S. and Barlow, W.K. A dewpoint hygrometer for water potential measurement. Agric. Meteorol. 12:113-121, 1973.

5. Brown, R.W. and Collins, J.M. A screen-caged thermocouple psychrometer and calibration chamber for measurements of plant and soil water potential. Agron. J. 72:851-854, 1980.

208

6. Savage, M.J., Wiebe, H.H. and Cass, A. In situ field measurement of water potential using thermocouple psychrometers. Plant Physiol. 73:609-613, 1983.

7. Savage, M.J., Wiebe, H.H. and Cass, A. Effect of cuticular abrasion on thermocouple psychrometric in situ measurement of leaf water potential. J. Exp. Bot. 35:36-42, 1984.

8. Campbell, G.S. and Campbell, M.D. Evaluation of a thermocouple hygrometer for measuring leaf water potential in situ. Agron. J. 60:24-27, 1974.

APPENDIX

Figure A1. C-52 Sample Chamber.

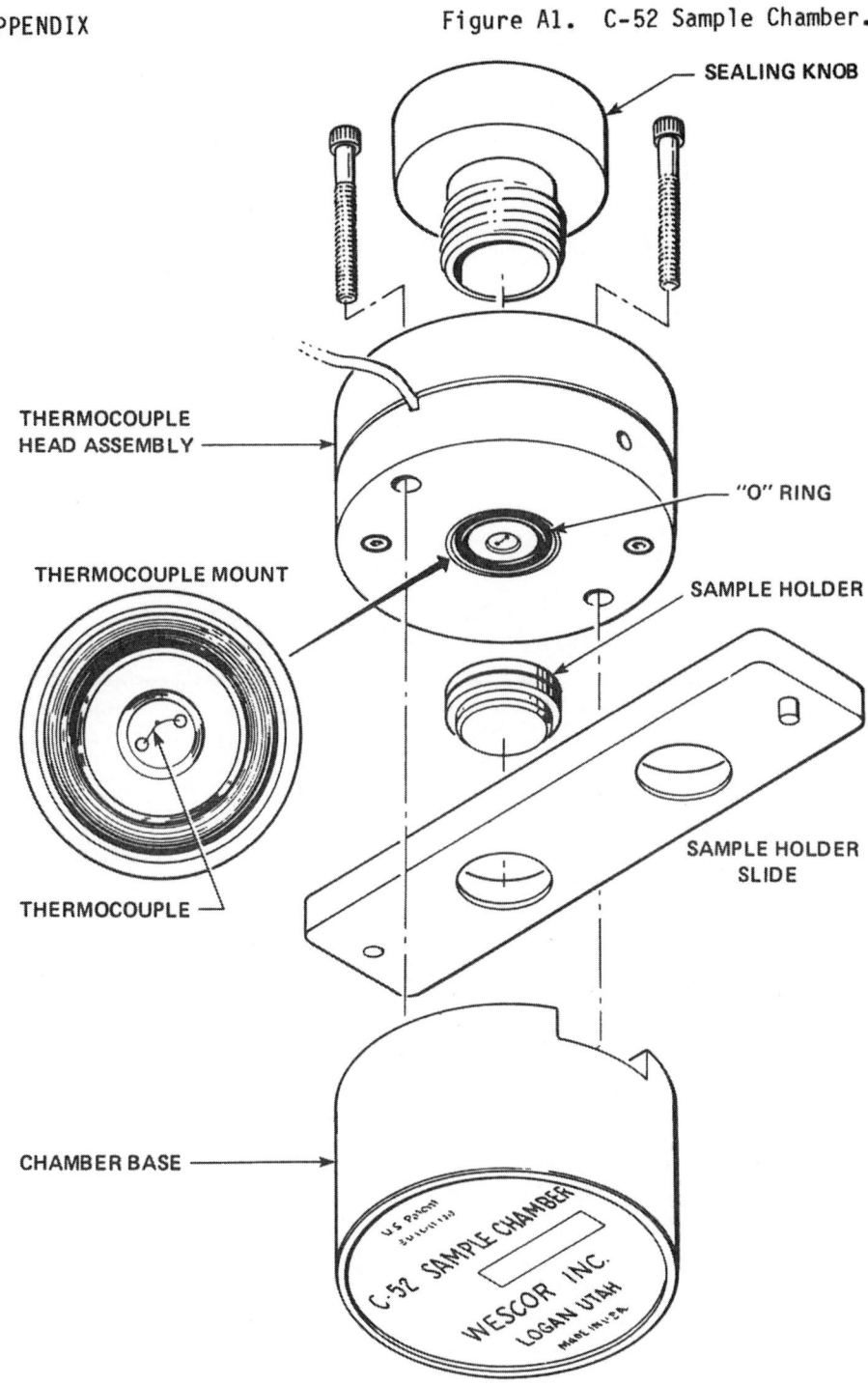

SEALING KNOB

THERMOCOUPLE
HEAD ASSEMBLY

"O" RING

THERMOCOUPLE MOUNT

SAMPLE HOLDER

THERMOCOUPLE

SAMPLE HOLDER
SLIDE

CHAMBER BASE

U.S. Patent
3810010 SA

C-52 SAMPLE CHAMBER

WESCOR INC.
LOGAN UTAH
MADE IN U.S.A.

RADIATION THERMOMETRY IN AGRICULTURE

Charles E. Everest

Everest Interscience, Inc.

INTRODUCTION

Many parts of the world are totally dependent on irrigation to produce plentiful crops. Some areas use irrigation as a supplement to rainfall and others use irrigation only to assure that crop yields won't be affected by a lack of rainfall for a few weeks.

The supply of water available for irrigation is decreasing. At the same time, the number of irrigation systems are increasing. The cost of the energy required to pump and move the water required for these irrigation systems is rising quickly and steeply.

These facts have led researchers from the United States Department of Agriculture (USDA) and agronomists from various universities to turn their attention to the science of irrigation and irrigation scheduling in particular.

According to Dr. Ray Jackson, "...irrigation-scheduling techniques fall into three categories: soil based, meterologically based, and plant based." (1) We will be concentrating on the plant based irrigation scheduling in our discussions.

Portable IR Thermometers in Agriculture

During the 1960's, infrared (IR) thermometers became commercially available as infrared technology advanced. The first instrument consisted of a lightweight sensing head connected to an electronic console by a cable. These instruments required external power sources.

The original sensing head contained a controlled temperature blackbody of about 40°C, which the sensor was exposed to alternately with radiation from the target. In the summer months, in the Southwest, the ambient air temperature was high and the sensing head was exposed to solar radiation. The reference temperature of 40°C was exceeded, causing erroneous temperature readings.

Subsequent designs of the infrared sensor made it portable by including a battery operated console. A number of lightweight, hand-held, battery operated IR thermometers have become available during the past two decades. Many of these instruments were made for use in buildings or other enclosed areas where there would be only slight changes of temperature. Some of these instruments failed in the field because of the changing air temperature. That is when several companies began to address this problem and significant progress has been made toward the solution. Temperature compensation circuits have been added as an integral part of instruments which are to be used outdoors.

In 1971 Hiler and Clark (2) and in 1974 Hiler et. al. (3) developed what they called a Stress Day Index (SDI) based on plant parameters. The plant stress factor and the crop susceptability factor yielded the SDI. The plant stress factor is an indication of the plant's water deficiency. The crop susceptability relates to the growth stage and crop species.

Several plant measurements, including leaf temperature, were measured and plant water potential was selected as a stress indicator. In his work in 1973, Ehler took vapor pressure into consideration and found leaf temperature to be a good indicator of stress. (4)

Due to the drawbacks inherent in making these leaf temperature measurements with thermocouples -- such as time involved and the lack of accuracy -- researchers turned to the use of infrared (IR) thermometers. These instruments offer the possibility of measuring the entire crop canopy or individual leaves quickly with much greater accuracy than with other conventional forms of temperature measurement.

In 1972, Aston and Van Bavel proposed that by measuring the visible and thermal radiant heat loads upon plant leaves the soil water depletion could be remotely detected as the soil surface dried. They suggested that various locations in the field would become stressed before others because of the inherent heterogeneity of soils. They suggested that locations in a field where the plants were stressed would have a canopy temperature with greater variability than the areas where there had been well-watered conditions. They proposed that this variability could be used to signal the onset of water deficits. (5)

In 1974 Stone and Horton (6) installed an IR thermometer above a field of grain sorghum. They measured both the canopy temperature and the air temperature to evaluate evapotransporation (ET). They used this method to obtain regional ET estimates. Then in 1976, Blad and Rosenburg (7) mounted an instrument over an alfalfa crop and compared the resulting surface temperature measurements with leaf thermocouples and aircraft temperature scanners. They, too, determined the ET comparisons between corn and alfalfa, based upon these temperature measurements.

Stress Degree Day (SDD)

In 1977, Idso et. al. (8) and Jackson et. al. (9) used infrared thermometers to measure canopy temperatures every day throughout a complete wheat growing season. They were trying to develop techniques for remotely evaluating crop water stress with a minimum of measurements. They defined a Stress Degree Day (SDD) as the difference between the canopy temperature T_C and the air temperature T_A. They used the SDD to predict final yields on crops making measurements for several wheat plots, and they plotted the accumulative SDD through maturity as a function of time. Each of these wheat plots received different amounts of irrigation water. They were able to demonstrate that canopy temperature data can be used to find crop water stress and the effects of the stress on the final yields of the crop.

In 1979, Walker and Hatfield (10) applied the SDD concept to red kidney beans. They, too, found that the final crop yield was inversely related to the SDD. They proposed that the less water available for transporation, the higher the SDD, proving the transpiration and canopy temperatures are related.

The SDD was evaluated by Jackson et. al. (11) as a possible irrigation scheduling tool. Measuring water depletion with a neutron moisture meter, they accumulated all of the positive values of SDD. They took the SDD to be 0 whenever it was negative. They concluded that irrigation should be given when or before the positive SDD's reached a value of 10. In 1982, Geiser et. al. (12) aligned canopy-air temperature difference with net radiation and vapor pressure data to use as an irrigation scheduling tool. He concluded that the use of irrigation water was less when this temperature differential was used as a scheduling criterion.

Early 1980 studies tested the deviation of mid-day canopy temperature as an irrigation scheduling tool. They found that in fully irrigated crops of corn, there were standard deviations of 0.3°C in fully irrigated plots of corn. The deviations were as great as 4.2°C in non-irrigated plots. In this work they concluded that plots exhibiting the standard deviation above or under 0.3°C were in need of irrigation. (13)

In 1981 Gardner et. al. (14) suggested that canopy-air temperature differences may be climate, crop and soil specific. Because plant water status has been determined by plant water potential, it was decided to compare such plant measurements with canopy temperature data. Depending upon the plant size and the species, either whole plants or parts of plants can be measured. Because of the entire canopy, considerable variation was expected. Measurements of the plant water potential were taken simultaneously with canopy temperatures using IR thermometers. Data received from measuring six differently irrigated plots showed a considerable amount of scatter in the data. From this data it was concluded that canopy temperatures and plant water potential are correlated, but not linearly. Their conclusion was that a quantitative measure of plant water stress in field conditions has not been achieved as yet.

Also, canopy temperatures can be dramatically affected with changing cloud conditions, increasing data variability. Leaf-temperature changes of up to 6°C have been reportedly caused by changing solar insulation. Being shaded by a cloud, a leaf requires 30-45 seconds to reach a new equilibrium. (15)

Quantification of Crop Water Stress

With regard to the three factors that are used to signal irrigation needs, the meterological and soil factors indicate when plants may be stressed but the plant factors indicate when they are stressed. These factors, such as water potential, require numerous samples to characterize a field, as they are point measurements. Making crop canopy measurements can minimize this problem. However, using crop canopy temperature measurements for quantifying plant water stress has not been entirely satisfactory.

Jackson reports that three temperature indices have been proposed in the literature:

> The SDD, which is the canopy air temperature difference measured post-noon during the time of maximum heating; the TSD, which is the difference in canopy temperatures between a stressed crop and a non-stressed (well-watered) crop; and the CTV, which is the range of temperatures encountered when measuring a plot during a particular measurement period. (16)

Remember, with SDD the plants were considered stressed if the value was positive and not stressed if negative. Some experiments have shown though that this arbitrary division of SDD = 0 does not prove to be appropriate for all environmental conditions.

With the Temperature Stress Day (TSD), the reference plot that is not stressed must be in very close proximity to a field that is stressed. The TSD compensates for environmental effects such as vapor pressure deficit in air temperature with the use of a well-watered plot as a reference. (17)

The degree of variability in soil properties inherent within a field may influence the Critical Temperature Variability (CTV).

Surface temperature as a function of net radiation and vapor pressure deficit have led to a temperature-based stress index that might prove to be a reliable means of quantifying crop water stress. (18)

The temperature of a plant has been recognized as a water availability indicator for some time now. Temperature measurements were made with contact sensors on or embedded in leaves before infrared thermometers were available. Infrared technology has progressed significantly in the past decade. Now infrared thermometers are available with accuracy of ±0.5°F or °C.

To quote Jackson et. al.:

> The rapidity of measurement and the ability to
> average the temperature of all plant parts within
> the field of view (FOV) of the instrument make
> these devices ideal for crop canopy tempera-
> ture measurements. (19)

Crop Water Stress Index and Soil Water

Canopy temperature has been used to quantify crop stress using a crop water stress index (CWSI). One common cause of plant stress is an inadequate supply of water in the soil. This causes the plants to transpire at a rate less than the evaporative demand of the atmosphere. It is desirable to minimize the amount of stress a plant undergoes in order to maximize its production in most cases.

The recent method of measuring plant temperature to characterize plant stress has been based on the fact that transpiration cools the leaf as water is evaporated from the leaf surface. When the amount of water becomes limited, the transpiration decreases, causing the leaf temperature to increase.

Even though a temperature measurement of individual leaves of a plant provide a good indication of the plant stress, this measurement is still only a single point measurement. These single point measurements can be very time consuming.

Infrared thermometry minimizes this necessity for multiple single point temperature measurements. Rapid measurement of large numbers of plants and measurement of plant temperatures over large areas are now available.

CWSI is based on the measurement of the foliage-air temperature differential ($T_F - T_A$ in °C) as a function of the vapor pressure deficit (VPD) of the atmosphere. This linear relationship is called the lower baseline. It has been developed for many crops under sunlit and completely shaded conditions.

The lower baseline for a crop can be derived very quickly by taking dry-bulb and wet-bulb temperatures as often as every ten minutes over a six- to eight-hour period centered about solar noon. This generally results in a fairly wide range of VPD over short periods of time.

First the relationship between the well-watered and potentially transpiring crop is defined. Then the upper limit baseline of the foilage-air temperature differential is calculated. This differential is independent of the vapor pressure deficit but dependent on air temperature. (20) Dry-bulb and wet-bulb temperatures are usually measured one to two hours past solar noon when the crop is most stressed.

The CWSI is the distance the measured $T_F - T_A$ (point B on Figure 1) is above the lower baseline at the measured VPD divided by the distance between the lower and upper baselines at the same VPD (AC). The index basically ranges between zero (no stress) and one (maximum stress). See Figure 1.

Using the Crop Water Stress Index (CWSI), Jackson found there were day-to-day changes in the CWSI in addition to the scatter that he expected due to errors in the canopy temperature measurement and the wet- and dry-bulb air temperature measurements on wheat crops. Both the CWSI and the fraction of extractable water used increased with time. The CWSI was roughly in parallel with the extractable water used. However, CWSI did not drop to its lowest value immediately after irrigation. It required five to six days minimum. This implies that the stressed wheat requires some time to recover. This may be because leaves need to rehydrate or roots that were previously in dry soil need time to develop new root hairs. The degree of stress experienced, the plant species and the age all have a part to play in the length of recovery period.

The fact that there is a recovery period for the temperature base index indicates that a unique relationship does not exist between plant temperatures and soil moisture. The relationship between irrigation scheduling and crop water stress must be es-

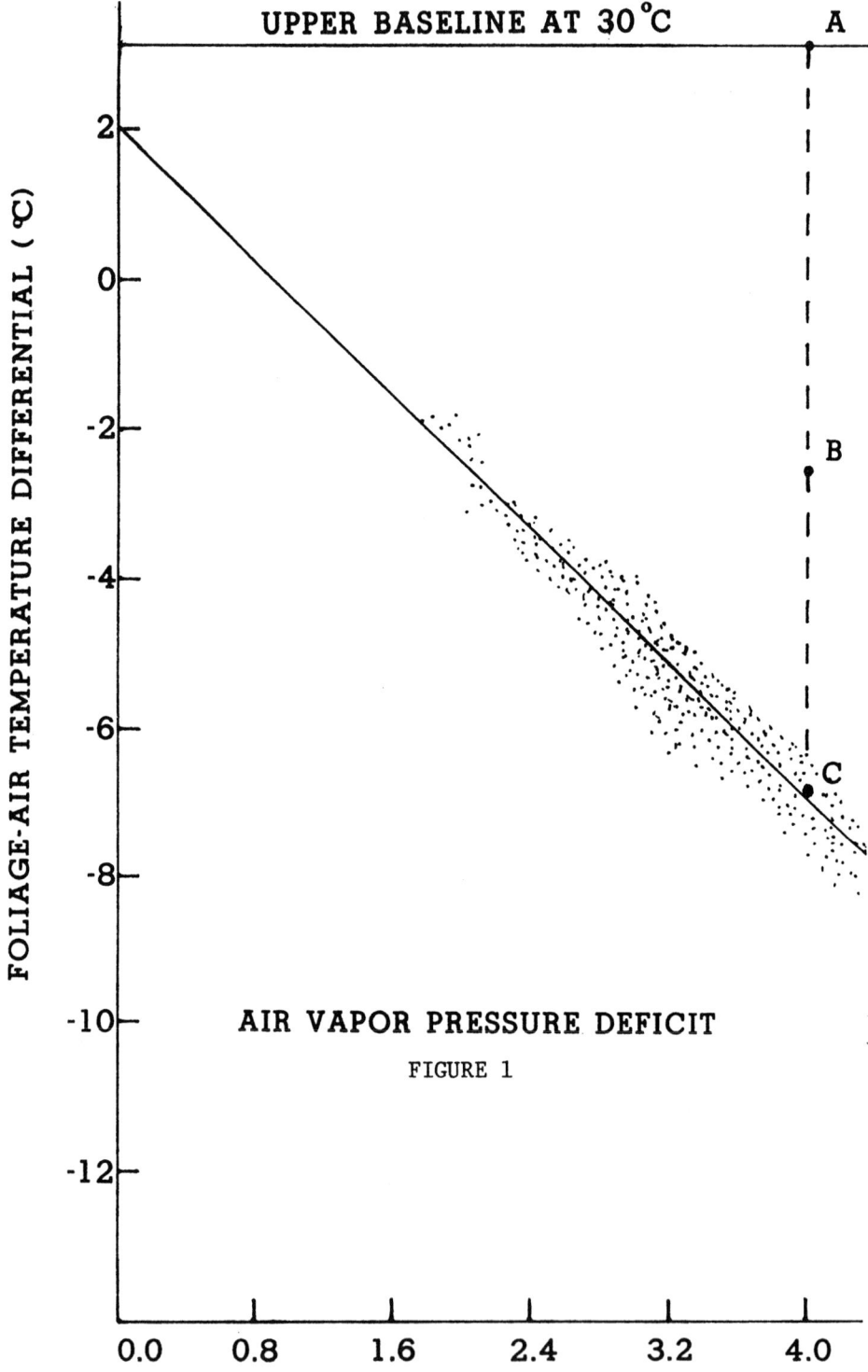

UPPER BASELINE AT 30°C

FOLIAGE-AIR TEMPERATURE DIFFERENTIAL (°C)

AIR VAPOR PRESSURE DEFICIT

FIGURE 1

tablished to make CWSI a usable irrigation scheduling tool. Although the statement, "Leaf temperatures are always warmer than the air," has been heard for some time, it is now rather obvious that canopy temperatures may be either warmer or cooler than the air. Some of the factors that would indicate this are environmental factors. Canopy temperatures will be near to or higher than air temperatures in humid climates. However, in arid climates temperatures may be more than 10°C below air temperature and have a variance of up to 15°C. Of course, it is in the arid areas where irrigation is practiced that there is the most need for temperature techniques.

However, care must be taken to avoid the complication of filling the field of view of the IR thermometer with anything other than the vegetation, especially when the vegetation is sparse.

Findings to Date

In his work, Jackson has found, "Its potential (IR) for use in the timing of irrigation appears promising." (21)

"... water is the greatest single use of energy in Nebraska agriculture," according to William Splinter who is Head of Nebraska University's Agricultural Engineering Department. (22) Irrigation scheduling can save money by reducing the amount of water that has to be pumped.

At the present time, even though more than one million acres utilize scheduling, there are seven million acres still unscheduled. In 1973 it cost farmers $31 million to pump irrigation water. Since then acreage has doubled and energy costs have increased <u>five times</u>. (23)

Also at Nebraska University, Blaine Blad, meterologist and climatologist, has been studying remote sensing using a hand-held infrared thermometer for years. "This approach... has potential for direct use by individual farmers." (24)

With the infrared thermometer, Blad's department found that there was a direct correlation between plant temperature and the amount of plant moisture. To quote Blad, "We think this is a tool that's going to be useful for plant breeders who are working on drought tolerance." (25)

His feeling is that on-farm use of infrared thermometers lies someplace in the future. Technology transfer is just beginning to take place. Infrared thermometers, however, are already being used by innovative farms and growers such as:

Crop Care
Salyer Land
Carian Sales
Hawaiian Sugar Planters Association
U & I, Inc.
Dekalb Ag Research
Ag Tech Co.
Westlands Water District

In our discussion of infrared thermometry my goal will be to educate you in the theory of infrared thermometry. A comprehensive list of references will be provided in order that you can find previous studies that relate to your specific area. Our discussion concentrates on an overview of infrared thermometry, emissivity problems, problems to avoid with infrared thermometry, breakthroughs in the state-of-the-art and the current status of research work.

Successful Agronomy Applications

Infrared thermometers have been found useful for a variety of applications with regard to the agronomy field. Their initial and most successful use to date has been optimizing irrigation scheduling. However, after learning of their existence, researchers have thought of many other types of applications.

For instance, they have been used to locate underground rivers from a helicopter and to locate breeding grounds of insects before they have wings.

Plants have been found to have an elevated temperature when they are stressed for water or by disease. A diseased plant can be detected by making a surface temperature measurement before there is any visual indication of a problem. And, soil temperatures can be measured without touching the soil. This means that the infrared thermometer does not disturb or change the temperature of the soil. However, keep in mind that the infrared thermometer measures only the surface temperature of the soil and does not penetrate the soil.

Infrared thermometers are very useful for measuring the surface temperature of any fragile materials or targets whose thermal equilibrium would be disturbed by other conventional forms of thermometry. The non-contact infrared thermometer poses no threat to such fragile materials as the crop canopy or individual leaves and stems.

THEORY OF INFRARED THERMOMETRY

Contact vs. Non-contact Thermometry

Infrared thermometers not only provide surface temperature measurements that are faster, safer and simpler to make, they also provide for more precise measurements. With most other forms of temperature measurement, a small portion of the quantity being measured must be consumed or diverted in the process of making the temperature measurement. In doing so, the magnitude of the target being measured is altered and an erroneous temperature reading results. This is true of all temperature measuring instruments other than infrared thermometers.

A probe must touch the object being measured with other conventional forms of thermometry. When the probe touches the object, heat transfer occurs between the probe and the object until equilibrium occurs if the object is at a different temperature from the probe. When this equilibrium occurs the object being measured is either cooler or warmer than it was previously.

This intrusion error does not apply to infrared thermometers because there is no contact between the infrared thermometer and the target. Since the object is radiating infrared whether or not the thermometer is present, its radiation characteristics and hence its temperature are not disturbed by the presence of the infrared thermometer.

Infrared thermometry is a non-invasive way to make surface temperature measurements. It is not necessary for the thermometer to make contact with the object being measured. All objects emit infrared radiation in all directions, whether or not the infrared thermometer is present. This infrared radiation is broadband electromagnetic radiation and hence is akin to neighboring visible light at shorter wavelengths and millimeter microwaves at longer wavelengths.

The broadband nature of this radiation implies that an object of a given temperature, T, will radiate wavelengths of varying amplitudes across this entire part of the spectrum. However, the energy content of wavelengths very far removed from the central radiating wavelength fall off very rapidly.

Emission of Radiation with Temperature

I'm sure you are aware that an object will emit light or visible radiation if it is sufficiently hot. This phenomenon is called "incandescence." Obvious examples of this phenomenon are light bulb filaments and smoldering embers. These incandescent objects also emit a tremendous amount of "invisible" infrared radiation.

Nature of Radiation

The difference between infrared and visible radiation is the wavelength of the electromagnetic wave. For instance, red light has a longer wavelength than blue light. Infrared radiation has longer wavelengths than both red and blue light. This is the only difference between these radiations; otherwise they are the same. They are composed of elementary packets of energy called photons. These photons travel in a straight line at the speed of light. These photons can be reflected by use of appropriate mirrors and they can be focused by the use of appropriate lenses.

The photons will dissipate their energy as heat when they are absorbed. The energy of a photon is inversely proportional to its wavelength.

EMISSIVITY

Blackbody Radiation

A "blackbody" is the strongest emitter at a given temperature. The reflectance and transmittance are zero and the emittance is unity.

A blackbody radiator is an idealized object that is the un-obtainable "perfect" radiator. All objects in the real world are known as "greybodies," as they fall short of this perfect per-formance. The ways and reasons that they fall short will be dis-cussed subsequently, but let's start with the ideal case and show how it deteriorates.

Consider a generalized object in space which we will call our blackbody radiator, which is composed of an infinite number of elements of mass at the same temperature, (T), each emitting energy uniformly in all directions. This energy is propagated from element to element of material in the blackbody until it reaches the last element of material at the blackbody surface. How this energy then transcends this surface determines whether the body is grey or black. This is profoundly important if ac-curate infrared temperature measurements are to be obtained.

The Concept of Emissivity

In a blackbody, there is a perfect impedance match between the body material at the surface and the propagating medium beyond. All the incident energy passes on unimpeded into space as easily as it passed from element to element within the body. We say that such a surface has an emission efficiency, or emissivity, of 1.0 (100%).

Following the principle of linear superposition, electromagnetic energy of a given kind will pass equally well in either direction across the blackbody surface. All incoming radiation energy will cross the blackbody surface and become absorbed. It will become indistinguishable from the internal energy transfer process. Therefore, a blackbody surface is at once a perfect radiator and a perfect absorber.

If there is an impedance mismatch at the surface, the body is said to be a greybody. Some of the energy is reflected back into the neighboring elements within the body and retained. The fraction of the energy that is reflected back is subtracted from 1.0 to yield the emissivity of the greybody surface. The fraction of the original energy that escapes is the numerical value of the emissivity of the greybody surface.

Lambertian Surfaces

Consider a surface mass element, δ, of our emitting body, and some of its identical neighbors. Each element radiates with equal intensity in all directions. However, only the radiation heading toward the air escapes. The rest hits its neighbors and is reabsorbed. Therefore, each surface element radiates uniformly into a 180° hemisphere. A flat surface composed of such elements is called a Lambertian surface. This uniformity of directional radiation is another ideal characteristic of a blackbody surface. Again, however, most real world infrared target surfaces have a directional dependence of radiation intensity.

Quantifying Blackbody (and Greybody) Radiation

We are indebted to Planck for providing us with the relationship between radiated power from a blackbody source, the temperature of the source, and the wavelength of the radiation. The power in watts can be measured directly as with a radiometer and if the "window" of wavelengths through which the radiation is allowed to pass is known, the absolute temperature of the blackbody will result. This relationship is Planck's Law, where:

W = the radiant power emitted into a hemisphere by the blackbody in power per unit area per wavelength interval. (Watts per sq. cm. per micron of bandwidth`

λ = the radiation wavelength in microns

e = the Napierian Logarithm base 2.71828...

T = the temperature of the blackbody in degrees Kelvin (°K = °Centigrade + 273)

C_1 = a constant = 3.7405×10^4 when area is square centimeters and wavelength in microns

C_2 = a constant = 1.43879×10^4 when square centimeters and microns are used

The radiant power emitted by a greybody therefore follows as:

$$W_\lambda = \varepsilon \times \left[\frac{C_1}{\lambda^5 \left(e^{(c_2/\lambda T)} - 1 \right)} \right]$$

Where:

ε = the emissivity of the greybody surface

If we differentiate Planck's equation and set the result equal to zero, we can determine the wavelength of peak intensity for a given temperature T:

$$\lambda_{Max} = \frac{2897.8}{T} \text{ Microns}$$

This is known as Wien's displacement law. Now, plugging the known values for λ max and T into Planck's equation, we obtain the value for the radiation power per micron of bandwidth at that wavelength:

$$W_{\lambda\ Max} = 1.288 \times 10^{-15} T^5$$

(watts per sq. cm. per micron)

Now we can see how rapidly the power per unit bandwidth falls off as we move away from the peak wavelength with our "window." The higher the temperature, the shorter the wavelength at which the peak occurs and W_λ max at the peak varies as T^5 .

By integrating Planck's equation with respect to wavelength from zero to infinity, we obtain the famous Stefan-Boltzmann (T^4) law:

$$W_{Tot} = 5.6697 \times 10^{-12} T^4 \text{ Watts per sq. cm.}$$

Note, however, that in infrared thermometers we limit the bandwidth "window" to a rather narrow range. Our signal power of interest is obtained by integrating only over the narrow range which produces various temperature exponents, depending on the window, which are always considerably less than four.

Consequently, Planck's law is the driving equation of interest in infrared thermometry. Wien's displacement law is second.

Tables of Emissivity are provided by most companies manufacturing infrared thermometers and most instruments have the capability of having an adjustable Emissivity setting. Remember, however, that plants and crop canopies are excellent radiators and that instruments are sometimes provided with preset emissivities. See Appendix C for a listing of some common emissivity values.

Emissivity Measurement

By comparing the reading on an infrared thermometer to a standard contact thermometer on the same surface, the Emissivity can be measured. A miniature, thin-film platinum resistance sensor is preferred for the contact sensor because it has a planar surface that makes better physical contact with the surface than a conventional probe. The contact sensor should be as small as possible and unhoused for the best results.

Attach the contact sensor to the surface under test and then elevate the temperature and stabilize near the upper end of the infrared thermometer's span for approximately one hour for thick sections and less for thin sections.

Aim the infrared thermometer at the target after stabilization, bringing it as close as possible to the contact sensor. If you have intra-optical light sighting on your instrument, focus the instrument until a ring configuration appears on the light spot, and position the contact sensor in the dead center section of the ring. Correct the Emissivity switch up or down until the infrared thermometer readout is the same as the contact thermometer. Record this reading for future use as the resulting Emissivity switch reading is equal to the Emissivity of the target.

Determination of a Greybody Surface Emissivity Value

A simple method of determining emissivity values of greybody surfaces up to approximately 400°C is to mask off a portion of the base surface with masking tape or scotch tape and pass the area adjoining the tape edge through a candle flame or "smokey" torch to deposit hydrocarbon soot on the area to effect a complete coverage. This soot has a high, well-defined emissivity of 0.955 ± 0.01, and has better thermal conductivity and a thinner section than e.g. paint so as not to affect the thermal balance of the substrate. If the tape is now pulled off of the target, a sharp line will divide the soot-covered area from the bare area to be evaluated.

The procedure is to heat the object to an appropriate temperature, and using an instrument with the Everest Interscience Lasite light beam sighting system with an emissivity setting of 0.95, measure the temperature of the soot area just adjacent to the bare material line. Next, move the instrument over to where the lite spot is just on the bare surface adjacent to the boundary and adjust the emissivity control to produce the same temperature reading. The value now residing in the thumbwheel is the desired emissivity value of the greybody surface.

PROBLEMS TO AVOID WITH INFRARED THERMOMETRY

The problems in industry are not the same as those in the agricultural field. In your area your main concern is that of being able to measure the temperature of fragile targets such as crop canopies or individual leaf or stem temperatures. The targets are sometimes small in spot size compared to industrial targets. Or, the target you might want to measure the temperature of may be very large, such as a crop canopy.

Other problems with providing precision infrared thermometers for use in the field are as follows:

> Low-temperature targets provide a greater challenge
> in infrared thermometry.
>
> Wind gusts can cause temperature variations.
>
> If an operator is trying to measure a very small
> target it may be difficult to tell where the
> thermometer is looking.
>
> Emissivity settings may or may not be a problem.
>
> If continuous operation is required, it is necessary
> to have an instrument that won't drift with time.

Low-Temperature Capability

As I stated before, all objects are emitting infrared radiation whether or not an infrared thermometer is present. The instrument is totally passive. High temperature objects give off a great amount of this radiation; however, low-temperature targets and those with temperatures close to the temperature of the instrument itself, give off very little infrared radiation. That is why the surface temperature measurement of low-temperature targets provides such a great challenge. Of course, agricultural applications fall into this category.

In the early 1970's the United States Department of Agriculture (USDA) challenged someone to design an instrument that would take temperatures in the range of -30 to +100°C with accuracy of ±0.5°C. They also requested that this instrument be capable of giving the temperature reading of the difference between the ambient air temperature and the temperature of the target automatically. This option is called Temperature Differential -- an option which is patented by Everest Interscience.

It took years of development, but the USDA's challenge has been met. Instruments to meet these specifications are available for use in the agronomy field. Technology is just now beginning to shift and researchers are passing their findings on to innovative farmers and growers.

One such progressive grower, Robert Carian said:

> Some investors had been losing money a couple of years on the land. I think the gun paid off in a couple of weeks. I was able to find the weak spots, the places that weren't getting enough water before I even had enough time to bring my own men in to check pressures and clean the lines. We picked 410 boxes (of grapes) to the acre there. Our neighbors didn't pick any more than 200. Without proper (water) management it would have produced like everyone else's vineyards in the area. (26)

Researchers, too, extol the advantages of using an infrared thermometer to optimize irrigation scheduling. Dr. Robert Reginato of the United States Water Conservation Laboratory in Phoenix, Arizona says, "I'm not saying the gun is the ultimate answer, but it's a good tool to help farmers make better irrigation decisions." (27)

Wind Gust and Ambient Air Temperature Problems

When an operator is taking temperature measurements outdoors, there is always the possibility that wind gusts will cause variations in the measurements over the field. This problem has been eliminated by the addition of a Data Averaging feature on some of the infrared thermometers available on the market today. This feature allows the operator to change the instrument from its normal instantaneous reading mode to a data averaging mode at the flip of a switch. The response time of the instrument is slowed down and the readings are then averaged over time.

Infrared thermometers for use in the field require particular precautions. Most portable infrared thermometers were made for use in buildings or other enclosed areas where only small temperature changes occur. When these instruments are used outdoors, the instruments give erroneous readings because components change in temperature.

Jackson suggests that:

> It is highly recommended that a potential purchaser determine whether an adequate ambient temperature compensation circuit is an integral part of any instrument that is intended for field use. (28)

Readability of Displays

Many of the original instruments for use in the natural environment used light emitting diode (LED) readouts. These were fine for the industrial instruments that were normally used indoors, but the readings washed out when they were in direct sunlight. The shift to liquid crystal display (LCD) readouts has corrected this problem and instruments are now available with standard LCD's or plug-in modules to convert the LED readouts to LCD's. Keep in mind that using the plug-in to convert to an LCD readout negates the possibility of having continuous monitoring through the output jack. Also, the use of the plug-in module may cause calibration problems.

Sighting Capabilities

The problem of knowing where an infrared thermometer is taking a temperature measurement is one that has been prevalent and addressed for some time. Various solutions have been offered but some of these solutions can cause as much of a problem as having no sighting system at all.

Many manufacturers have used a scope on top of the instrument for sighting. The cross-hairs of the scope are meant to show the exact center of the target. This may or may not be true. If it is true, it may be showing the exact place where a temperature is not being taken. In the dual-mirror systems common to portable infrared thermometers of every make, unless the instrument is focused at its exact focal plane, the center may well be the place where the temperature measurement is not being taken.

Also, scopes can get knocked out of alignment very easily. If this happens, the operator thinks he knows the exact focus of the instrument and where he is taking a temperature measurement, but he really doesn't. This is worse than having no sighting tool at all because of the false sense of security the operator experiences.

Laser beams have been used, also. The same problems inherent with the scopes are inherent with the laser sighting systems. Additionally, the power needed for the laser adds bulk and other problems with this system of sighting.

A third approach to the sighting problem is a visible light which is projected directly through the infrared optics. This light illuminates the exact and entire area where a temperature measurement is being taken. The light shares the same path as the infrared optics. Even if the instrument is dropped and the optical housing is knocked out of alignment, the light sighting and the optics are still focused on the same target. Also, to an experienced operator, the intensity of the infrared radiation being emitted can be visually noted and the places on the target with the greater radiation can be seen.

Normally this sighting system is used for close focus, small target spot sizes because the light does dissipate at greater distances.

Emissivity Settings

Normally, when an operator is taking the temperature of plants, leaves or stems, the Emissivity of the target is quite high and produces no problem. However, if soil temperature measurements are required, an Emissivity setting that can be adjusted is a must. Please refer to the previous section on Emissivity.

Here I would only like to make the point that if an Emissivity problem does exist, to be sure readings are taken at the correct setting, the control should be visible to the operator at all times. Some instruments have a keypad to enter the Emissivity setting, but the actual number is not visible at all times. This can lead to erroneous readings and any operator should be cautioned about the importance of taking the readings at the correct setting.

Continuous Operation

Another time when an operator should be cautious is when continuous operation is desirable. Most portable infrared thermometers on the market today cannot be run continuously. They begin to "drift" when they are run for more than a few minutes. They need to be turned off occasionally in order to "re-zero."

Minute temperature differences within the field of view of the detector, but not within the field of view of the instrument, cause "Zero Drift." This point is illustrated in Figure 2 . The detector element responds to all object temperatures within its field-of-view (θ) that are different from its own exact temperature.

The field-of-view (θ) is considerably smaller and within α . Ideally, this is all that we want the detector to see. Any temperature change that is within α but outside of θ , is indistinguishable by the detector from a change within θ, but is, in fact, an error signal which is referred to as "drift."

Newer state-of-the-art infrared thermometers have an optical chopper that automatically corrects for this "drift." This is what is meant by a re-zero feature. This optical chopper is a gold-plated shutter that periodically closes in such a way as to reflect back the detector's view onto itself.

Any residual signal is defined as "zero drift" and is artificially eliminated in this part of the cycle. When the optical chopper opens up the instrument's field-of-view, the resulting signal is defined as the true signal amplitude.

Now there are instruments with automatic re-zero. These instruments provide temperature measurements on a continuous basis with no drift in temperature measurement accuracy over time.

If your application calls for continuous operation, be sure that you have an instrument that has an automatic re-zero feature. With this feature continuous operation for months at a time is possible.

CONCLUSIONS

Recent breakthroughs in the state-of-the-art of infrared thermometry have brought space age electronics to the agricultural researchers. The capability of measuring surface temperatures with 0.1°C precision and ±0.5°C accuracy is a reality. Also,

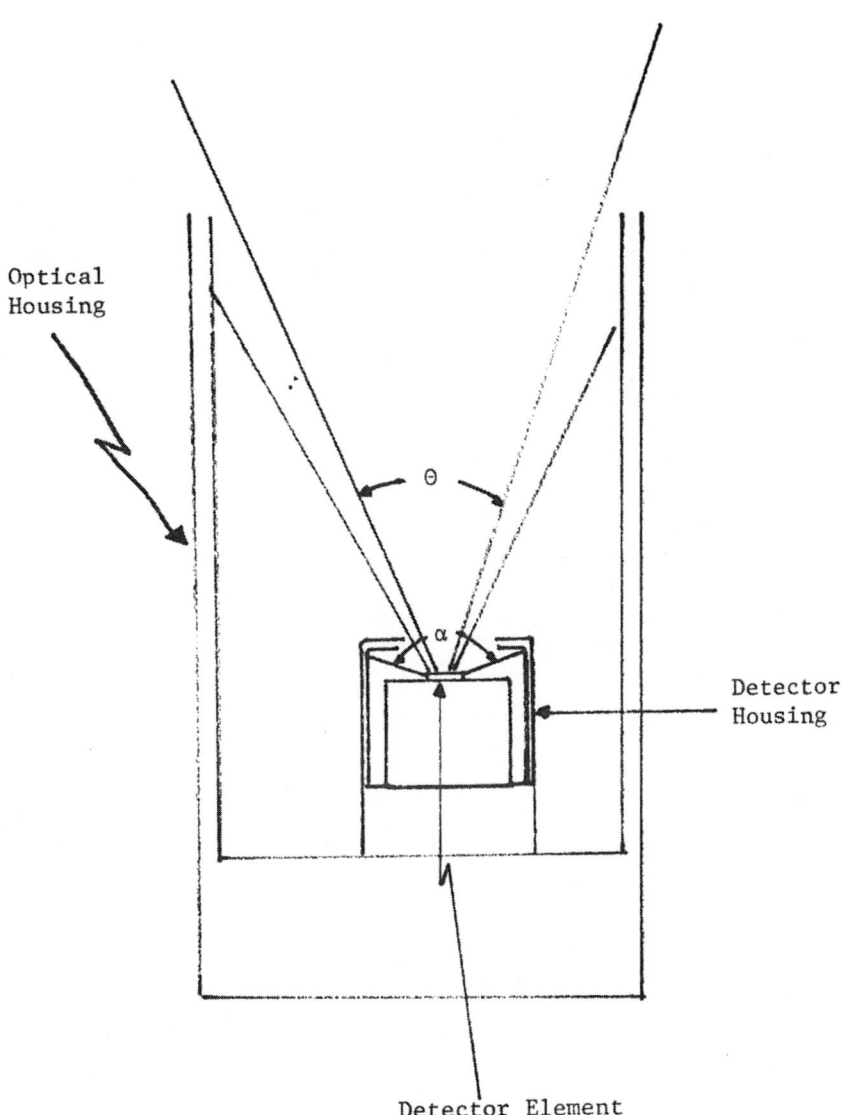

FIGURE 2

a new light sighting system, coupled with microscopic capabilities, has opened the doors of infrared thermometry to many applications for agronomists, plant pathologists and entomologists. Also, to enhance the precision of the instruments, the effects of unwanted sky radiation have been explored.

Precision and Accuracy

Accuracy refers to the fact that an instrument takes a temperature reading of a known standard and is within a stated percentage or number of degrees of that known traceable standard. The accuracy of the instrument is then said to be traceable to that standard, such as a National Bureau of Standards thermometer.

Precision refers to the repeatability of an instrument. If a target is measured and it reads 97.1°C, the instrument should read 97.1°C the next time a reading is taken if the target has remained stable from one reading to the next. If it does, the instrument can be labelled as being "precise."

Electronic infrared thermometers have been perfected that maintain an unprecedented absolute accuracy of ±0.5°C over the entire range of -30 to +100°C. Infrared thermometers are now available that can respond in a fraction of a second with 0.1°F precision. This fast response time makes it very easy for a researcher to quickly scan the canopy of a crop.

Temperature Differential

The Temperature Differential feature also has been perfected. This feature is used extensively to determine optimum irrigation scheduling. The leaf-temperature/air-temperature differential has been shown to be a crucial element for measuring and evaluating leaf energy status. Using the Temperature Differential feature, researchers can determine optimum irrigation scheduling using the Stress Degree Day (SDD) concept. Temperature Differential measurements of a crop canopy are made at the same location for a number of days at the same time. As a crop becomes stressed for water, the SDD figure rises. Thus, the researchers are able to determine the optimum time to irrigate. This not only saves water, but it saves energy required for pumping the water. Therefore, crop yields are enhanced by optimizing irrigation.

Using the Temperature Differential feature, a precision thermistor is housed in the end of the antenna which telescopes out of the optical housing about six inches. This removes the thermistor from the thermal effects of the instrument and the operator's hand. The end user simply touches the sensitive trigger and the difference between the absolute temperature of the target and the ambient air temperature registers on the large liquid crystal dis-

play. The antenna is then stored in the housing when it is not
in use.

Environmentally Sealed Units

Electronic infrared thermometers are usually configured as
hand-held, self-contained units when used by agronomists. However,
there are environmentally sealed units available that can be
fixed-mounted in the field and run continuously. These units can
be mounted up to five miles from the remote readout box with the
digital temperature information being transmitted by fiber optics,
greatly reducing interference noise.

Lasite Light Sighting

When using infrared thermometers, it is necessary that the
target fill the infrared optics' field of view. If the target
does not fill the field, the instrument will average the temperature of the first object it sees. For instance, if an operator
wants to measure an individual tiny leaf, and that leaf is smaller
than the smallest spot size of the infrared thermometer, an erroneous reading will result. The instrument will average the temperature of the leaf and anything else it sees, such as the soil in
the background.

To meet the challenge of this sighting problem, a Lasite
Intra-Optical Light Sighting System has been developed. A pulsating light is directed through the infrared optics. This light
illuminates the exact and entire field of view, thus eliminating
the guesswork about where the instrument is "looking." It gives
an indication of the exact place where a temperature measurement
is being taken. (29)

Sky Spy

Sky conditions can determine whether or not accurate temperature measurements are being taken in a field. To solve this problem, a second detector is positioned in a window on top of E.I.'s
instruments to detect sky radiation conditions. This feature,
which we call the Sky Spy, addresses the sky radiation error possibility. Circuitry is added that helps to correct the accuracy
of the surface temperature reading which otherwise would have been
affected by the unwanted radiation on the target.

Microscopic Capabilities

Coupling the Lasite Intra-Optical Light Sighting capabilities
with advances in optical focusing has provided models with microscopic capabilities. Now it is possible to measure the temperature
of targets as minute as 125 microns at safe, comfortable working

distances. We will not discuss these microscopic instruments in detail as they are not generally used in the agronomy applications. (30)

Breakthrough in State-of-the-Art Technology

The advancement in the state-of-the-art in infrared technology during the past few years has brought about the production of lightweight, hand-held IR thermometers that can be used in measuring plant canopy temperatures rapidly. Although many of these instruments have just become available to aricultural researchers, some have already been purchased by farmers and growers. There should be an explosion of knowledge about canopy temperatures in the next few years. There should be a similar explosion in knowledge about how they can be used in irrigation management. To quote Dr. Jackson, "The data accumulated so far strongly suggests success." (31)

APPENDIX A. GUIDE TO COMMON TERMS IN ELECTRO OPTICS

The terms and definitions given here are typical of those that might be used in the description of an optical or electro-optical system or device.

ABSORPTANCE – the ratio of absorbed radiant flux to incident radiant flux.

A/D CONVERTER – a device that converts an analog signal, such as raw video output, to a digital signal. This is usually done for easier signal processing.

AFOCAL – an optical system with object and image points at infinity.

ATMOSPHERIC ATTENUATION – the reduction in flux resulting from absorption and scattering as light travels through the atmosphere.

BANDPASS FILTER – an optical device that transmits radiation above or below a specified frequency, or between two specified frequencies.

BANDWIDTH – the range of frequencies over which an optical system will transmit radiation.

BEAM SPLITTER – an optical device that divides a beam of light into two or more separate beams.

BLACKBODY – (1) an object that completely absorbs all radiant energy striking it, (2) a device containing a radiation source with an emissivity of unity. Radiators with emissivity of less than unity are referred to as graybodies.

CASSEGRAIN OPTICS – an optical system that uses two mirrors to fold the image path back on itself. It has the advantage of shortening the physical length of the optical system.

COMPOUND LENS –   a composite lens made of two or more elements which may or may not be cemented together. A compound lens is used to surmount the aberrations inherent in a single lens.

CONCAVE LENS –   a lens with a spherical surface that curves inward. Also known as a diverging lens.

CONVEX LENS –   a lens with a spherical surface that curves outward. Also known as a converging or convergent lens.

DETECTOR –   (1) a device that converts incident radiation into another form such as electrical, visual, or photographic, (2) a device that alters an electrical flow in response to incident radiation. Typical devices include photoconductive, photojunction, and photovoltaic cells, phototransistors, phototubes or thermal devices such as bolometers and thermocouples.

EMISSIVITY –   the ratio of the radiance emitted by a source to the radiance emitted by a blackbody radiator at the same temperature and wavelength.

EXTREME INFRARED –   radiation with a wavelength greater than 15 micrometers.

FAR INFRARED –   radiation with a wavelength between 6 and 15 micrometers.

FIBER OPTICS –   (1) a branch of optics concerned with the transmission of light through long thin fibers of transparent material and (2) any long thin material capable of transmitting light through a series of internal reflections.

FIELD OF VIEW -   the angular measurement of the volume of space a system views in normal operation.

FILTER -   a device used to attenuate or transmit a selected wavelength of radiation.

FOCAL LENGTH -   the distance between the second principal point and the second focal point. Sometimes referred to as the equivalent focal length.

FOCAL PLANE -   a plane through the focal point which is at right angles to the principal axis of the system. The best image of the system is formed on the focal plane.

FRESNEL LENS -   a lens that resembles a surface cut into rings and flattened out. Fresnel lenses are often used in large area optics or where weight or thickness may be a constraining factor.

INFRARED (IR) -   an invisible portion of the spectrum that lies between approximately 0.75 micrometers and 1000 micrometers.

KELVIN -   the SI unit of temperature. Absolute zero is 0 degrees on the Kelvin scale. Each degree Kelvin is equal to 1 degree Centigrade. Water freezes at about 273.16K. No degree symbol is used in Kelvin system.

LENS -   a transparent optical component with one or more curved surfaces (usually, but not limited to spherical). Lenses serve to converg or diverge incident optical radiation.

MICRON -   a metric measurement equal to 10^{-6} meters. Often used instead of micrometers when refering to wavelength.

MIDDLE INFRARED –
(Mid IR)

the portion of the infrared spectrum between 3.0 and 6.0 micrometers.

NEAR INFRARED –
(Near IR)

infrared radiation from about .75 to 3.0 micrometers in wavelength.

NOISE –

random fluctuations that distort a signal. There are two major types of noise in electro-optics: electronic noise generated by detectors, amplifiers, and other electronic sources and optical noise generated by fluctuation in the background radiation.

NOISE EQUIVALENT
TEMPERATURE DIFFERENCE –

the temperature difference in two adjacent elements in a scene that will give a signal equal to the system noise.

OBJECTIVE LENS –

the optical element that receives the light from an object or scene and forms the first or primary image.

OPTICAL CHOPPER –

a device used to modulate light output. Usually a spoked wheel rotating at a constant rate placed between the emitter and receiver.

VISIBLE RADIATION –

radiation between 0.4 and about 0.75 micrometers in wavelength, also called light or visible light because radiation between these points can be seen with the unaided eye.

WAVELENGTH –

the physical distance covered by one cycle of a sinusoidal wave.

APPENDIX B. EMISSIVITY OF TOTAL RADIATION FOR VARIOUS METALS

Materials	°C	Emissivity
Alloys		
Nickel–Chromium		
20Ni–25Cr–55Fe, oxidized	200	0.90
60Ni–12Cr–28Fe, oxidized	270	0.89
80 Ni–20Cr	100	0.87
Aluminum		
Polished	50– 500	0.04–0.06
Rough Surfaces	20– 50	0.06–0.07
Strongly Oxidized	55– 500	0.2 –0.3
Oxidized	200	0.11
Asbestos Board	20	0.96
Bismuth, Oxidized	25	0.048
Brass		
Dull, Tarnished	200	0.61
Oxidized @ 600°C	200	0.61
Unoxidized	25	0.035
Polished	200	0.03
Rolled Sheet	20	0.06
Bronze, Polished	50	0.1
Carbon		
Filament	1000–1400	0.53
Graphite	0–3600	0.7 – 0.8
Lamp Black	20– 400	0.96
Soot applied to solid	50–1000	0.96
Soot with water glass	20– 200	0.96
Unoxidized	100	0.81
Chromium, Polished	50	0.1
Cobalt, Unoxidized	500	0.13
Columbium, Unoxidized	1500	0.19
Copper		
Calorized	100	0.26
Calorized, Oxidized	200	0.18
Commercial, scoured to a shine	20	0.07
Oxidized	50	0.6 –0.7
Polished	50– 100	0.02
Unoxidized	100	0.02
Unoxidized, liquid	---	0.15

continued

Material	°C	Emissivity
Fire Brick	1000	0.75
Glass	20- 100	0.94-0.91
Gold		
Carefully polished	200- 600	0.02-0.03
Unoxidized	100	0.02
Enamel	100	0.37
Graphite	0-3600	0.7 -0.8
Gypsum	20	0.93
Iron, Cast		
Oxidized	200	0.64
Strongly Oxidized	40	0.95
Unoxidized	100	0.21
Unoxidized, liquid	---	0.29
Rusted	25	0.65
Wrought, Dull	100	0.05
Lamp, Black	25	0.94
Lead	20- 400	0.96
Oxidized	200	0.05
Unoxidized	200	0.63
Mercury, Unoxidized	25	0.10
Molybdenum	600-1000	0.08-0.13
Monel Metal, Oxidized	200	0.43
Nichrome Wire		
Clean	50	0.65
Oxidized	50- 500	0.95-0.98
Nickel		
Industrial,Polished	200- 400	0.07-0.09
Oxidized	200	0.37
Oxidized @ 600°C	200- 600	0.37-0.48
Unoxidized	25	0.045
Platinum		
Clean, Polished	200- 600	0.05-0.1
Unoxidized	25	0.037
Wire	50- 200	0.06-0.07
Porcelain, Glazed	20	0.92
Rubber		
Hard	20	0.95
Soft, gray, rough	20	0.86

continued

Materials	°C	Emissivity
Silica Brick	1000	0.80
Silver		
Clean, Polished	200- 600	0.02-0.03
Unoxidized	100	0.02
Skin	32	0.98
Soil		
Dry	20	0.92
Wet	20	0.95
Soot applied to solid surface	50-1000	0.91-0.94
Soot with water glass	20- 200	0.96
Steel		
Alloyed (8% Ni, 18% Cr)	500	0.35
Aluminized	50- 500	0.79
Dull Nickel Plated	20	0.11
Flat, rough surface	50	0.95-0.98
Cast, polished	750-1050	0.52-0.56
Sheet, ground	50	0.56
Oxidized	200- 600	0.8
Calorized, oxidized	200	0.52
Sheet with shiny layer oxidized	20	0.82
Strongly oxidized	50	0.88
Unoxidized	100	0.08
Unoxidized, liquid	---	0.28
Tantalum, Unoxidized	1500	0.21
Tungsten, Unoxidized	25	0.024
Varnish	40- 100	0.8 -0.95
Dull Black	40- 100	0.96-0.98
Glossy Black sprayed on iron	20	0.87
Water	20	0.96
Ice	10	0.96
Snow	10	0.85
Zinc		
Polished	200- 300	0.04-0.05
Unoxidized	300	0.05

APPENDIX C.   TEMPERATURE CONVERSION CHART

°C	°F	°C	°F	°C	°F
-40	-40.0	13	55.4	80	176
-38	-36.4	14	57.2	85	185
-36	-32.8	15	59.0	90	194
-34	-29.2	16	60.8	95	203
-32	-25.6	17	62.6	100	212
-30	-22.0	18	64.4	105	221
-28	-18.4	19	66.2	110	230
-26	-14.8	20	68.0	115	239
-24	-11.2	21	69.8	120	248
-22	- 7.6	22	71.6	125	257
-20	- 4.0	23	73.4	130	266
-19	- 2.2	24	75.2	135	275
-18	- 0.4	25	77.0	140	284
-17	+ 1.4	26	78.8	145	293
-16	3.2	27	80.6	150	302
-15	5.0	28	82.4	155	311
-14	6.8	29	84.2	160	320
-13	8.6	30	86.0	165	329
-12	10.4	31	87.8	170	338
-11	12.2	32	89.6	175	347
-10	14.0	33	91.4	180	356
- 9	15.8	34	93.2	185	365
- 8	17.6	35	95.0	190	374
- 7	19.4	36	96.8	195	383
- 6	21.2	37	98.6	200	392
- 5	23.0	38	100.4	205	401
- 4	24.8	39	102.2	210	410
- 3	26.6	40	104.0	215	419
- 2	28.4	41	105.8	220	428
- 1	30.2	42	107.6	225	437
0	32.0	43	109.4	230	446
+ 1	33.8	44	111.2	235	455
2	35.6	45	113.0	240	464
3	37.4	46	114.8	245	473
4	39.2	47	116.6	250	482
5	41.0	48	118.4	255	491
6	42.8	49	120.2	260	500
7	44.6	50	122.0	265	509
8	46.4	55	131.0	270	518
9	48.2	60	140.0	275	527
10	50.0	65	149.0	280	536
11	51.8	70	158.0	285	545
12	53.6	75	167.0	290	554

Temperature Conversion Chart (Cont'd)

°C	°F	°C	°F
295	563	600	1112
300	572	650	1202
305	581	700	1292
310	590	750	1382
315	599	800	1472
320	608	850	1562
325	617	900	1652
330	626	950	1742
335	635	1000	1832
340	644	1050	1922
345	653	1100	2012
350	662	1150	2102
355	671	1200	2192
360	680	1250	2282
365	689	1300	2372
370	698	1350	2462
375	707	1400	2552
380	716	1450	2642
385	725	1500	2732
390	734	1550	2822
395	743	1600	2912
400	752	1650	3002
405	761	1700	3092
410	770	1750	3182
415	779	1800	3272
420	788	1850	3362
425	797	1900	3452
430	806	1950	3542
435	815	2000	3632
440	824	2050	3722
445	833	2100	3812
450	842	2150	3902
455	851	2200	3992
460	860	2250	4082
465	869	2300	4172
470	878	2350	4262
475	887	2400	4352
480	896	2450	4442
485	905	2510	4550
490	914	2566	4650
495	923	2593	4700
500	932		
550	1022		

CONVERSION METHOD:

$°F = 9/5°C + 32°$
$°C = 5/9 (F° - 32°)$

FOOTNOTES

(1).   R. D. Jackson, "Canopy Temperature and Crop Water Stress."
        Advances in Irrigation, Volume 1, page 44.

(2).   E. A. Hiler and R. N. Clark (1971). "Stress Day Index
        to Characterize Effects of Water Stress on Crop Yields."
        Trans. ASAE 14, pages 757-761.

(3).   E. A. Hiler, T. A. Howell, R. B. Lewis and R. P. Boos (1974)
        "Irrigation Timing by the Stress Day Index Method." Trans.
        ASAE 17, pages 393-398.

(4).   W. L. Ehrler (1973). "Cotton Leaf Temperatures as Related
        to Soil Water Depletion and Meteorlogical Factors."
        Agronomy Journal 65, pages 404-409.

(5).   A. R. Aston and C. H. M. Van Bavel (1972). "Soil Surface
        Water Depletion and Leaf Temperature." Agronomy Journal 64,
        pages 368-373.

(6).   L. R. Stone and M. L. Horton (1974). "Estimating Evapo-
        transpiration Using Canopy Temperatures: Field Evaluation."
        Agronomy Journal 66, pages 540-545.

(7).   B. L. Blad and N. J. Rosenberg (1976). "Measurement of Crop
        Temperature by Leaf Thermocouple, Infrared Thermometry, and
        Remotely Sensed Thermal Imagery." Agronomy Journal 68,
        pages 635-641.

(8).   S. B. Idso, R. D. Jackson, and R. J. Reginato (1977).
        "Remote Sensing of Crop Yields." Science 196, pages 19-25.

(9).   R. D. Jackson, R. J. Reginato, and S. B. Idso (1977).
        "Wheat Canopy Temperature: A Practical Tool for Evaluating
        Water Requirements." Water Resources Res. 13, pages 651-
        656.

(10).   G. K. Walker and J. L. Hatfield (1979). "Test of the
        Stress-Degree-Day Concept Using Multiple Planting Dates of
        Red Kidney Beans." Agronomy Journal 71, pages 967-971.

(11).   R. D. Jackson, S. B. Idso, R. J. Reginato, and P.J. Pinter,
        Jr. (1981). "Canopy Temperature as a Crop Water Stress
        Indicator." Water Resource Res. 17, pages 1133-1138.

(12).   K. M. Geiser, D. C. Slack, E. R. Allred, and K. W. Stange
        (1982). "Irrigation Scheduling Using Crop Canopy-Air
        Temperature Difference." Trans. ASAE (in press).

(13). B. L. Blad. B. R. Gardner, D. G. Watts, and N. J. Rosenberg
       (1981). "Remote Sensing of Crop Moisture Status." Remote
       Sensing Quarterly 3, pages 4-20.

(14). B. R. Gardner, B. L. Blad, D. P. Garrity, and D. G. Watts
       (1981). "Relationships Between Crop Temperature, Grain
       Yield, Evapotranspiration and Phenological Development in
       Two Hybrids of Moisture Stressed Sorghum." Irrigation
       Science 2, pages 213-224.

(15). C. L. Wiegand and L. N. Namken (1966). "Influences of Plant
       Moisture Stress, Solar Radiation, and Air Temperature on
       Cotton Leaf Temperature." Agronomy Journal 58, pages 552-
       556.

(16). Op. Cit. #1, page 65.

(17). Ibid.

(18). Ibid.

(19). Ibid.

(20). Ibid, page 48.

(21). Ibid, page 45.

(22). William Splinter. "Scheduling Saves Energy in NU Test."
       Irrigation Age, April 1984, page 26.

(23). Ibid.

(24). B. L. Blad. "Two Irrigation Scheduling Schemes Becoming
       More Sophisticated." Irrigation Age, April 1984, page 29.

(25). Ibid.

(26). Robert Carian. "Thermometer Gun Shoots Grape Yields To
       High Levels." Irrigation Age, December 1983, pages 24D,E.

(27). Ibid.

(28). Op. Cit. #21.

(29). C. E. Everest. "Aiming Technique Takes Guesswork Out Of
       Infrared Thermometry." Industrial Research & Development,
       January 1982.

244

(30). C. E. Everest. "Infrared Thermometers Provide Small-Target Accuracy." <u>Design News</u>, February 1984.

(31). Op. Cit. #21.

245

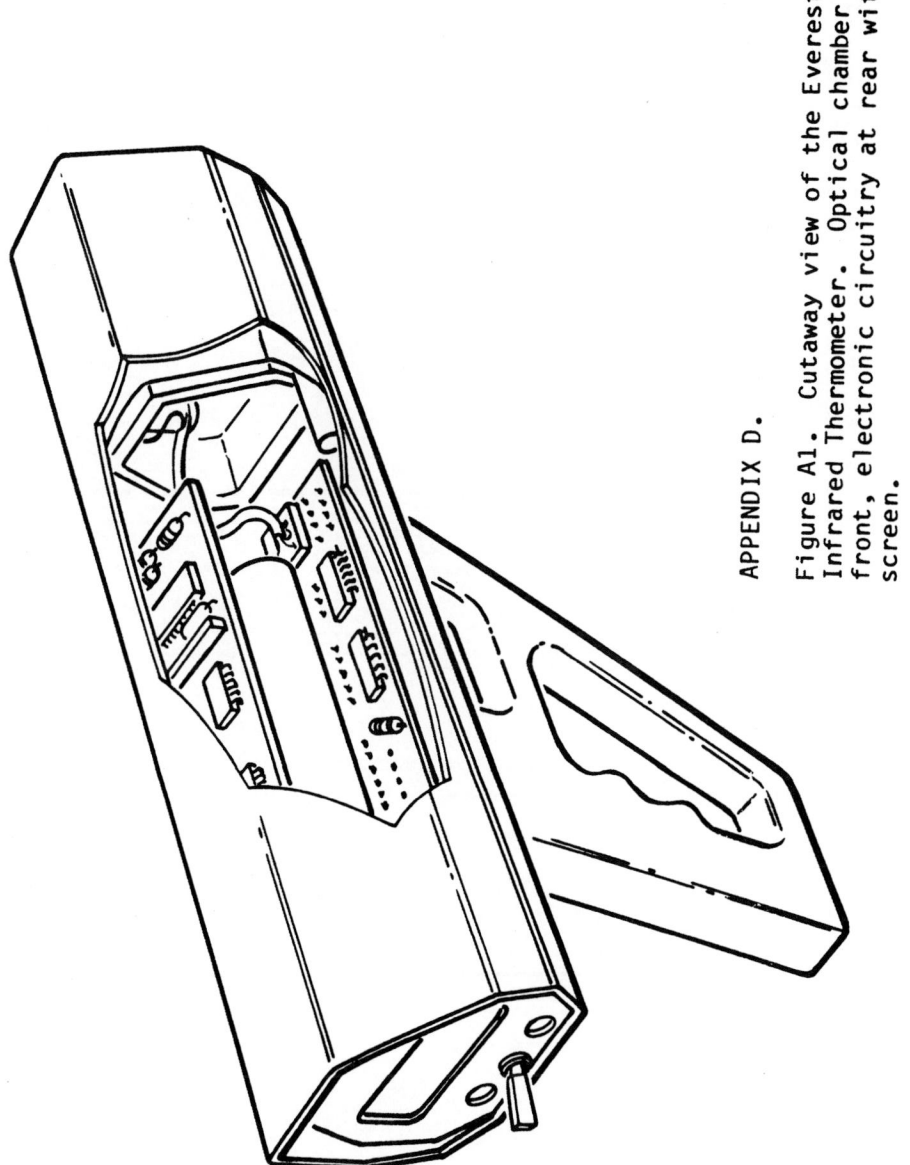

APPENDIX D.

Figure A1. Cutaway view of the Everest Portable Infrared Thermometer. Optical chamber at the front, electronic circuitry at rear with viewing screen.

THEORETICAL FRAMEWORK

INSTRUMENTATION USE IN A COMPREHENSIVE DESCRIPTION OF PLANT-ENVIRONMENT INTERACTIONS

John M. Norman

Department of Agronomy
203 KCR Building
University of Nebraska
Lincoln, NE 68583-0817 USA

1. SUMMARY

The definition of the word instrument is stretched to include use of a leaf-environment model to aid in monitoring leaf status in the field. A set of equations that couple photosynthesis, stomatal conductance, the leaf energy balance, and leaf and soil water status are briefly described for a typical C_3 plant. The Basic language code for these equations, which was developed on a Radio Shack, Model 100 microcomputer, is included with documentation including input and output values of all important variables along with a sample session. Some representative output from the model is discussed along with limitations of the equations.

2. INTRODUCTION

Instruments extend our senses to reveal consistent relationships among factors in our environment. Since the topic of this volume is Advanced Agricultural Instrumentation, one expects to find discussions of electronic, mechanical or electrochemical gadgets that will help us probe into the inner workings of plants. I should like to take the liberty of using a very broad definition of the word "instrument," albeit the last definition in my American College Dictionary - "A device for measuring the present value of the quantity under observation." Thus, an instrument is used to measure-- to estimate the relative amount or value of something by comparison with some standard. This process of comparison with some standard may require a model or representation of whatever is under investigation to determine the appropriate standard or even the appropriate measurement. Certainly the interpretation of any measurement implies a model of what is being observed. This model may be physical, chemical, intuitive, or just plain "common sense," and may not always be apparent. It is interesting to note that even we humans incorporate "models" so that the brain can interpret information fed to it by our senses. Psychologists spend much time trying to reveal such "models." Thus, the development of models to interpret data amounts to using an image of ourselves to create an understanding of the world in our minds.

This broad definition for the word instrument does not imply any upper level of complexity for the model, which is used to determine the appropriate standard or interpret the observations. Thus a comprehensive soil-plant-atmosphere model may be part of an instrument for observation of quantities that cannot yet be observed by simpler instruments, which use simpler models for interpretation.

One example of such a complex model being an essential part of an instrument is a device to indirectly and nondestructively measure the leaf area index and leaf angle distribution of a canopy. To use such an instrument, a one, two or three dimensional canopy radiative transfer model must be used as a "kernel" in the integral inversion of the radiation measurement. A direct frontal attack on this problem of measuring canopy architecture with mechanical devices can be quite frustrating because of rapid leaf movements in the wind, heliotropic leave movements, leaf rolling or wilting from water stress, leaf surface irregularities, and stems and petioles, to name just a few of the confounding factors. All of these factors can be incorporated in the indirect measurement. Clearly without this complex radiative transfer model, we do not have an instrument for indirectly measuring canopy architecture. Further, a model part of this canopy architecture instrument is indistinguishable in character from the very simple model of a mechanical spring movement that is implicit in the old dial volt meters, which we rarely encounter in this modern electronic age of digital voltmeters.

The development of our understanding of plant processes is closely tied to the instruments that we have for observations. Unfortunately a whole ensemble of complex processes are intimately coupled with each other and all the environmental factors, and our simple instruments relate to only one or perhaps a few of these essential processes. The interconnectedness of all these processes makes it impossible to observe one while maintaining all other invariant. We attempt to reduce confounding effects of natural environmental variations by conducting experiments in controlled environments, but we then add another confounding factor in that our plants are adapting to this artificial environment and not the natural environment that is our main interest (1). Perhaps we can consider that detailed plant-environment models are to field research what controlled environments are to laboratory research: a method for monitoring many factors that would be impractical to measure in detail. Thus, a physically-based, plant-environment model may permit us to track detailed processes under field conditions while requiring only a few simple routinely measured inputs along with some more easily measured checks on the model outputs. Figure 1 contains a diagramatic representation of the use of such an "instrument."

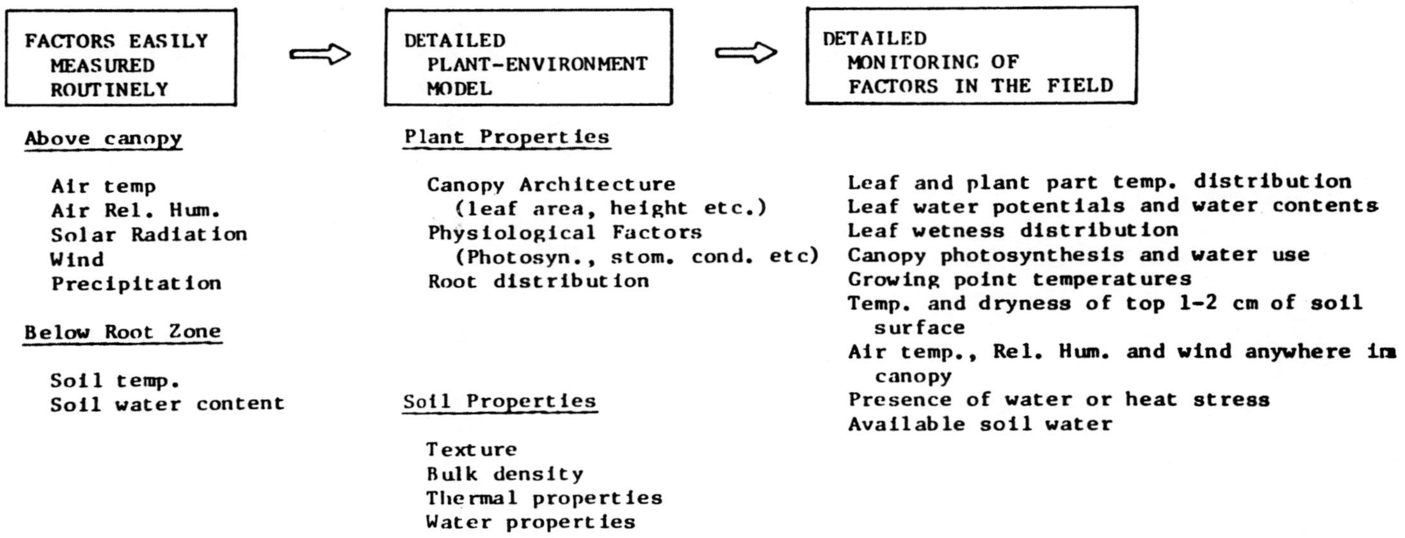

FACTORS EASILY
MEASURED
ROUTINELY

⇨

DETAILED
PLANT-ENVIRONMENT
MODEL

⇨

DETAILED
MONITORING OF
FACTORS IN THE FIELD

Above canopy

Air temp
Air Rel. Hum.
Solar Radiation
Wind
Precipitation

Below Root Zone

Soil temp.
Soil water content

Plant Properties

Canopy Architecture
  (leaf area, height etc.)
Physiological Factors
  (Photosyn., stom. cond. etc)
Root distribution

Soil Properties

Texture
Bulk density
Thermal properties
Water properties

Leaf and plant part temp. distribution
Leaf water potentials and water contents
Leaf wetness distribution
Canopy photosynthesis and water use
Growing point temperatures
Temp. and dryness of top 1-2 cm of soil
  surface
Air temp., Rel. Hum. and wind anywhere in
  canopy
Presence of water or heat stress
Available soil water

Figure 1. Illustration of the flow of information in a plant environment model from typical inputs
to typical outputs.

An "instrument" such as that depicted in Figure 1 has been used to study the water balance of an irrigated corn crop (14), the interaction between microenvironment, plant and pest (15), the influence of frost and environmental amelioration in orchards (17), or the remote sensing of canopy architecture (6).

Numerous instruments are described in this volume, and in this chapter we shall attempt to draw some of these measurements together and hopefully begin to form a more coherent view of a few aspects of plant environment relations. The following topics in the volume relate directly or indirectly to the discussion in this chapter:

Plant and soil nutrient status
Psychrometry (leaf total, leaf osmotic and soil)
Stem diameter
Soil water content
CO_2 and water gas exchange on leaves
Kinetics of photosynthetic reactions
Water flow in plants
Fluorescence
Leaf spectral properties
Leaf temperature
Radiation thermometry
Multiband radiometry

A diagramatic representation of these effects is shown in Figure 2.

3. LEAF MODEL DESCRIPTION

The leaf model of photosynthesis and water use has four main parts: (a) photosynthesis equations, (b) equations that couple stomatal conductance to photosynthesis and other plant or environmental factors, (c) leaf energy balance equations, and (d) water stress considerations. A representation of these interrelationships is diagrammed in Figure 3.

3.1 Photosynthesis

The photosynthesis equations used in this model are taken from Farquhar and von Cammerer (4). They divide the Rubisco enzyme kinetics into two regions and estimate assimilation from the equation describing the region most likely to be limiting photosynthesis. Therefore, the CO_2 assimilation is either limited by the RuBP saturated rate of carboxylation

$$P1 = VC \frac{CI - GS}{CI + KK} - RD \tag{1}$$

where KK = KC (1 + O/KO), or it is limited by RuBP regeneration

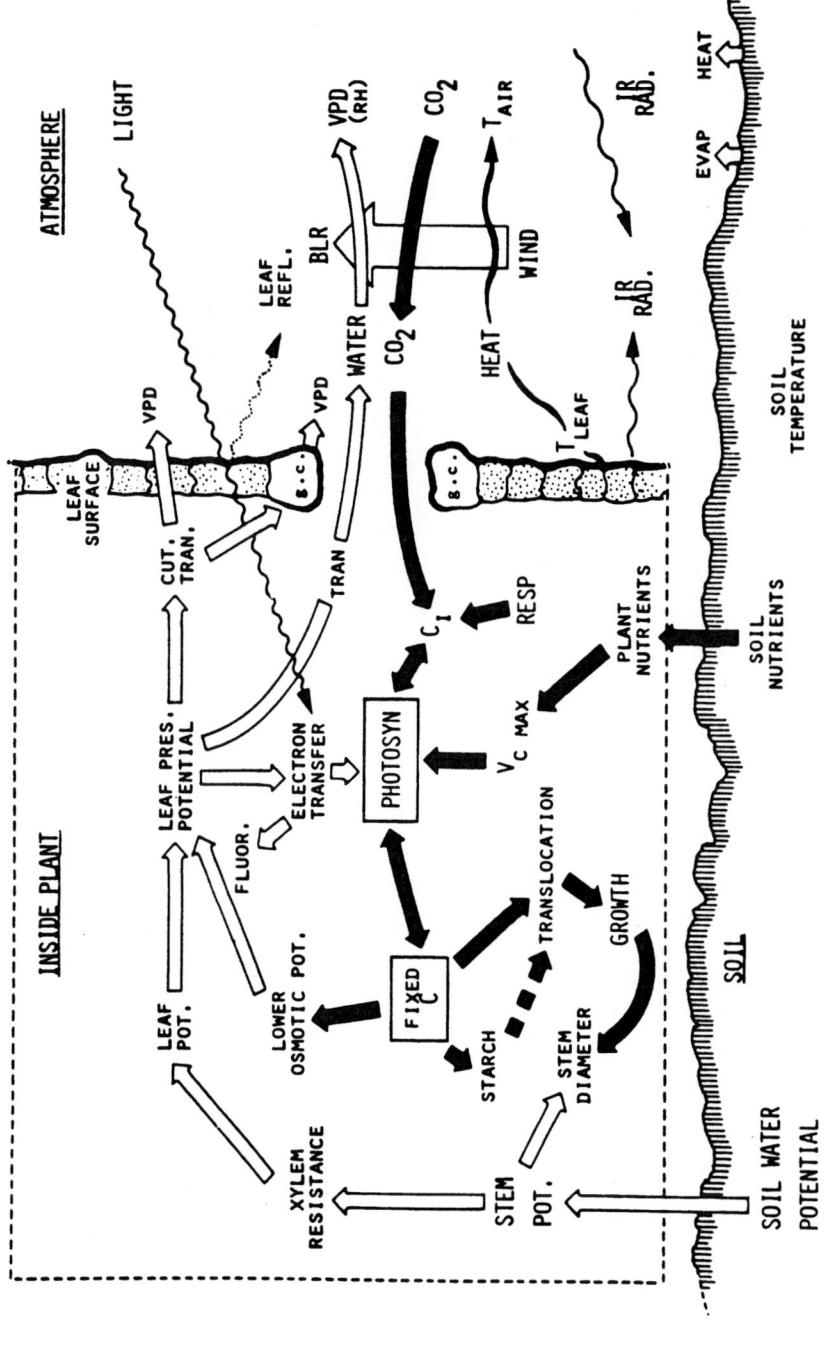

Figure 2. Sketch showing a few of the many related factors pertinent to the study of leaf-environment relations.

254

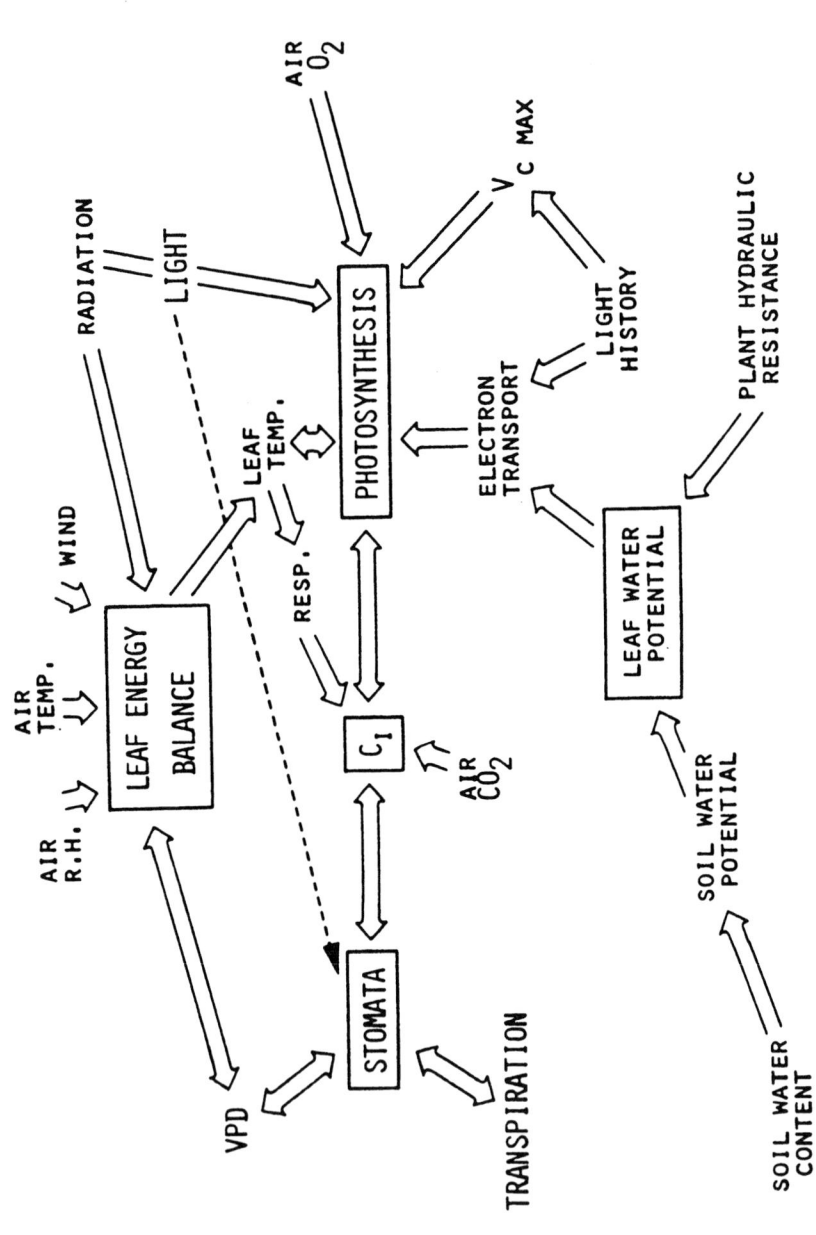

Figure 3. Illustration of the flow of information in the model Culeaf, which is described in this chapter.

$$P2 = \frac{J \; (CI - GS)}{4.5 \; CI + 10.5 \; GS} - RD \tag{2}$$

where

$$J = \frac{JM \; E}{E + 2.1 \; JM} \tag{3}$$

so that the photosynthetic rate is given by

$$P = \min \; (P1, \; P2) \tag{4}$$

Equation (1) is appropriate at low CI and Equation (2) at high CI. The CO_2 photocompensation point is given by GS

$$GS = 0.105 \; 0 \; KC/KO \tag{5}$$

and is the CO_2 compensation point in the absence of day, dark respiration. From Farquhar and von Cammerer (4), VC, KO, KC and RD have similar temperature dependencies. Using KC as an example

$$KC = K1 \; \exp \; (A1 \; x \; K2) \tag{6}$$

where

$$A1 = (TL - TO) \; / \; [(TO + 273) \; R \; (TL + 273)]$$

and the coefficients K1 and K2 are obtained from Farquhar et al.(3). The temperature dependence of potential electron transport rate (JM) is quite different:

$$JM = J2 \; \exp \; (E1 \; . \; A1)/B \tag{7}$$

where

$$B = 1 + \exp \; [(S(TL + 273) - H) \; / \; (R(TL + 273)]$$

Again the coefficients are obtained from Farquhar et al. (3).

Other factors can affect VC and JM. Some results of measurements made on P. Vulgaris exposed to a history of low light (2) indicate both VC and JM may be affected. In this model, we have taken some liberty with the effect of leaf exposure to a history of low light by reducing JM and VC by (0.8 FE + 0.2), where FE is a dimensionless number between 0 and 1 representing the fraction of full sunlight that the leaf has been exposed to for an extended period. The dark respiration is reduced by a lesser amount (0.6 FE + 0.4). Further, the effect of water stress is assumed to alter JM when leaf pressure potential (PP) is reduced below a threshold pressure (PT); JM is just reduced by PP/PT for PP < PT. Such effects of water stress are suggested by chlorophyll fluorescence studies (7,8) but are hardly well established. These functions for altering VC and JM are consistent with the literature but chosen quite arbitrarily and thus should not be used without caution.

3.2 Coupling of Stomatal Conductance and Photosynthesis

A very simple method for coupling stomatal functioning with photosynthesis can be inferred from results presented by Wong et al. (19). They suggest that CI/CA is approximately constant over a wide range of conditions near a value of 0.72 for C_3 species and 0.4 for C_4 species. Although the constancy of CI/CA will not always hold, as is shown by the more detailed coupling model of Farquhar and Wong (5), this is a useful starting place for coupling stomata and photosynthesis.

Two conditions when constancy of CI/CA will surely fail are (1) when stomata are not in equilibrium with the photosynthetic system (environmental conditions changing rapidly) and (2) when stomata are affected independently of photosynthesis (direct vapor pressure deficit effect on stomata through cuticular or peristomatal transpiration; 10). Although essential processes in stomatal functioning are known to operate independently of photosynthesis (10,20), until more of the essential relationships are better understood, assuming CI/CA to be constant can provide much useful insight into photosynthesis-transpiration relations.

If CI/CA is assumed constant, then stomatal conductance to water vapor can be computed from

$$KS = \left[\frac{CA - CI}{1.6\ P} - Rh\right]^{-1} \tag{8}$$

where the 1.6 is the ratio of the diffusivity of water vapor to CO_2, Rh is the boundary layer resistance of the leaf when it is in the chamber that was used to measure P. Of course, Rh depends on whether leaves are hypostomatus or amphistomatus. The right hand side of Equation (8) may become negative if the net photosynthetic rate becomes small or negative. Clearly KS can never be negative so it is set to a limiting value of zero so the leaf conductance is the cuticular conductance (CE) and CI/CA is no longer constant.

Any factor that affects the stomata independently of photosynthesis will cause a change in the ratio CI/CA. In our simple model the vapor pressure deficit between the leaf and the environment (VP) can cause direct stomatal closure if above some threshold value, D1. When this occurs, the altered KS is calculated from the known dependence of KS on VP, and CI is obtained by setting a rearranged form of Equation (8) (solved for P) equal to either Equation (1) or Equation (2), whichever is appropriate as defined by Equation (4). This estimate of CI obtains from the solution of a quadratic equation and the largest root is used. Whether the solution with Equation (1) or Equation (2) is appropriate is determined by the largest value of CI; thus both equations are used and the largest value of CI used as the final solution.

3.3 Leaf Energy Balance Equations

The coupling between stomata and photosynthesis implies a very direct relation between transpiration and photosynthesis. However, transpiration is affected by factors other than stomatal conductance such as wind speed, air temperature, and available short- and long-wave radiation. Thus a solution of the leaf energy budget must be combined with the photosynthesis and stomatal conductance models. The key to solving the leaf energy budget is a solution for the leaf-air temperature difference (DT) in terms of the other variables. This is most easily done by linearizing the saturation vapor pressure and black body radiation relations with temperature:

$$DT = (EN - 1825 \ KL \ \ DE - T5) \ / \ (1825 \ KL \ \ SA + R2 + T6) \qquad (9)$$

where KL is the leaf conductance to water vapor, DE is the air vapor pressure deficit, SA is the slope of the saturation vapor pressure versus temperature relation, RZ = 2400/RX (RX = leaf boundary layer resistance), T5 is the thermal radiation emitted by the leaf, T6 is the slope of black body radiation versus temperature relation and EN is the net incoming radiation absorbed by the leaf. The net incoming radiation absorbed by a leaf exposed to the sun and sky is given by

$$EN = A_p \ EP + 1.15 \ A_n \ EP + QE \qquad (10)$$

where QE is the thermal radiation from the cool, clear sky and surroundings assumed to be at air temperature (9), EP is the PAR radiation and 1.15 EP is the approximate near infrared radiation for clear sky conditions (16), A_p is absorbtivity of the leaf to PAR radiation (0.85) and A_n is near infrared absorbtivity of the leaf (0.15). The net allwave absorbed radiation for the leaf is

$$RN = EN - T5 + T6 \ \ DT. \qquad (11)$$

These leaf energy budget equations are described in more detail in Norman (11). The transpiration rate (TR) can be calculated easily after the leaf temperature has been obtained from Equation (9):

$$TR = 1825 \ KL \ (DE + SA \ DT). \qquad (12)$$

The leaf sensible heat (Q) is given by

$$Q = 2400 \ DT/RX \qquad (13)$$

and the energy balance is checked by

$$RN = TR + Q \qquad\qquad (14)$$

because the leaf is assumed to have negligible heat storage and other energy sources or sinks, such as photosynthesis, are small.

3.4 Water Stress Considerations

The limited availability of water can have a profound effect on photosynthesis and transpiration. For many years researchers have struggled to determine whether water stress affects mainly the stomata or the photosynthetic mechanisms. Some results suggest that water stress leads to stomatal closure and photosynthesis then is reduced because of the large diffusion resistance. This control by the stomata is most likely to occur with rapidly imposed water stress such as occurs with plants growing in small pots under controlled environment conditions. When stomatal closure "shuts down" the photosynthetic system, CI decreases substantially. If water stress is slowly imposed, perhaps over weeks or more as in field conditions, then CI tends to remain near a value for an unstressed plant (19). Under these conditions, stomata of stressed plants often appear to behave similarly to unstressed plants. Perhaps both stomata and photosynthesis are being affected similarly but independently by water stress, but that remains to be determined. The response of stomata to slowly imposed water stress can be simulated by reducing photosynthesis in response to low leaf turgor and allowing the stomata to adjust so that CI/CA remains constant. This is a defendable approach but far from being proven. The pressure potential is used to indicate the onset of stress because leaf osmotic potentials can vary throughout a day, and stomatal closure seems to correlate more closely with turgor. In fact, in maize leaves, osmotic potentials can decrease 10 bars during the day, presumably from photosynthate temporarily stored in the photosynthesizing leaves. Shade leaves on the same plant will close stomata at much higher total potentials than the sunlit leaves, which have osmoregulated.

The effect of water stress can be simulated for this individual leaf by choosing a hydraulic conductance (KH) that produces a reasonable relation between transpiration and leaf water potential. This is somewhat more artificial for this single leaf than for a whole plant made up of many leaves. The leaf water potential (PL) is a function of the soil potential (PS) and the mass transpiration rate (TM) as

$$PL = PS - TM/KH. \qquad\qquad (15)$$

Further, $PL = PO + PP$ and thus PL is a function of leaf osmotic (PO) and pressure potentials (PP). If PP is less than some threshold turgor pressure (PT), then the photosynthetic rate can be reduced by PP/PT. This is a very simple method for simulating the effect

of low soil potentials or high transpiration rates on photosynthesis.
Simple as this method is, iteration to a solution of all the relevant equations can present numerical problems.

4.   PROGRAM DESCRIPTION

The Basic computer code listing for Culeaf (Appendix A) is
included so that the reader can obtain all the details associated
with the equations, which have been described only briefly in this
chapter. As one who has written many programs, I make no guarantee
that this listing is error-free; such a claim would be foolish.
Also it may not be written in the most computationally efficient
form; primarily because clarity is more important than execution
efficiency, provided that execution times are reasonable. The
program speed can be increased by numbering statements sequentially
in increments of one, but such code is very difficult to add statements to. Programs are available for all microcomputers that
resequence statement numbers for any program.

The Radio Shack Model 100 lap microcomputer is convenient but
slow in execution of numerical computations, particularly exponen-
tials or log functions. Therefore, exponential or power functions
that are executed in a loop many times have been linearized to speed
execution. The screen plotting subprogram is unique to the Model
100 and probably not easily transferred to other microcomputers.

One inefficiency apparent in the program listing is that many
variables are defined by two names; a two-letter name related to
the meaning of the name and a variable array [V1(I)]. This two-
letter name allows variables to be identified much more readily in
the program code and the array facilitates manipulating variables
for input and output.

The program listing in Appendix A contains many comments that
document the code. The reader is referred to this listing for
detailed information. Here we will just outline the most general
features of this program.

60-340	Data and declaration statements
1000-1100	Fixed constants in program
1110-1900	Setting up input changes from default
1960-2500	Setting up independent and dependent variables for execution and plotting
2520-3490	Equivalence between 2-letter names and V1(I), prelimi- nary calculations and initial values of variables to be iterated
3495-3990	Solution of the leaf energy budget with photosynthesis and stomatal conductance subprogram
7000-8300	Equivalence statements and preliminary output
10000-11430	Photosynthesis and stomatal conductance subprogram
15000-16000	Setting up leaf energy balance
40000-41000	Plotting subprogram

An example session with this program is included in Appendix B. Since the Model 100 has only a 40-character display, only the added comments in this example session go beyond 40 columns. The graphs included in this example session are character renditions of the Model 100 graphical output. Of course the graphics on the Model 100 are much better than these forgeries. Also the characters on the Model 100 LCD display are much larger than these printed letters making the output display remarkably good. However, only eight lines of output can be viewed at one time as the display scrolls.

Anyone who might wish to implement this program will need to do so from the listing in Appendix A. This listing has been taken directly from the computer listing and not retyped so it should be accurate. However, to facilitate accurate entry of the Basic code, an extensive output from the example session is included in Appendix C.

The only portion of the program that is marginally satisfactory is that part concerning iteration when water stress occurs. The procedure used here is very simple but slow, and under certain peculiar conditions can result in a pressure potential that may be somewhat in error. Faster and more accurate convergence algorithms are available, but the code for these is more complex and quite difficult to decipher. This probably is the first area for the ambitious modeler to improve.

5. MODEL APPLICATION AND LIMITATIONS

The model Culeaf has been used to predict various quantities that relate to leaf functioning. In this section only a few outputs are highlighted to indicate the kind of insights that can be gained concerning leaf-environment relations from even such a simplified model.

The effect of stomatal closure from low leaf turgor on leaf-air temperature difference is quite clear from Figure 4a. Clearly high air temperatures may not necessarily cause damage to a leaf unless transpiration is limiting for some reason.
The leaf water status that may be responsible for stomatal closure is a combination of reduced soil potential (PS = -6 bars in this stress case) and transpiration, because at 25°C, the reduced soil potential does not affect stomatal conductance.

Dark respiration and photosynthetic rate also are a strong function of leaf temperature without the presence of any water stress in the leaf (Figure 4b).

The affect of reduced soil water potential on leaf pressure potential is shown in Figure 4c. Although the reduced pressure potential (threshold pressure potential, PT = 4 bars) below the threshold reduced photosynthesis and subsequently stomatal conductance, transpiration continues to increase down to a pressure potential of 2 bars because of increasing vapor pressure deficit with temperature.

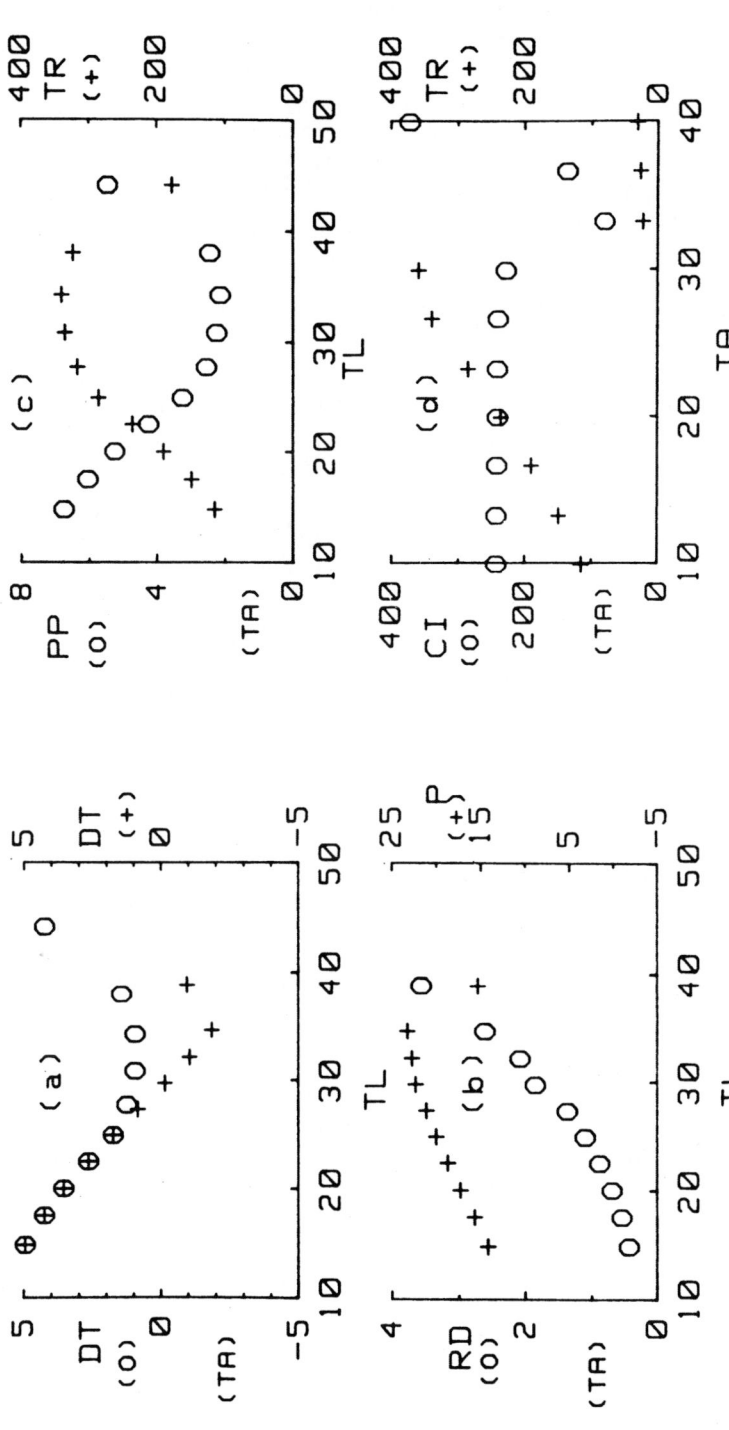

Figure 4. Representative output from the model Culeaf. (a) Temperature difference between leaf and air (DT=TL-TA) as a function of leaf temperature (TL) with (PS=-6,0) and without (+) water stress. (b) Dark respiration (0) and photosynthesis (+) as a function of TL. (c) Leaf pressure potential (0) and transpiration rate (+) versus TL. (d) Internal CO$_2$ concentration (0) and transpiration rate (+) as a function of air temperature (TA) without low, bulk, leaf pressure potentials, but with stomatal closure from direct stomatal response to vapor pressure deficit. See Appendix B for values of other variables.

As the temperature continues to increase so that photosynthesis decreases because of high temperature, the high temperature effect may override the water stress effect and leaf turgor may be restored with stomatal closure. Whether such a sequence of events ever actually occurs in a leaf is difficult to determine.

If stomata should close in response to atmospheric humidity because of cuticular or peristomatal transpiration, then transpiration can be nearly stopped even when no water deficit occurs in the bulk leaf (Figure 4d). When stomata close in response to direct vapor pressure deficit, the internal CO_2 concentration (CI) decreases. In the example shown in Figure 4d, the lowest value of CI occurs at 33°C when stomata are closed, and CI increases at higher temperatures because of reductions in photosynthesis from increased temperatures.

The dependence of photosynthesis and ratio of internal to external CO_2 concentration (CR = CI/CA) is shown in Figure 5a. As expected, photosynthesis saturates as light intensity increases and negative net photosynthetic rates at low light can produce high CI values. Because of assumptions used in this model, leaf conductance (to water vapor) has a dependence on light that is similar to photosynthesis (Figure 5b). Further, the leaf-air temperature difference is dependent on the incident radiation (Figure 5b).

The relation between photosynthesis and internal CO_2 concentration, which is similar to that of Farquhar and von Cammerer (4), depends on the previous light history of a particular leaf (Figure 5c). The stomatal conductance also depends on the internal CO_2 concentration; however, the dependent shown in Figure 5d does not agree well with data for low CI. The KS versus CI relation in Figure 5d is reasonable for air CO_2 concentrations above the normal 300 ppm; below such ambient concentrations, data suggest stomata continue to open and do not close again, as predicted by this model, at very low CI. The model of Farquhar and Wong (5) predicts KS versus CI relationships that compare more closely with measured results at low CI. This model, thus, is limited to normal atmospheric concentrations of CO_2 or higher. The dependence of CR (CR = CI/CA) on internal CO_2 concentration (CI) is similar to that predicted by Farquhar and Wong (5).

This relatively simple model of leaf photosynthesis and water use can provide some useful insights into leaf-environment relations, but it has some obvious limitations.

The equations in this chapter are appropriate for a single leaf. Predictions for a canopy of leaves can be obtained by integrating the collective effect of many single leaves that may be functioning at different light levels, temperatures, relative humidities, etc. If this integration over single leaves to obtain canopy prediction is done in detail, the procedure can be somewhat involved (11,12,18). However, some simplified methods (12,13) provide results that compare so well with more rigorous treatments that the more complex approaches may be beyond the point of diminishing returns.

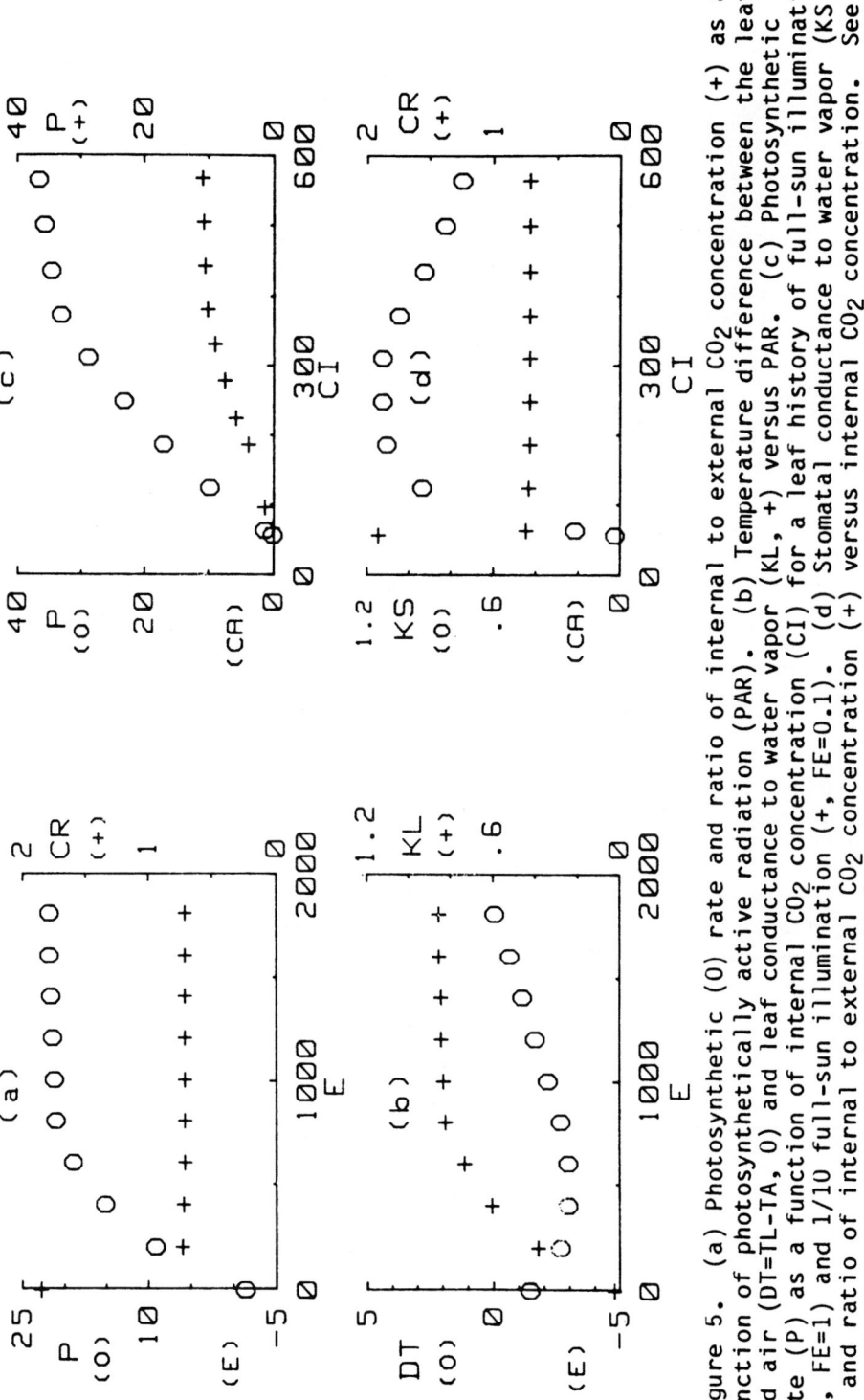

Figure 5. (a) Photosynthetic (0) rate and ratio of internal to external CO_2 concentration (+) as a function of photosynthetically active radiation (PAR). (b) Temperature difference between the leaf and air (DT=TL-TA, 0) and leaf conductance to water vapor (KL, +) versus PAR. (c) Photosynthetic rate (P) as a function of internal CO_2 concentration (CI) for a leaf history of full-sun illumination (0, FE=1) and 1/10 full-sun illumination (+, FE=0.1). (d) Stomatal conductance to water vapor (KS, 0) and ratio of internal to external CO_2 concentration (+) versus internal CO_2 concentration. See Appendix B for values of other variables.

A basic assumption made in this chapter is that photosynthesis and stomata are in equilibrium. In canopies of leaves this may not be a reasonable assumption because leaves can be moved about by the wind, and sunflecks are continually moving because of changing sun angles. When photosynthesis and stomata are not in equilibrium, CI/ CA can be highly variable (10). When a sunfleck moves over a previously shaded portion of leaf, some significant time passes before the stomata and photosynthesis are adjusted to this new environment. This may occur because of delays in stomatal opening or photoactivation of enzymes or both. In any event, the delay in starting up photosynthesis appears to be longer than the delay in reducing it as the sunfleck passes on to another portion of the leaf leaving our region of concern in the shade. The net result of these moving sunflecks is that static methods of integration described previously (11,18) tend to overestimate photosynthesis. Corrections for such overestimates are straightforward if the dynamics of photosynthesis and stomatal functioning are known. The model presented in this chapter is limited to normal ambient atmospheric CO_2 concentrations (340 ppm by volume) and above.

In this chapter we have not discussed leaf boundary layer resistance (RX). Usually RX is small so that uncertainties in "constants" relating RX to leaf size are of minor concern. However, under some conditions, stomatal conductances can exceed 2 cm/s, and uncertainties in the dependence of RX on leaf size (L) and wind speed (V) become a major concern.

The effect of water stress in this model is to reduce photosynthesis. Although this produces reasonable predictions for slowly-imposed water stress as might develop under field conditions over days or weeks, it is far from an established fact. The occurrence of water stress causes so many changes in the plant that much research remains to be done before we can make confident predictions.

Undoubtedly every person who looks at the equations in this chapter will find flaws. In spite of the shortcomings, perhaps this chapter might serve as a useful starting point for the uninitiated in leaf-environment relations.

ACKNOWLEDGEMENTS

I wish to thank Bill Gensler for his leadership in providing the opportunity to participate in the NATO Advanced Studies Institute, and for his persistance, patience, and encouragement throughout the preparation of this chapter. The Agriculture Research Division of the University of Nebraska also provided support for this activity.

REFERENCES

1. Beardsell, M.F. Effects of routine handling on maize growth. Aust. J. Plant Physiol. 4:857-861, 1977.

2. von Cammerer, S. and Farquhar, G.D. Some relationships between the biochemistry of photosynthesis and the gas exchange of leaves. Planta 153:376-387, 1981.

3. Farquhar, D.G., von Cammerer, S. and Berry, J.A. A biochemical model of photosynthetic CO_2 assimilation in leaves of C_3 species. Planta 149:78-90, 1980.

4. Farquhar, G.D. and von Cammerer, S. Modeling of photosynthetic response to environmental conditions. In: Physiological Plant Ecology II (O.L. Lange, P.S. Nobel, C.B. Osmond and H. Zeigler, eds.) Encycl. Plant Physiol. New Ser. Vol. 12B, Springer-Verlag, Berlin, pp. 549-588, 1982.

5. Farquhar, G.D. and Wong, S.C. An empirical model of stomatal conductance. Aust. J. Plant Physiol. 11:191-210, 1984.

6. Goel, N.S. and Thompson, R.L. Inversion of vegetation canopy reflectance models for estimating agronomic variables V: Estimation of LAI and average leaf angle using measured canopy reflectances. Remote Sensing Environ. 16:69-85, 1984.

7. Govindjee, W.J., Downton, S., Fork, D.C. and Armond, P.A. Chlorophyll. A fluorescence transient as an indicator of water potential of leaves. Plant. Sci. Letters 20:191-194, 1981.

8. Havaux, M. and Lannoye, R. Chlorophyll fluorescence induction: A sensitive indicator of water stress in maize plants. Irrig. Sci. 4:174-151, 1983.

9. Idso, S.B. A set of equations for full spectrum and 8-14 μm and 10.5 - 12.5 μm thermal radiation for cloudless skys. Water Resour. Research 17:295-304, 1981.

10. Jarvis, P.G. and Morison, J.I.L. The control of transpiration and photosynthesis by stomata. In: Stomatal Physiology (P.G. Jarvis and T.A. Mansfield, eds.) Cambridge University Press, pp. 247-279, 1981.

11. Norman, J.M. Modeling the complete crop canopy. In: Modification of the Aerial Environment of Plants (B.J. Barfield and J. Gerber, eds.) ASAE Monograph, Amer. Soc. Agric. Engr., St. Joseph, MI, pp. 249-277, 1979.

12. Norman, J.M. Interfacing leaf and canopy light interception models. In: Predicting Photosynthate Production and Use for Ecosystem Models (J.D. Hesketh and J.W. Jones, eds.) CRC Press, Boca Raton, FL, pp. 49-67, 1980.

13. Norman, J.M. Simulation of microclimates. In: Biometeorology in Integrated Pest Management (J.L. Hatfield and I.J. Thomason, eds.) Academic press, NY, pp. 65-99, 1982.

14. Norman, J.M. and Campbell, G.S. Application of a plant-environment model to problems in irrigation. In: Advances in Irrigation, Vol. II (D. Hillel, ed.) Academic Press, NY, pp. 155-188, 1983.

15. Toole, J.L., Norman, J.M., Holtzer, T.O. and Perring, T. Simulating Banks Grass Mite population dynamics as a subsystem of a crop canopy-microenvironment model. Environ. Entomol. 13:329-337, 1984.

16. Weiss, A. and Norman, J.M. Partitioning solar radiation into direct and diffuse, visible and near-infrared components. Agric. Meterorol. 34:205-213, 1985.

17. Welles, J.M., Norman, J.M. and Martsolf, J.D. A model of foliage temperatures in a heated orchard. J. Amer. Soc. Hort. Sci. 104:602-610, 1979.

18. deWit, C.T. Photosynthesis of leaf canopies. Agr. Res. Rep. No. 663, Central Agr. Publ. Doc., Wageningen, 1965.

19. Wong, S.C., Cowan, I.R. and Farquhar, G.D. Stomatal conductance correlates with photosynthetic capacity. Nature 282:424-426, 1979.

20. Zeigler, E., Bloom, A.J. and Hepler, P.K. Ion transport in stomatal guard cells: A chemiosmotic hypothesis. Whats New Plant Physiol. 9:29-32, 1978.

```
          APPENDIX  A  -  PROGRAM LISTING

                Program Name - Culeaf

1    '
2    '
3    '
4    '
5    '
6    '
9    'LEAF MODEL PROGRAM OF PHOTOSYNTHESIS AND AND WATER USE AS PART OF
10   'NATO ADVANCED STUDIES INSTITUTE ON ADVANCED AGRICULTURAL INSTRUMENTATION.
15   'IL CIOCCO (PISA),ITALY,1984.
20   'PROGRAMMED BY JOHN M NORMAN,DEPT OF AGRONOMY,203 KCR BLDG,UNIV OF
25   'NEBRASKA,LINCOLN,NE 68683-0817. PLEASE WRITE WHEN MAJOR ERRORS ARE FOUND.
30   '
34   'OPEN RAM FILE FOR OUTPUT IF DESIRED. IF TOO LITTLE MEMORY, USE LPRINT.
35   IO%=0 'IF IO%=1 WRITE OUTPUT
36   IF IO%=1 THEN OPEN"RAM:DATOUT"FOR OUTPUT AS 1
37   '
40   '
45   'THIS PROGRQAM WAS DEVELOPED AND RUN ON A RADIO SHACK TRS-80 MODEL 100 LAP
50   'MICROCOMPUTER.THE REMAINING MEMORY WITH THIS PROGRAM INSTALLED IS ABOUT 7K
55   'ON A 24K MACHINE.
60   DEFSNG A-Z
100  '
150  '
200  DIM X(20),Y(20),W(20),V1(42),V$(42),U1(20),U2(20),U3(20)
210  '
240  'VBL NAMES FOR VBLS TO BE PLOTTED
245  '
250  DATA "EP","TA","RH","V","CA","P","KS","TL","EA","DT","RN","TR","Q","RS"
```

```
251 DATA "RX","FS","L","FE","RD","DE","PO","PP","PS","KC","KO","R3","V1","JM"
252 DATA "J1","J","GS","G","CO","CI","D1","D2","E","KL","PL","KH","CR","VP"
260 '
290 'DEFAULT VALUES OF EP(WM-2)TA(C)RH(FRAC)V(MS-1)CA(PPM)PS&PO(BAR)D1&D2(MB)
291 'FS(.5-1)L(M)FE(FRAC)KH(10-3GM-2S-1BAR-1)
295 '
300 DATA 395,30,0.4,1,340,-0,-15,70,90,.5,.1,1,20
305 '
310 'PHOTOSYNTHESIS EQ COEFF. T0(C)K1(10-6BAR)K2(JMOL-1)K3(10-3BAR)K4(JMOL-1)
311 'K5(JMOL-1)K6(JMOL-1)O(10-3BAR)J2(10-6EQM-2S-1)PT(BAR)VM(10-6MOLM-2S-1)
315 '
320 DATA 25,460,59356,330,35948,66400,58520,210,250,4,100
325 '
330 'E1(JMOL-1)H(JMOL-1)S(JMOLK-1)RO(10-6MOLM-2S-1)
335 '
340 DATA 37000,220000,710,1.1
350 '
360 '
370 '
1000 ' MAIN PROG SETS UP PLOTTING VBLS AND VBL TO BE SCANNED ON,DOES
1005 'PRELIMINARY CALC TO SPEED UP SUBPROG OF PHOTOSYNTHESIS AND STOM COND,
1010 'SOLVES LEAF ENERGY BALANCE WITH WATER STRESS AND SETS UP OUTPUT.
1011 '
1012 '
```

```
1013 'FIXED CONSTANTS
1014 '
1015 SI=5.67E-8    'STEF-BOLTZ (WM-2K-4)
1020 R=8.314    'GAS CONST (JMOL-1K-1)
1025 KE=.0084    'EPIDERM OR CUT COND WATER.0002 MS-1*42=.0084 MOLM-2S-1
1100 NVZ=42    'NO OF VBLS FOR PLOTTING
1110 NPZ=5    'NO OF POINTS ON PLOT
1115 'INPUT 'TO CHNG NO. OF POINTS PLOTTED IF DESIRED.
1120 INPUT "NPZ=5 OK";NPZ
1130 'READ IN VBL NAMES FOR OUTPUT.
1150 FOR I=1 TO NVZ
1200 READ V$(I)
1250 NEXT I
1260 'READ IN DEFAULT VALUES OF VBLS THAT CAN BE CHANGED BEFORE RUN.
1300 FOR I=1 TO 5
1350 READ V1(I) 'EP,TA,RH,V,CA
1400 NEXT I
1405 READ V1(23)    'PS SOIL POT BARS
1410 READ V1(21)    'PO LEAF OSMOTIC POT BARS
1415 READ V1(35)    'D1 THRESHOLD OF STOM CLOSURE FROM VPD
1420 READ V1(36)    'D2 VPD AT WHICH STOM AREE CLOSED
1425 READ V1(16)    'FS=1 HYPOSTOM, FS=.5 AMPHISTOM VALUES .5 OR 1.
1430 READ V1(17)    'L  LEAF WIDTH  (M)
1435 READ V1(18)    'FE AVE RADIATION ENVIRON THAT LEAF EXPOSED TO AS FRAC FULL SUN
1440 READ V1(40)    'KH PLANT HYDRAULIC RESIS (MGM-2S-1BAR-1)
1445 'READ IN PHOTOSYNTHESIS COEFFICIENTS.
1450 READ T0,K1,K2,K3,K4,K5,K6,O,J2,PT,VM
1455 READ E1,H,S,R0
```

```
1460  '
1465  'PRINT OUT DEFAULT VALUES AND LET USER CHANGE THEM IF DESIRED.
1470  'PROGRAM RECYCLES BACK TO HERE WHEN DONE WITH PREVIOUS RUN.
1475  '
1490  PRINT USING "EP=### TA=##.# RH=#.# V=##.# ";V1(1);V1(2);V1(3);V1(4);
1491  PRINT USING "CA=### FS=#.# L=#.### ";V1(5);V1(16);V1(17);
1495  PRINT USING "FE=#.# PS=###.#PO=###.#D1=## ";V1(18);V1(23);V1(21);V1(35);
1496  PRINT USING " D2=## KH=##";V1(36);V1(40)
1500  INPUT "NAME,VAL";XD$,X3  'LOOP
1550  IF XD$=X1$ AND X3=X4 THEN GOTO 1950
1600  X1$=XD$
1650  X4=X3
1700  FOR I=1 TO 5
1750  IF X1$=V$(I) THEN V1(I)=X4
1800  NEXT I
1810  IF X1$=V$(16) THEN V1(16)=X4
1820  IF X1$=V$(17) THEN V1(17)=X4
1830  IF X1$=V$(18) THEN V1(18)=X4
1840  IF X1$=V$(23) THEN V1(23)=X4
1850  IF X1$=V$(21) THEN V1(21)=X4
1860  IF X1$=V$(35) THEN V1(35)=X4
1870  IF X1$=V$(36) THEN V1(36)=X4
1880  IF X1$=V$(40) THEN V1(40)=X4
1900  GOTO 1490
```

```
1950  '
1955  X1$="ZZ"
1960  'PRINT OUTPUT VARIABLES AND SELECT VARIABLES TO BE PLOTTED.
1970  '
2100  FOR I=1 TO NV%
2120  PRINT USING "\      \";V$(I);
2140  NEXT I
2160  PRINT
2200  INPUT "X,Y FOR PLOT";X$,Y$  'LOOP
2202  W$=""  'IF W$="" THEN YOU CAN SAVE 3RD 4TH & 5TH VBLS FROM THE PREVIOUS RUN
2205  INPUT "2nd  Y";W$
2206  IF W$=""THEN GOTO 2210
2207  INPUT "3rd Y";U1$:INPUT "4th Y";U2$:INPUT "5th Y";U3$
2210  IX%=0:IY%=0:IZ%=0:IW%=0:I1%=0:I2%=0:I3%=0
2220  FOR I=1 TO NV%
2240  IF X$=V$(I) THEN IX%=I
2260  IF Y$=V$(I) THEN IY%=I
2265  IF W$=V$(I) THEN IW%=I
2266  IF U1$=V$(I)THEN I1%=I
2267  IF U2$=V$(I)THEN I2%=I
2268  IF U3$=V$(I)THEN I3%=I
2270  NEXT I
2275  IF IW%=0 AND I1%=0 THEN W$=""
2280  IF IX%=0 OR IY%=0 THEN GOTO 2200
2400  INPUT "VBL TO BE SCANNED";Z$
2420  FOR I=1 TO NV%
2440  IF Z$=V$(I) THEN IZ%=I
2460  NEXT I
```

```
2462 ITZ=1    'IF SCANNED VBL IS TA THAN ITZ=0 AND VBL DEP ON TA ARE CALC.
2464 IF IZZ=2 THEN ITZ=0
2470 IF IZZ=0 THEN GOTO 2400
2480 INPUT "MIN,MAX";ZN,ZX
2490 '
2500 '
2510 '
2520 'MAIN CALCULATION LOOP OVER NUMBER OF OUTPUT VALUES TO BE PLOTTED.
2530 '
2540 '
3120 FOR I=1 TO NPZ                                    'BEGIN CALCULATION LOOP
3140 IF NPZ>1 THEN V1(IZZ)=ZN+(I-1)*(ZX-ZN)/(NPZ-1)
3150 IF NPZ=1 THEN V1(IZZ)=ZN
3160 '
3170 'RESET VALUES FOR EQUIVALENCE BETWEEN 2-LETTER NAMES AND V1 ARRAY.
3180 '
3230 EP=V1(1)    'PAR W m-2
3232 E=4.6*EP:V1(37)=E    'PAR 10-6MOLQ M-2S-1
3240 TA=V1(2)    'AIR TEMP  C
3250 RH=V1(3)    'AIR REL HUM  FRAC
3259 'ITZ=1 IF TA SAME AS LAST TIME
3260 IF ITZ=1 AND I>1 THEN GOTO 3270
3265 ES=6.108*10^(7.5*TA/(237.3+TA))
3270 EA=ES*RH
3272 DE=ES-EA    'VAPOR PRESS DEFICIT OF AIR
3274 V1(20)=DE
3280 V1(9)=EA
```

```
3290 V=V1(4)       'WIND SPEED  m s-1
3294 L=V1(17)      'LEAF WIDTH  (M)
3295 RX=180*SQR(L/(V+.02))  'BOUNDARY LAYER RESIS THE .02 ADJ FOR FREE CONV
3296 IF RX>500 THEN RX=500  'S M-1
3300 CA=V1(5)      'AIR CO2  ppm
3310 FS=V1(16)     '1=HYPO,  .5=AMPHI
3320 IF FS<>.5 THEN FS=1
3340 PO=V1(21)  'LEAF OSMOTIC POT (BAR)
3350 PS=V1(23)  'SOIL POT (BAR)
3360 D1=V1(35):D2=V1(36)  'VPD STOM CLOSURE THRESHOLDS (10-3 BAR)
3370 FE=V1(18)  'MEAN LIGHT ENVIRON LEAF CURRENTLY ADJUSTED TO(FRAC)
3375 KH=V1(40)
3380 IF I=1 THEN PP=PT   'INITIAL TURGOR POT ESTM.
3390 IF I=1 THEN DT=0 'INITIAL TL
3391 TL=TA+DT
3392 VP=DE+SA*DT
3393 '
3395 'CALC KC,KO ETC AT TAIR FOR LINEARIZATION IN PHOTO SUBPROG
3397 '
3400 AZ=(TA-T0)/((T0+273)*R*(TA+273))
3405 H0=R*(TA+273)*(TA+273)
3410 KV=K1*EXP(AZ*K2)  'KC
3415 H1=K2/H0
3420 KW=K3*EXP(AZ*K4)  'KO
3425 H2=K4/H0
3430 KX=R0*EXP(AZ*K5)  'R3 OR RD
3435 H3=K5/H0
3440 KY=VM*EXP(AZ*K6)  'V1 OR VC
```

```
3445 H4=K6/H0
3480 GOSUB 10000
3482 TL=TA
3485 GOSUB 15000
3490 INZ=1:PP=PT
3491 '
3492 'SOLUTION OF LEAF ENERGY BALANCE WITH WATER STRESS.
3493 '
3495 RS=1/KS
3500 DT=(EN-1825*KL*DE-T5)/(1825*KL*SA+R2+T6)
3520 TN=TA+DT 'PRINT "TL,TN,DT";TL;TN;DT 'SAME AS TL
3540 VP=DE+SA*DT
3560 TR=1825*KL*VP
3580 TM=TR/2.45 '10-3GM-2S-1
3600 PL=PS-TM/KH
3620 IF PL<PO THEN PL=PO
3640 PN=PL-PO 'SAME AS PP
3641 'PRINT"PN,PP,INZ,TN,TL";PN;PP;INZ;TN;TL
3650 IF ABS(TN-TL)<0.5 THEN GOTO3652 ELSE TL=TN:GOTO 3705
3652 TL=TN:IF PN>=PT AND INZ=1 THEN PP=PN:GOTO 3840
3653 IF PN<PT AND INZ=1 THEN PZ=PT/2:PP=PZ:TL=TN:GOTO 3700
3655 IF ABS(PN-PP)<.1 THEN GOTO 3840
3657 PZ=PZ/2:IF PZ<.1 THEN PZ=.1:
3658 IF INZ>20 THEN GOTO 3840
3660 TL=TN
```

```
3670 PP=PP+SGN(PN-PP)*PZ
3671 PRINT@ 281,"IN PP";INZ;PP;:PRINT@241,""
3680 DT=TL-TA .
3690 '
3700 INZ=INZ+1
3705 GOSUB 10000
3710 GOTO 3495
3840 RN=EN-(T5+T6*DT)'2*.95*SI*(TL+273)^4
3880 Q=R2*DT
3990 '
3991 '
3992 'SETTING UP EQUIVALENCE BETWEEN 2-LETTER VBL NAMES AND V1 ARRAY FOR PLOT.
3993 '
7000 V1(6)=P    'PHOTOSYN RATE 10-6 mol CO2 m-2 s-1
7020 V1(7)=KS*100 'STOM COND   cm s-1
7040 V1(8)=TL  'TEMP LEAF  C
7060 V1(10)=DT 'TLEAF-TAIR  (C)
7080 V1(11)=RN 'LEAF NET RAD (W M-2)
7100 V1(12)=TR 'LEAF TRANSPIRATION (W M-2)
7120 V1(13)=Q  'LEAF SENSIBLE HEAT (W M-2)
7140 V1(14)=RS 'STOMATAL RESISTANCE (S M-1)
7160 V1(15)=RX 'BOUNDARY LAYER RESIS (S M-1)
7180 V1(19)=RD 'DARK RESP (10-6MOLM-2S-1)
7200 V1(20)=DE 'VPAOR PRES DEF OF AIR(MB)
7220 V1(22)=PP 'LEAF TURGOR PRES (BAR)
7240 V1(24)=KC 'MICHAELIS CONST FOR CO2(10-6 BAR)
7260 V1(25)=KO 'MICHAELIC FOR O2(10-3BAR)
```

276

```
7280 V1(26)=R3  'DARK RESP NOT ALTERED BY CURRENTLY ADJUSTED LIGHT ENVIRON
7300 V1(27)=V1  'MAX CARBOYX VEL NOT ALTERED BY CURRENTLY ADJ LIGHT ENVIRON
7320 V1(28)=JM  'LIGHT SAT MAX RATE OF ELECTRON TRANSPRT UNALTERED(10-6EQM-2S-1)
7340 V1(29)=J1  'JM REDUCED FOR WATER STRESS AND LIGHT HISTORY
7360 V1(30)=J   'POTENTIAL RATE OF ELECTRON TRANSPORT
7380 V1(31)=GS  'CO2 COMP POINT W/O DARK RESP (PPM)
7400 V1(32)=G   'CO2 COMP POINT(PPM)
7420 V1(33)=CO  'KNEE OF P VS CI CURVE(PPM)
7440 V1(34)=CI  'INTERNAL CO2 CONC(PPM)
7460 V1(37)=E   'PAR(10-6MOLQ M-2S-1)
7480 V1(38)=KL*100 'LEAF COND(MOLM-2S-1 IN CALC AND CM S-1 IN OUTPUT)
7500 V1(39)=PL  'LEAF WATER POT (BAR)
7520 V1(41)=CI/CA  'CR
7540 V1(42)=VP  'VPD OF LEAF(MBAR)
8000 X(I)=V1(IX%)
8010 Y(I)=V1(IY%)
8020 IF W$ <> "" THEN W(I)=V1(IW%)
8030 IF W$="""THEN W(I)=0
8035 IF W$<>"" THEN U1(I)=V1(I1%)
8040 IF W$<>"" THEN U2(I)=V1(I2%)
8045 IF W$<>"" THEN U3(I)=V1(I3%)
8049 '
8050 'OUTPUT TO FILE OR PRINTER IF DESIRED
8052 IF IO%<>1 THEN GOTO 8090
8055 LPRINT USING"I = ## ";I
8057 FOR N=1 TO 42 STEP 7
```

```
8060 LPRINT USING"\    \=-####.##  ";V$(N);V1(N);V$(N+1);V1(N+1);V$(N+2);V1(N+2);
8061 LPRINT USING"\    \=-####.##  ";V$(N+3);V1(N+3);V$(N+4);V1(N+4);V$(N+5);V1(N+5);
8062 LPRINT USING"\    \=-####.##  ";V$(N+6);V1(N+6)
8065 NEXT N
8070 '
8085 'PRINT X Y & W VALUES ON SCREEN AS THEY ARE COMPUTED FOR MONITORING.
8090 IF I<=2 THEN PRINT@241,""
8100 IF I>2 THEN PRINT@201,""
8110 PRINT USING "## ####.### ####.### ####.##";I;X(I);Y(I);W(I)
8120 IF I=1 THEN GOTO 8190
8130 IF I<NPZ THEN PRINT@ 201,""
8190 NEXT I '                            END CALCULATION LOOP
8300 GOSUB 40000
8400 GOTO 1490
8500 '
8510 '
8520 '
8530 '
8540 '
9997 '
9998 '
9999 '
10000 'SUBPROGRAM PHOTOSYNTHESIS AND STOMATAL CONDUCTANCE*********************
10010 '
10030 '
10190 'EVALUATE PHOTOSYN TEMP FUNCTIONS USING LINEARIZED FORMS TO SAVE TIME.
10200 '
```

```
10205 MO=(TL+273)/(273*44.5)
10210 'KS=KS*MO:KL=KL*MO 'MS-1 TO MOLM-2S-1
10215 '
10220 A1=(TL-TO)/((TO+273)*R*(TL+273))
10240 'KC=K1*EXP(A1*K2)
10246 KC=KV*(1+H1*DT)
10260 'KO=K3*EXP(A1*K4)
10262 KO=KW*(1+H2*DT)
10270 'R3=RO*EXP(A1*K5)
10272 R3=KX*(1+H3*DT)
10275 'V1=VM*EXP(A1*K6)
10277 V1=KY*(1+H4*DT)
10280 '
10300 JM=J2*EXP(E1*A1)/(1+EXP((S*(TL+273)-H)/(R*(TL+273))))
10305 'TB=H/(S+R*LOG(H/E1-1))
10310 PA=PP/PT:IF PA>1THEN PA=1:IF PA<0 THEN PA=0
10320 J1=JM*PA*(.8*FE+.2)   'ADJUST FOR TURGOR LOSS AND LOW LIGHT ON JMAX
10340 J=J1*E/(E+2.1*J1)
10350 GS=.5*.21*O*KC/KO
10360 KK=KC*(1+O/KO)
10380 V2=V1
10390 RE=R3/V2
10400 G=(GS+KK*RE)/(1-RE)
10420 V3=J/(4.5*V2)
10440 CO=(KK*V3-2.33333*GS)/(1-V3)
10445 CI=.72*CA
10450 IF CI<0 THEN CI=0
```

```
10500 '
10520 P1=V2*(CI-GS)/(CI+KK)-R3
10540 P2=J*(CI-GS)/(4.5*CI+10.5*GS)-R3
10560 IF P1<=P2 THEN P=P1
10580 IF P2<P1 THEN P=P2
10585 'SUBTR BLR OF CHAMBER USED FOR P.S. MEAS FROM CALC OF KA
10590 IF P<1.E-5 THEN P=1.E-5
10620 KA=1/((CA-CI)/(1.6*P)-30*FS*MO)
10640 IF KA<0 THEN KA=KE ELSE KA=KA+KE
10700 VC=V2*(.8*FE+.2)
10720 RD=R3*(.6*FE+.4)   'REDUCE RESP LESS THAN VCMAX IS REDUCED BY LOW LIGHT
10730 C4=CI
10940 '
10960 RE=RD/VC
10980 G=(GS+KK*RE)/(1-RE)
10985 KS=KA
10986 'CALC EFFECT OF DIRECT VAPOR PRESS DEFICIT ON STOM COND.
10990 IF VP<D1 THENGOTO11005'
10995 IF VP>D1 AND VP<D2 THEN KS=(KA-KE)*(D2-VP)/(D2-D1)+KE
11000 IF VP>D2 THEN KS=KE
11001 'SOLVE FOR CI USING KS FROM ABOVE AND PHOTOSYNTHESIS PARAMETERS.
11002 'THIS IS SOLUTION OF EQUALITY BETWEEN P CALC FROM CI,CA & KS WITH
11003 'FARQUHAR EQUATIONS FOR PHOTOSYNTHESIS.
11005 RS=1/KS
11015 KL=1/(RS+FS*RX/42)
11140 A2=KL/1.6:A3=A2*CA
11160 A4=A3+RD-A2*KK-VC
```

```
11180 A5=(A3+RD)*KK+VC*GS
11200 A6=SQR(A4*A4+4*A2*A5)
11220 A7=(-A4-A6)/(-2*A2)
11230 AB=2.3333*GS:AC=J/4.5:V3=AC/VC
11240 CO=(KK*V3-AB)/(1-V3)
11250 '
11260 A8=A3+RD-A2*AB-AC
11270 A9=(A3+RD)*AB+AC*GS
11280 AA=SQR(A8*A8+4*A2*A9)
11285 AD=(-A8-AA)/(-2*A2)
11290 IF AD>A7 THEN CI=AD ELSE CI=A7
11320 P1=VC*(CI-GS)/(CI+KK)-RD
11330 P2=J*(CI-GS)/(4.5*CI+10.5*GS)-RD
11340 IF P1<=P2 THEN P=P1
11360 IF P2<P1 THEN P=P2
11370 'CONVERT KS AND KL FOR OUTPUT.
11400 KS=KS*MO'(TL+273)/(273*44.5) 'STOM COND IN MS-1
11420 KL=KL*MO'(TL+273)/(273*44.5) 'LEAF COND MS-1)
11430 'PRINT P;CI;GS;TL;IR%
14000 RETURN
14001 '
14002 '
14003 '
```

```
15000 'SUBPROGRAM TO SET UP LEAF ENERGY BALANCE*********************************
15005 '
15010 '
15100 'LEAF ENERGY BALANCE
15110 '
15220 IF ITZ=1 AND I>1 THEN GOTO 15510    'ITZ IS VBL THAT IS 1 IF TA NOT CHANGED
15221 'FROM LAST TIME THIS ROUTINE WAS CALLED
15250 T2=TA+273
15260 T3=T2^3:T4=T2^4:T5=1.077E-7*T4:T6=4.308E-7*T3
15300 SA=5313*ES/T2^2    'SLOPE OF SAT V.P. VS TA CURVE
15500 QE=5.95E-5*EXP(1500/T2)    'TEMP PART OF SKY THERMAL RAD.
15505 QE=.95*(1.7+EA*QE)*SI*T4    'SKY AND GROUND THERMAL RAD
15510 EN=1.15*EP+QE
15530 R2=2400/RX
16000 RETURN
16010 '
16020 '
16030 '
40000 'PLOTTING SUBPROGRAM***************************************************
40002 '
40004 '
40009 GOTO 40065
40010 '
40015 FOR M=1 TO NPZ:IF M<=2 THEN PRINT@241,"":IF M>2 THEN PRINT@201,""
40020 PRINT USING"## ####.### ####.### ####.##"; M; X(M); Y(M); W(M)
40025 IF M=1 THEN GOTO 40035
40030 IF M<NPZ THEN PRINT@201,""
```

```
40035 SOUND 0,25 :NEXT M
40065 PRINT USING "VBLS FOR PLOT \ \ \ \ \ \ ";V$(IX%);V$(IY%);V$(IWZ)
40070 PRINT USING "VBL SCANNED \ \ RANGE ####.# ####.#";V$(IZ%);ZN;ZX
40075 INPUT "XMIN,XMAX,YMIN,YMAX";X1,X2,Y1,Y2
40076 IF IWZ<>0 THEN INPUT "Y2MIN,Y2MAX";W1,W2
40080 IF W$="""" THEN W1=0:IF W$=""""THEN W2=0
40085 '
40090 SY$="("
40095 SZ$=")"
40100 CLS
40120 PRINT USING "####.##                    ####.##";Y2;W2
40160 PRINT@ 83,Y$;:PRINT@ 105,W$;:PRINT USING " \\";U1$;U2$;U3$;W$;
40170 PRINT
40180 PRINT USING "                    XMAX=####.##";X2;
40200 PRINT USING "                    XMIN=####.##";X1;
40210 PRINT:PRINT
40212 PRINT USING "####.##                  ####.##";Y1;W1;
40220 PRINT@ 295,X$;
40300 LINE (45,0)-(145,63),1,B
40330 OP%=0
40400 FOR I=1 TO NP%
40420 XP%=45+100*(X(I)-X1)/(X2-X1)
40430 IF XP%>145 OR XP%<45 THEN GOTO 40600
40440 YP%=63*(Y(I)-Y2)/(Y1-Y2)
40445 IF IWZ<>0 THEN WP%=63*(W(I)-W2)/(W1-W2)
40450 IF YP%>62 OR YP%<1 THEN GOTO 40470
40460 LINE (XP%-1,YP%-1)-(XP%+1,YP%+1),1,B
```

```
40465 GOTO 40480
40470 OPZ=OPZ+1
40480 IF IWZ=0 THEN GOTO 40520
40490 IF WPZ>62 OR WPZ<1 THEN GOTO 40600
40500 PSET (XPZ,WPZ)
40502 PSET (XPZ+1,WPZ)
40504 PSET (XPZ-1,WPZ)
40506 IF WPZ < 63 THEN PSET (XPZ,WPZ+1)
40508 IF YPZ > 0  THEN PSET (XPZ,WPZ-1)
40520 GOTO 40620
40600 OPZ=OPZ+1
40620 NEXT I
40630 PRINT@ 201,SY$;V$(IZZ);SZ$
40640 IF OPZ>0 THEN PRINT@ 66,"     OFF PLOT=";OPZ
40650 XOZ=45+100*(-X1)/(X2-X1)
40660 YOZ=63*(-Y2)/(Y1-Y2)
40665 IF IWZ<>0 THEN WOZ=63*(-W2)/(W1-W2)
40670 FOR I=1 TO 110 STEP 2
40680 II=35+I
40690 IF YOZ>=1 AND YOZ<=62 THEN PSET (II,YOZ)
40700 NEXT I
40710 FOR I=1 TO 63 STEP 2
40720 IF XOZ>=40 AND XOZ<=140 THEN PSET (XOZ,I)
40730 NEXT I
40735 IF IWZ=0 THEN GOTO 40810
40740 FOR I=1 TO 110 STEP 3
```

```
40750 II=45+I
40760 IF WO%>=1 AND WO%<=62 THEN PSET(II,WO%)    :NEXT I
40810 LINE (25,25)-(27,27),1,B
40820 IF IW%=0 THEN GOTO 40900
40840 PSET(155,25):PSET(156,25):PSET(154,25):PSET(155,26):PSET(155,24)
40900 PRINT@ 267," ";
40905 LINE INPUT DU$
40910 'SOUND 1,250
40920 IF DU$="" THEN GOTO 40000
40930 Z1$="0" 'IF VBL LETTERS TYPED IN THEN PRINT OUT VBL VALUE
40940 FOR I=1 TO NV%
40945 IF DU$=U1$ THEN W$=U1$:IW%=I1%:CLS:FORM=1TONP%:W(M)=U1(M):NEXT M:Z1$="1"
40946 IF DU$=U1$ THEN GOTO 40010
40950 IF DU$=U2$ THEN W$=U2$:IW%=I2%:CLS:FORM=1TONP%:W(M)=U2(M):NEXT M:Z1$="1"
40951 IF DU$=U2$ THEN GOTO 40010
40955 IF DU$=U3$ THEN W$=U3$:IW%=I3%:CLS:FORM=1TONP%:W(M)=U3(M):NEXT M:Z1$="1"
40956 IF DU$=U3$ THEN GOTO 40010
40960 IF DU$=V$(I) THEN PRINT@ 268,V$(I);V1(I)
40965 IF DU$=V$(I) THEN Z1$="1"
40970 NEXT I
40980 IF Z1$="0" THEN GOTO 41000
40990 GOTO 40900
41000 RETURN
```

APPENDIX B - EXAMPLE SESSION

NPZ=5 OK?CR (NOTE: CR INDICATES CARRIAGE RETURN-ON MODEL 100 HIT ENTER)

```
EP=395  TA=30.0 RH=0.4  V = 1.0 CA=340
FS=0.5  L =0.100 FE= 1.0 PS=  0.0PO=-15.
D1=70   D2=90   KH=20
```

NAME,VAL?PS,-6

```
EP=395  TA=30.0 RH=0.4  V = 1.0 CA=340
FS=0.5  L =0.100 FE= 1.0 PS= -6.0PO=-15.
D1=70   D2=90   KH=20
```

NAME,VAL?CR

```
EP TA RH V  CA P  KS TL EA DT
RN TR Q  RS RX FS L  FE RD DE
PO PP PS KC KO R3 V1 JM J1 J
GS G  CO CI D1 D2 E  KL PL KH
CR VP
```

X,Y FOR PLOT?TL,P
2ND Y?TR

```
3RD Y?PP
4TH Y?DT
5TH Y?RD

VBL TO BE SCANNED?TA
MIN,MAX?10,40

1   14.930   14.125   112.946
5   44.196    3.310   174.776
IN PP 1  3.75  (NOTE: This appears on display if water stress affects
                photosynthesis. IN=iteration no. and PP is press poten)


VBL FOR PLOT TL  P   TR
VBL SCANNED TA  RANGE   10.0    40.0
XMIN,XMAX,YMIN,YMAX?10,50,-5,25
Y2MIN,Y2MAX?0,320

25.00!     +           ! 320.00
     !      o          !   OFF PLOT= 1
   P !   o             !TR PP DT RD TR
   o ! o +       o     +!+ XMAX= 50.000
     !                 !   XMIN= 10.000
(TA) !  +           o!
     ..!.............! C
  -5.00!    TL       ! 0.00


(NOTE: C is where the cursor is, OFF PLOT is no. of points out of range
and thus not plotted, (TA) indicates the name of the variable scanned or
independent variable, and the dotted line is the zero line and it extends
beyond the appropriate axis.
```

If EP is typed then display becomes)

```
25.00!              +        ! 320.00
      !        o             !      OFF PLOT= 1
  P   !                      !TR PP DT RD TR
      !  o    o     +        !+ XMAX= 50.000
    o !  o +        o        !      XMIN= 10.000
      !                      !
 (TA) !       +          o   !
      !..!...............!  EP 395
 -5.00!           TL_         ! 0.00
```

(NOTE: If DT is typed)

```
1    14.930   14.125    4.930
5    44.196    3.310    4.196
VBLS FOR PLOT TL  P    DT
VBL SCANNED TA  RANGE    10.0    40.0
XMIN,XMAX,YMIN,YMAX?CR
Y2MIN,Y2MAX?0,10
```

```
25.00!                         ! 10.00
     !                         !
  P  !            o            !DT PP DT RD DT
  o  !       o         o       !+ XMAX= 50.000
     !       o         +       !  XMIN= 10.000
     !              +!         !o!
(TA) !     +                    C
-5.00!..!........+..+......!      0.00
              TL
```

(NOTE:If QQ, or any letter not a vbl name, is typed)
EP=395 TA=30.0 RH=0.4 V = 1.0 CA=340
FS=0.5 L =0.100 FE= 1.0 PS= -6.0PO=-15. D1=70   D2=90   KH=20
NAME,VAL?CR

```
EP TA RH V  CA P  KS TL EA DT
RN TR Q  RS RX FS L  FE RD DE
PO PP PS KC KO R3 V1 JM J1 J
GS G  CO CI D1 D2 E  KL PL KH
CR VP
```

X,Y FOR PLOT?TL,CI
2ND Y?CR   (NOTE: this will keep PP,DT,RD from previous run but TR lost.

VBL TO BE SCANNED?TA
MIN,MAX?10,40

etc....

APPENDIX C - EXAMPLE SESSION EXTRA OUTPUT

```
I = 1
EP= 395.00 TA= 10.00 RH=   0.40 V =   1.00 CA= 340.00 P =  14.13 KS=   0.64
TL=  14.93 EA=  4.91 DT=   4.93 RN= 322.89 TR= 112.95 Q = 209.94 RS= 155.67
RX=  56.36 FS=  0.50 L =   0.10 FE=   1.00 RD=   0.41 DE=   7.37 PO= -15.00
PP=   6.69 PS= -6.00 KC= 190.30 KO= 196.74 R3=   0.41 V1=  41.93 JM= 151.04
J1= 151.04 J = 128.59 GS=  21.33 G =  25.39 CO= 685.88 CI= 241.34 D1=  70.00
D2=  90.00 E =1817.00 KL=   0.54 PL=  -8.31 KH=  20.00 CR=   0.71 VP=  11.38
I = 2
EP= 395.00 TA= 17.50 RH=   0.40 V =   1.00 CA= 340.00 P =  17.50 KS=   0.83
TL=  20.75 EA=  8.00 DT=   3.25 RN= 336.94 TR= 198.43 Q = 138.51 RS= 120.26
RX=  56.36 FS=  0.50 L =   0.10 FE=   1.00 RD=   0.71 DE=  12.00 PO= -15.00
PP=   4.95 PS= -6.00 KC= 312.93 KO= 263.00 R3=   0.71 V1=  68.42 JM= 198.17
J1= 198.17 J = 161.24 GS=  26.24 G =  32.43 CO= 490.24 CI= 239.85 D1=  70.00
D2=  90.00 E =1817.00 KL=   0.68 PL= -10.05 KH=  20.00 CR=   0.71 VP=  16.09
```

```
I = 3
EP= 395.00  TA=   25.00  RH=    0.40  V =    1.00  CA= 340.00  P =  20.17  KS=   0.99
TL=  26.31  EA=   12.67  DT=    1.31  RN=  361.31  TR= 305.71  Q =  55.61  RS= 101.03
RX=  56.36  FS=    0.50  L =    0.10  FE=    1.00  RD=   1.26  DE=  19.01  PO= -15.00
PP=   2.75  PS=   -6.00  KC=  519.57  KO=  355.88  R3=   1.26  V1= 112.77  JM= 257.05
J1= 176.72  J =  146.75  GS=   32.19  G =   41.89  CO= 230.45  CI= 237.92  D1=  70.00
D2=  90.00  E =1817.00   KL=    0.78  PL=  -12.24  KH=  20.00  CR=   0.70  VP=  21.48

I = 4
EP= 395.00  TA=   32.50  RH=    0.40  V =    1.00  CA= 340.00  P =  14.24  KS=   0.69
TL=  33.39  EA=   19.56  DT=    0.89  RN=  376.89  TR= 338.96  Q =  37.93  RS= 144.74
RX=  56.36  FS=    0.50  L =    0.10  FE=    1.00  RD=   2.31  DE=  29.34  PO= -15.00
PP=   2.13  PS=   -6.00  KC=  892.97  KO=  493.49  R3=   2.31  V1= 192.33  JM= 271.53
J1= 144.25  J =  123.64  GS=   39.90  G =   55.86  CO= 103.54  CI= 241.48  D1=  70.00
D2=  90.00  E =1817.00   KL=    0.58  PL=  -12.92  KH=  20.00  CR=   0.71  VP=  31.82

I = 5
EP= 395.00  TA=   40.00  RH=    0.40  V =    1.00  CA= 340.00  P =   3.31  KS=   0.16
TL=  44.20  EA=   29.50  DT=    4.20  RN=  353.44  TR= 174.78  Q = 178.66  RS= 611.55
RX=  56.36  FS=    0.50  L =    0.10  FE=    1.00  RD=   5.24  DE=  44.25  PO= -15.00
PP=   5.43  PS=   -6.00  KC= 1863.34  KO=  775.63  R3=   5.24  V1= 397.33  JM=  79.28
J1=  79.28  J =   72.62  GS=   52.97  G =   85.33  CO= -28.59  CI= 251.96  D1=  70.00
D2=  90.00  E =1817.00   KL=    0.16  PL=   -9.57  KH=  20.00  CR=   0.74  VP=  61.03
```

Definitions of Variables used in Program Code
and Mathematical Equations

A1	10220	Temperature dependent exponent of Arrhenius temperature functions for photosynthesis used in photosynthesis subroutine.
A2	11140	Dummy variables used to speed up calculation of "knee" of A vs C_i curve. Transition from RuP_2 saturated region to RuP_2 regeneration limited zone.
.		
.		
.		
A9	11270	
AA	11280	
AB	11230	
AC	11230	
AD	11285	
AZ	3400	Same as A1 but in main program.

CA   300,3300 CO_2 concentration or partial pressure of the air (10^{-6} liter liter^{-1}, ppm, 10^{-6} bar)

CI   10445,11290   Leaf internal CO_2 concentration (10^{-6} liter liter^{-1}, ppm, 10^{-6} bar).

CO   10440,11240   Knee of photosynthesis versus internal CO_2 concentration curve (10^{-6} liter liter^{-1}, ppm, 10^{-6} bar).

CR   7420   Ratio of leaf internal to external CO_2 concentration (CI/CA).

DE   3272   Vapor pressure deficit of air, ES-EA (10^{-3} bar).

DT   3390,3500   Temp difference between leaf and air, TL-TA (C).

DU$ 40920   String variable to allow user to request the value of any variable while current plot is on screen and this process does not disturb plot on screen. A null here allows plot to be redone without executing program again.

D1	300,1415 3360	Threshold for beginning stomatal closure from VPD (mb).
D2	300,1420 3360	VPD where stomatal closure from VPD is complete (mb).
E	3232	Same as EP (10^{-6} mol quanta $m^{-2} s^{-1}$).
EA	3270	Actual vapor pressure of the air (10^{-3} bar).
EN	3500,15510	Net incoming radiation on leaf. This is net absorbed shortwave plus incoming thermal from environment; does not include thermal radiated from leaf ($W\ m^{-2}$).
EP	300,3230	Photosynthetically active radiation ($W\ m^{-2}$).
ES	3265	Saturation vapor pressure of the air (10^{-3} bar).
E1	340,1455 10300	Parameter in temp dependence of J2, max. rate of electron transport ($J\ mol^{-1}$).
FE	300,1435 3370	Average incidence radiation flux density that leaf is exposed to as a fraction of full sun (0.1 - 1.0).
FS	300,1425 3310	Amphistomatus leaves FS = 0.5, hypostomatus FS = 1.0.

294

Variable	Value	Description
G	10400	CO_2 compensation point (10^{-6} liter liter^{-1}, ppm, 10^{-6} bar).
GS	10350	CO_2 compensation point without dark respiration (ppm, 10^{-6} bar).
H	340,1455 10300	Parameter in temp dependence of J2, max. rate of electron transport (J mol^{-1}).
HØ	3405	Denominator of Arrhenius temperature functions computed once to save time (J mol^{-1} C).
H1	3415,10246	Slope for linearization of KC vs leaf temp.
H2	3425,10262	Slope for linearization of KO vs leaf temp.
H3	3435,10272	Slope for linearization of RO vs leaf temp.
H4	3445,10277	Slope for linearization of VC vs leaf temp.
I	3120	Loop dummy variable.
IR	10920	Number of iterations.
IN%	3490,3700	Number if iterations in leaf-temp-water-stress loop.
IT%	2462,3260	Flag, if IT% = 0 then TA is scanned variable and functions dependent on TA must be recalculated. If IT% = 1 only calculate functions dependent on TA once.

IW%	2265	Subscript value of V$ (I) So V$ (IW%) = W$ on plot.
IX%	2240	Subscript value of V$ (I) So V$ (IX%) = X$ on plot.
IY%	2260	Subscript value of V$ (I) So V$ (IY%) = Y$ on plot.
IZ%	2440	Subscript value of V$ (I) So V$ (IZ%) = Z$ for scanned variable.
I1%	2266	Subscript value of V$ (I) So V$ (I1%) = U1$ on plot.
I2%	2267	Subscript value of V$ (I) So V$ (I2%) = U2$ on plot.
I3%	2268	Subscript value of V$ (I) So V$ (I3%) = U3$ on plot.
J	10340	Potential rate of electron transport (10^{-6} Eq m^{-2} s^{-1}).
JM	10300	Light saturated maximum rate of electron transport not altered by water stress but dependent on temp. (10^{-6} Eq m^{-2} s^{-1}).
J1	10320	JM reduced by water stress (10^{-6} Eq m^{-2} s^{-1}).
J2	320,1450 10300	Light saturated potential rate of electron transport at temp to (10^{-6} Eq m^{-2} s^{-1}).

296

KA	10620	Stomatal conductance to water vapor with no effects of vapor pressure deficit on stomata (Mol m^{-2} s^{-1}).
KC	10246	Michaelis constant for CO_2 at the leaf temperature from KV to increase program speed (10^{-6} bars).
KE	1025	Epidermal or cuticular conductance for water (2×10^{-4} m s^{-1} or 8.4×10^{-3} Mol m^{-2} s^{-1}).
KH	300,1440 3375	Internal plant hydraulic conductance (mg [m^2 leaf area]$^{-1}$ s^{-1} bar^{-1}).
KK	10360	Part of equation for RuBP saturated rate of carboxylation (P1).
KL	3500,11015 11420,7480	Leaf diffusion conductance for water vapor (mol m^{-2} s^{-1} m calculations between stm 10500 - 11360, and cm s^{-1} in output).

KO	10262	Michaelis constant for O_2 at leaf temperature computed from KX to increase program speed (10^{-3} bars).
KS	11400,7020	Stomatal diffusion resistance to water vapor ($m\ s^{-1}$, output in $cm\ s^{-1}$).
KV	3410	KC at air temperature (10^{-6} bars).
KW	3420	KO at air temperature (10^{-3} bars).
KX	3430	RO (dark respiration) at air temperature ($10^{-6}\ mol\ m^{-2}\ s^{-1}$).
KY	3440	VC (carboxylation velocity) at air temperature ($10^{-6}\ mol\ m^{-2}\ s^{-1}$).
K1	320,1450 3410	Coefficient in KC temperature dependence, KC at TO(10^{-6} bar).
K2	320,1450 3410	Coefficient in KC temperature dependence activation energy (J mol^{-1}).
K3	320,1450 3420	Coefficient in KO temperature dependence, KO at TO(10^{-3} bars).

K4	320,1450 3420	Coefficient in KO temperature dependence activation energy ($J\ mol^{-1}$).
K5	320,1450 3430	Coefficient in RO temperature dependence activation energy ($J\ mol^{-1}$).
K6	320,1450 3430	Coefficient in VC temperature dependence activation energy ($J\ mol^{-1}$).
L	300,1430 3294	Leaf width (m).
M	40015	Dummy loop variable for plotting.
MO	10205,11400	Conversion of stomatal conductance from $mol\ m^{-2}\ s^{-1}$ to $m\ s^{-1}$.
NV%	1100	No. of variables for plotting (42).
NP%	1110,40015	No. of points on output plot.
OP%	40470,40640	No. of points assigned to be plotted that were off the plot and so could not be plotted.
O	320,1450	Oxygen partial pressure (10^{-3} bars).
P	11340	Leaf photosynthetic rate ($10^{-6}\ mol\ m^{-2}\ s^{-1}$).
PA	10310	PP/PT ratio of pressure potential to threshold pressure potential for water stress reduction of electron transport.
PL	3600	Total leaf water potential (bars).

PN   3640   Leaf pressure potential used in iteration of leaf energy budget; nearly equal to PP at end of iteration (bars).

PO   300,1410   Leaf osmotic potential (bars).
      3340

PP   3670   Leaf pressure potential (bars).

PS   300,1405   Soil water potential (bars).
      3350

PT   320,1450   Threshold leaf pressure potential for reduction of electron transport by water stress (bars).

PZ   3653,3657   Intermediate value of leaf pressure potential used in iteration (bar).

P1   10800,11320   Net photosynthesis based on the RuBP-saturated (rate of carboxylation $(10^{-6}$ mol m^{-2} s^{-1}).

P2   10820,11330   Net photosynthesis based on the electron transport/photophosphorylation limited rate of RuBP regeneration $(10^{-6}$ mol m^{-2} s^{-1}).

Q   3880   Leaf sensible heat flux density (W m^{-2}).

QE   15300   Thermal radiation as a function of air temp and vapor pressure.

300

R	1020	Universal gas constant (8.314 J mol^{-1} K^{-1}).
RD	10720	Day, dark respiration (10^{-6} mol m^{-2} s^{-1}).
RE	10390	R3/V2.
RH	300,3250	Air relative humidity (fraction).
RN	3840	Net all-wave radiation absorbed by the leaf (W m^{-2}).
RS	3495	Stomatal diffusion resistance to water vapor (S m^{-1}).
RX	3295	Boundary layer resistance (S m^{-1}).
RO	340,1455 3430	Day, dark respiration rate (10^{-6} mol m^{-2} s^{-1}).
R2	15530	2400/RX (W m^{-2} k^{-1}).

R3	10273	Day, dark respiration not altered by reduced light environment (10^{-6} mol m^{-2} s^{-1}).
S	340,1455 10300	Parameter in temp. dependence of J2, max rate of electron transport (J mol^{-1} K^{-1}).
SA	15500	Slope of saturation vapor pressure versus temperature relation at air temp. (10^{-3} bar K^{-1}).
SI	1015	Stephan Boltzman constant (5.67×10^{-8} W m^{-2} K^{-4}).
SY$	40090	String variable for "(".
SZ$	40095	String variable for ")".
TA	300,3240	Air temperature (C).
TB	10305	Optimum temperature of maximum rate of electron transport versus temperature (K).
TL	3391,3660	Leaf temperature (C).
TM	3580	Leaf transpiration rate in mass units (10^{-3} g m^{-2} s^{-1}).

TN	3520	Temporary value of leaf temperature used in iteration of leaf energy budget (C).
TO	320,1450 3400	Reference temperature for photosynthesis Arrenius temperature functions (C).
TR	3560	Leaf transpiration rate (W m^{-2}).
T2	15250	Air temp., TA+273 (K).
T3	15260	T2^3 (K^3).
T4	15260	T2^4 (K^4).
T5	15260	1.077×10^{-7} T4 (W m^{-2}) ($1.077 \times 10^{-7} = 2 \times 0.95 \times$ SI)
T6	15260	4.308×10^{-7} T3 (W m^{-2} K^{-1}).
U1(I)	8035	Additional array containing variable to be plotted on vertical axis if desired.

U2(I)	8040	Additional array containing variable to be plotted on vertical axis if desired.
U3(I)	8045	Additional array containing variable to be plotted on vertical axis if desired.
U1$	2207	Third y-variable string ID for output plot if desired on right vertical axis.
U2$	2207	Fourth y-variable string ID for output plot if desired on right vertical axis.
U3$	2207	Fifth y-variable string ID for output plot if desired on right vertical axis.
V	300,3290	Wind speed ($m\ s^{-1}$).
VC	10700,10800	Carboxylation rate of RuBP ($10^{-6}\ mol\ m^{-2}\ s^{-1}$).
VM	320,1450 3440	Maximum carboxylation rate of RuBP ($10^{-6}\ mol\ m^{-2}\ s^{-1}$).
VP	3392	Difference between saturation vapor pressure at leaf temperature and air vapor pressure ($10^{-3}\ bar$).
V$(I)	1200	String names of variables for plotting read in from data statement (stm 250).

304

V\$(1) = EP
V\$(2) = TA
V\$(3) = RH
V\$(4) = V
V\$(5) = CA
V\$(6) = P
V\$(7) = KS
V\$(8) = TL
V\$(9) = EA
V\$(10) = DI
V\$(11) = RN
V\$(12) = TR
V\$(13) = Q
V\$(14) = RS
V\$(15) = RX

V\$(16) = FS
V\$(17) = L
V\$(18) = FE
V\$(19) = RD
V\$(20) = PE
V\$(21) = PO
V\$(22) = PP
V\$(23) = PS
V\$(24) = KC
V\$(25) = KO
V\$(26) = R3
V\$(27) = V1
V\$(28) = JM
V\$(29) = J1
V\$(30) = J

V\$(31) = GS
V\$(32) = G
V\$(33) = CO
V\$(34) = CI
V\$(35) = D1
V\$(36) = D2
V\$(37) = E
V\$(38) = KL
V\$(39) = PL
V\$(40) = KH
V\$(41) = CR
V\$(42) = VP

V1   10277   Maximum carboxylation velocity not altered by reduced light environment (10^{-6} mol m^{-2} s^{-1}).

V2   10380   Carboxylation rate adjusted for low light history (10^{-6} mol m^{-2} s^{-1}).

V1(I)   1405   Numerical value of string variable V\$(I).

V3   10420   J/(4.5 v 2).

W(I)	8020	Array containing variable to be plotted on W axis (right vertical axis).
WP%	40445	Pixel count of second y-variable corresponding to W(I).
W$	2205	Second y-variable string ID (right vertical axis) for output plot.
WO%	40665	Pixel count of second y-variable (W-variable) origin for plot.
W1	40075	Minimum value of W-variable (second y-variable) for plot.
W2	40075	Maximum value of W-variable (second y-variable) for plot.
X(I)	8000	Array containing variable to be plotted on X axis (horizontal axis).
X$	2200	String ID of X-variable (horizontal axis) for plot.
XD$	1500	Dummy string input variable.
XP%	40420	Pixel count of X-variable corresponding to X(I).
X1	40075	Minimum value of X-variable for plot.
X2	40075	Maximum value of X-variable for plot.

X3	1500	Dummy numerical input variable.
X4	1550	Dummy numerical input variable.
X0%	40650	Pixel count of X-variable origin for plot.
X1$	1600	Dummy string input variable.
Y(I)	8010	Array containing variable to be plotted on y axis (left vertical axis).
Y$	2200	First y-variable string ID (left vertical axis) for plot.
Y0%	40660	Pixel count of y-variable origin for plot.
YP%	40440	Pixel count of y-variable corresponding to Y(I).
Y1	40075	Minimum value of y-variable for plot.
Y2	40075	Maximum value of y-variable for plot.

Variable	Line	Description
ZN	3100	Minimum value of variable to be scanned, which generates output.
ZX	3100	Maximum value of variable to be scanned, which generates output.
Z$	2400	Variable to be scanned to generate series of output values for the plot. This need not be the x-axis variable.
Z1$	40960	Flag used with DU$ to determine whether user wants to exit program. If DU$ is not equal to any allowed variable name, then Z1$ remains 0 and plotted routine is exited.

GROWTH AND NUTRIENT STATUS

CROP ESSENTIAL ELEMENT STATUS ASSESSMENT IN THE FIELD

J. Benton Jones, Jr.

Department of Horticulture
University of Georgia
Athens, Georgia 30602 USA

ABSTRACT: High crop yields of exceptional quality are dependent on
a specific and adequate essential element status in the producing
plants. When elemental levels in these plants are under or over
the sufficiency range, usually visual symptoms of stress appear
which provides one means of identification. Some form of chemical
testing of the plant is required to confirm an essential element
insufficiency or to determine if sufficiency is being achieved.
Such chemical tests performed in the laboratory are called "plant
analysis," while tests performed in the field on fresh tissue are
referred to as "tissue tests." Chemical tests of the growing
media, whether soil or some other substance, can also be conducted
to predict plant growth and probable crop response to applied
fertilizer. Currently, there are no electrical or electrochemical
means for assessing either the soil or plant essential element
status in situ. Heterogeneous distribution of elements in the
plant and rooting media makes for placement of measuring devices
difficult, if not impossible. Therefore, one is left with chemical
testing as the means of essential element assessment.

1. INTRODUCTION

Many believe that there is a world food crisis developing in
terms of both quantity as well as quality, the result of many
nations inability to keep pace with increasing population. Dwindling
world grain carryover supplies and a changing pattern of human
activity with fewer people engaged in food production are growing
concerns (1). Recent surveys show that the vast majority of soils
suitable for food production are requiring greater inputs of fertilizer and water to increase and/or maintain their productivity (2).
Only about 10% of the soils being cultivated today would be classed

as "moderately fertile," with the remaining 90% requiring treatment to make them suitable for profitable crop production.

One of the major requirements for successful crop production is essential element sufficiency, a demand met by the proper use of agricultural limestone and chemical fertilizers, or natural fertilizers such as, animal manures, sewage sludge, etc. It has been estimated that the world-wide demand for chemical fertilizers (mainly to supply the major essential elements: nitrogen, phosphorus, and potassium) will more than double before the turn of the century (3). How much fertilizer is needed in any particular crop production system is determined by a number of factors, the one most important being the soil itself measured in terms of its ability to supply the plant's essential element requirements. This measurement can only be done by means of a soil test (4). Another technique which determines the essential element status of the plant itself, is a tissue test (5), or plant analysis (6).

The common practice in the past, and still used extensively today, is to determine essential element needs based on a combination of factors, past experiences, known crop requirements and/or visual appearance of the growing crop. Most farmers are keen observers and rely to a considerable extent on their past experience as a means of determining those practices that give the best results. Unfortunately, the "best result" may be far from the maximum potential. New crop varieties, particularly products of the "green revolution" have high essential element requirements, requirements that must be satisfied in order to achieve their maximum yield potential. Better control of insects and diseases, as well as the increasing use of irrigation, place greater demands on the need for adequate essential element status to fully utilize these crop production advances.

This paper discusses the various commonly used procedures to assess essential element need based on visual observations, and the chemical assay of soils and plant tissues, procedures whose employment are required for success in any crop production system.

2. VISUAL APPEARANCE

Plants, stressed due to an insufficient quantity of an essential element, will visually appear abnormal, a description of these symptoms is given in Table 1 (7). However, visual appearance is not always a reliable means of assessing the plant's actual status. A plant may be insufficient in a particular essential element but not sufficiently so to produce the characteristic symptom. It is not unusual for a crop yield to be reduced 10 to 15 per cent by an essential element insufficiency, and yet, look perfectly normal. Plants may be stressed by lack of excess soil moisture, low or high air temperatures, restricted root development, and disease or insect infestation, and yet, give the appearance of an essential element insufficiency.

313

TABLE 1   CHARACTERISTIC VISUAL SYMPTOMS OF ESSENTIAL ELEMENT INSUFFICIENCY
Source: Davidescu and Davidescu (7)

Element	Manner of plant growth and development	Leaf coloration	Other signs	Location and phase of first appearance
Nitrogen	Insufficiency observed especially in the growth stage; stems and shoots short & weak; leaves small; poor stooling in Graminae; less branching; trees flower poorly; adventive roots form poorly on strawberries	Light green & later turning yellow; on cabbage, fruit tree, and shrub leaves appear orange and reddish shades	Early leaf loss; seconday shoots and leaves have a vertical position; fruits are strongly colored	On lower, more mature leaves; signs may appear in first phase of growth
Phosphorus	Accelerated growth; retarded development, especially blooming and ripening; adventive roots form poorly on strawberries	Dark blue-green; rose and purple shades appear in Cruciferae; bronzed; dried leaves have a dark, almost black color	Small fruit (apple); change of fruit color early leaf loss; appearance of rusty-brown spots in potato to tubers; thickening of cell walls	On older and lower leaves
Potassium	Slow growth; short internodes; slower stooling in Graminae, smaller # of shoots; plants with compound leaves have their lobes drawn in; supporting tissue is poorly developed	Dark blue-green; necrosis along leaf blade;yellow, brown or dead tissue between veins;'burning'of leaf margins; brown shades appear in leaves; chlorosis of clover leaves	Uneven growth of leaf blade, veins appear to be sunken into the leaf tissue; uneven ripening of tomato, apple fruits, etc.; premature death of upper stems in potato; leaves appear wilted, although plant is well supplied with water	On mature and lower leaves; symptoms often appear in the middle of the period of plant development

TABLE 1 (continued)

Element	Manner of plant growth and development	Leaf coloration	Other signs	Location and phase of first appearance
Magnesium	Development phases are retarded	Chlorosis, beginning especially at margin & center of the leaf blade, often leaving green strips between(cucumber,pear etc.);appearance of yellow,brown & red (cabbage) shades; 'marbled' chlorosis of leaves (radish, kohlrabi)	Incomplete ripening of fruit (apple); shrivelling & twisting leaf margin tobacco); brittleness of leaves (potato) due to high water content	On mature, lower leaves, proceeding higher; symptoms appear in later phases
Calcium	Damage & loss of buds & terminal roots; rosette formation of the small leaves; strong root branching	Chlorosis; appearance of white strips along leaf margins radish, kohlrabi, cabbage)	Leaf margins twist upwards(beet, potato, etc.); crooked leaf margins; loss of turgidity (tomato);potato tubers damaged (brown spots of dead tissue); appearance of manganese toxicity symptoms (on acid soils)	On younger leaves & plant parts

TABLE 1 (continued)

Element	Manner of plant growth and development	Leaf coloration	Other signs	Location and phase of first appearance
Boron	Dying of buds & terminal roots; intensive development of lateral shoots, giving plant a bushy look; wrinkling of upper leaves; poor blooming, dropping of flowers & fruits; leaves arranged in form of rosette	Yellowing of vegetative organs, especially young ones; suberification of tissues; deformation of fruits; various colored spots (brown, dark green); splitting of fruits	Hollow stems & roots; root rot(sugar beet); damaged fruit (tomato); dead tissue formation on & in fruit; poor development of cauliflower head; appearance of brown shades; potato tubers small & uneven; stem lignification (tomato)	On younger leaves & plant parts
Manganese	Vertical position of shoots and leaves	Chlorosis between leaf veins, leaves often spotted; leaf veins (down to the smallest) remain green, giving a striped appearance; grey-green & brown shades in oats; small dark brown spots along leaf veins & tissue death(potato);pale reddish color of beet leaves	Dying of leaf tissue; vertical position of shoots & secondary leaves, triangular-shaped leaves,twisted upwards(beet); brown or dry spots on inside surfaces of pea seed halves	On the younger parts of the plant (especially at the base of the leaf)

TABLE 1 (continued)

Element	Manner of plant growth and development	Leaf coloration	Other signs	Location and phase of first appearance
Copper	Poor growth & even cessation of growth	Chlorosis, whitening of leaf tips	Loss of turgidity in young leaves & stems	On the younger parts of plant
Zinc	Shortening of internodes; small leaves arranged in rosettes; poor fruiting	Yellowing or spotting of leaves(chlorosis) sometimes including veins; leaves take on shades of bronze	Thickened leaves (tobacco); expressed more strongly on trees in the summer	On older leaves in different phases of plant development
Sulphur	Stem increases slowly in diameter	Yellow-green coloration of leaf blade & oftentimes of veins, without tissue death	Symptoms similar to those of nitrogen insufficiency; poor nodule formation in legumes	On the younger parts of the plant
Iron	Cessation of growth	Non-uniform chlorosis between leaf veins; light green & yellow leaves without necrosis	Fruits strongly colored; twig loss when iron lack is accentuated (pear, plum, cherry)	On the younger parts of the plant

And finally, visual symptoms may be confusing even to the trained eye when distinguishing among some of the elemental deficiency symptoms as well as those caused by an excess of the suspected deficient element, or due to an imbalance among the elements.

Visual plant symptoms and their value for determining essential element needs can be misleading. Therefore, farmers are advised to confirm suspected insufficiencies by means of either a tissue test or plant analysis. A soil test may also be helpful, but should not be relied on alone without an assay of the plant itself.

Excellent color plates of typical essential element deficiency and excess visual symptoms may be found in several books (8,9,10,11).

3. SOIL TESTING

Soil testing has had a 40 year history of development and utilization, primarily in the more developed agricultural areas of the world. Many of the soil testing procedures still in use today had their origin in the 1940s and 50s. The value of the soil test was best summarized in a 1951 report (12) by the National Soil Test Workgroup when they said,

"there is good evidence that the competent use of soil tests can make a valuable contribution to the more intelligent management of the soil."

Fitts and Nelson (13) have given the objectives for soil testing as:
1) to group soils into classes for the purpose of suggesting fertilizer and lime practices;
2) to predict the probability of getting a profitable response to the application of plant nutrients;
3) to help evaluate soil productivity; and
4) to determine specific soil conditions which may be improved by addition of soil amendments or cultural practices.

Without a soil test result, the soil's fertility status can not be adequately evaluated, and therefore, any addition of lime or fertilizer to the soil would be indiscriminate and potentially hazardous to successful crop production free from elemental stress. With intensive cropping of even the best natively fertile soils, stress eventually occurs if proper procedures are not followed. Replacement of elements removed by crops, countering soil acidification, and keeping the proper balance of elements are required to sustain soil productivity. The soil test provides the guidance to satisfy all three requirements.

The technique of soil testing involves 4 steps: field sampling, sample preparation, laboratory analysis and interpretation. There is considerable uncertainty about how to best collect the soil sample from the field and considerable controversy regarding the soil test result interpretation.

3.1 Field Sampling

Most fields and orchards used for crop production usually are not homogeneous in depth or type. The challenge is to obtain a sample(s) that is representative of the area under test, the common procedure being to take a number of cores to form a composite for the laboratory assay. Peck and Melsted (14) have given the requirements for soil sampling required to ensure good representation. They recommend sampling plowed fields to the plow depth and orchards to a depth of 50 cm., collecting 8 to 16 cores per each 8 hectares to form one composite sample. Keeping samples from subsoil and surface soil depths separate, not mixing soil types together and not coring in uncharacteristic regions of the field are some of the additional instructions.

3.2 Sample Preparation

The field collected composite soil sample is air dried, gently crushed and then passed through a 10-mesh sieve. Oven drying is not recommended nor should the soil sample be pulverised. Wolf (15) has developed a soil testing procedure where the field composite sample is neither dried or sieved since the laboratory tests are performed on a volume of sample rather than using weighed aliquots.

3.3 Laboratory Analysis

Laboratory tests performed on the prepared soil sample will depend on the use of the test results, although the tests for water pH and level of extractable elements are the more commonly ones performed. If the soil is found to be acidic (ph <6.8), a lime requirement test is also made.

The methods for determining soil water pH and the lime requirements for acidic soils have been described by McLean (16); descriptions of the more commonly used procedures are given in Table 2 (17). The SMP Buffer procedure (18) is best suited for use on silt and clay loam soils of moderate organic matter content, while the Adams-Evans procedure (19) is only for use on sandy soils with low cation exchange capacity and low organic matter content.

The essential element status of a soil is determined by extraction procedures, frequently specific for each element or group of similar elements. Those procedures for the determination of phosphorus (P) have been described by Thomas and Peaslee (20), and more recently by Olsen and Sommers (21). A list of these procedures are given in Table 3 (17). The three most commonly used extraction procedures for P are Bray P1 (22) for acid soils of moderate cation exchange capacity, Mehlich No. 1 (23) for acid sandy soils, and Olsen (24) for alkaline soils.

TABLE 2 SOIL WATER AND LIME BUFFER PH BY THE SMP AND ADAMS-EVANS FUBBER PROCEDURES

Source: Jones (17)

Soil Test	Sample Size	Extraction Reagent	Volume Used	Shaking Time	Method of Determination
Soil Water pH (for SMP buffer)	5.0 g (4.25-cm^3)	Water	5 ml	10 min. with intermittent stirring.	Read pH to nearest 0.1 with glass electrode pH meter calibrated with pH 4.0 and 7.0 buffers
Soil Water pH (for Adams-Evans Buffer)	10-cm^3	Water	10 ml	10 min. with intermittent stirring.	Read pH to nearest 0.1 with glass electrode pH meter calibrated with pH 4.0 and 7.0 buffers
Soil pH in 0.01M CaCl$_2$	5.0 g (4.25-cm^3)	0.01M CaCl$_2$	5 ml	30 min. with intermittent stirring.	Read pH to nearest 0.1 with glass electrode pH meter calibrated with pH 4.0 and 7.0 buffers
SMP Buffer pH	5.0 g (4.25-cm^3) soil in 5 ml water	SMP Buffer	10 ml	Shake 10 min., intermittent stirring for 20 min.	Read pH to nearest 0.1 with glass electrode pH meter calibrated with pH 4.0 and 7.0 buffers
Adams-Evans Buffer pH	10-cm^3 soil in 10 ml water	Adams-Evans Buffer	10 ml	Shake 10 min., intermittent stirring for 20 min.	Read pH to nearest 0.5 with glass electrode pH meter calibrated to read pH 8.00 in mixture of 10 cc buffer reagent and 10 cc water

TABLE 3 COMMONLY USED METHODS FOR DETERMINING EXTRACTABLE PHORPHORUS (P) IN SOILS

Parameter	Mehlich No. 1	Bray P1	Method Mehlich No. 3	Olsen	AB-DTPA
Adaptability Limits	Sandy soils, acid, low in CEC	Acid soils with moderate CEC	All soils	Alkaline soils	Alkaline soils
Sample Size, Weight (volume)	5 g (4-cm^3)	2 g (1.70-cm^3)	2.5-cm^3	2.5 g (2-cm^3)	10 g (8.5-cm^3)
Volume of Extractant, ml	25	20	25	50	20
Extraction Reagent	0.05N HCl in 0.025N H$_2$SO$_4$	0.03N NH$_4$F in 0.025N HCl	0.2N HAc in 0.25N HN$_4$NO$_3$ in 0.015N NH$_4$F in 0.013N NHO$_3$ in 0.001M EDTA	0.5N NaHCO$_3$ at pH 8.5	1M NH$_4$HCO$_3$ 0.005M DTPA at pH 7.6
Shaking Time, minutes	5	5	5	30	15
Shaking Action and Speed	Reciprocating 180+ oscillations/min.	Reciprocating 180+ oscillations/min.	Reciprocating 180+ oscillations/min.	Reciprocating 180+ oscillations/min.	Reciprocating 180+ oscillations/min.
Method of P Determination in Extract	Molybdenum Blue	Molybdenum Blue	Molybdenum Blue	Molybdenum Blue	Molybdenum Blue

TABLE 3 (continued)

Parameter	Method				
	Mehlich No. 1	Bray P1	Mehlich No. 3	Olsen	AB-DTPA
Range in Soil P concn without Dilution, kg/ha	2-100	2-250	2-200	2-200	2-100
Sensitivity, kg/ha	1	1	1	1	2
Primary Reference	Mehlich (23)	Bray & Kurtz (22)	Mehlich (31)	Olsen, Cole, Watanabe & Dean (24)	Soltanpour & Schwab (32)

321

The exchangeable cations, calcium (Ca), potassium (K), magnesium (Mg) and sodium (Na), are extracted from the soil using neutral normal ammonium acetate (25), and for the sandy soils by the Mehlich No. 1 extractant (23). These extraction procedures, and others, are described by Doll and Lucas (26), and a list of these procedures are given in Table 4 (17).

The determination of the micronutrient status of soils is unique in that there are a number of extraction procedures, some for only one micronutrient element, such as the hot water extraction procedure (27) for boron (B) and ammonium oxalate extraction (28) for molybdenum (Mo), and multielement extraction of copper (Cu), iron (Fe), managanese (Mn) and zinc (Zn) using diethylenetriaminepentataacetic acid (DTPA) as the extraction reagent (29). Cox and Kamprath (30) discuss in some detail the soil testing procedures for determining the micronutrient status of soils which should be limited to only those soil-crop situations as described in Table 5.

More recent emphasis has been placed on the development and use of "universal" extraction reagents that can be used for the determination of all the major elements (P, K, Ca and Mg) and micronutrients (B, Cu, Fe, Mn and Zn) as one single procedure. There are currently three such extraction reagents, Mehlich No. 3 (31) and the Wolf modified Morgan Extractant (15) for use on most acid soils, and the AB-DTPA Extractant (32) for alkaline soils. The extract obtained by all three methods can be assayed for its elemental content using an ICP emission spectrometer (33).

Baker (34) has proposed an entirely different soil testing system which employs an equilibrium solution that interacts with the soil being tested. The method has considerable promise and has been developed into a diagnostic approach to soil testing (35). McLean (36) has also suggested a similar approach to soil testing based on chemical equilibration with soil buffer systems. Although both methods have considerable promise, neither method has been adopted for routine use by soil testing laboratories.

There are testing procedures for determining the nitrogen (N) and sulfur (S) content of soils, although most of these tests have limited application. Dahnke and Vasey (37) have reviewed those testing procedures for N, and Tabatabai (38) for S. The other determination that has wide use and application for many soil situations is the test for soluble salts (39).

Although all of these soil testing procedures have been well documented in the literature, slight modifications are frequently made to accommodate their use in the laboratory, modifications that can invalidate the interpretation. The need to specify the exact laboratory methodology required for commonly used soil test methods has resulted in the publication of the HANDBOOK ON REFERENCE METHODS FOR SOIL TESTING (17) as well as three other publications which detail procedures for soil tests used in particular regions of the United States (40,41,42).

TABLE 4 COMMONLY USED PROCEDURES FOR DETERMINING EXCHANGEABLE CATIONS (CA, MG, K AND NA) IN SOILS

Parameter	Method				
	Mehlich No. 1	1N NH$_4$OAc pH 7.0	Mehlich No. 3	Water	AB-DTPA[1]
Adaptability Limits	Sandy soils, acid, low in CEC	Wide range of soils	Wide range of soils	Wide range of soils	Alkaline soils
Sample Size	5 g (4-cm^3)	5 g (4.25-cm^3)	2.5-cm^3	5 g (4.25-cm^3)	10 g (8.5-cm^3)
Volume of Extractant, ml	25	25	25	25	20
Extraction Reagent	0.05N HCl in 0.025N H$_2$SO$_4$	1N NH$_4$OAc at pH 7.0	0.2N HAc in 0.25N NH$_4$NO$_3$ in 0.015N NH$_4$F in 0.013N HNO$_3$ in 0.001M EDTA	Pure water	1M NH$_4$HCO$_3$ 0.005M DTPA at pH 7.6
Shaking Time, minutes	5	5	5	30	15
Shaking Action and Speed	Reciprocating 180+ oscillations/min.	Reciprocating 180+ oscillations/min.	Reciprocating 180+ oscillations/min.	Reciprocating 180+ oscillations/min.	Reciprocating 180+ oscillations/min.
Method of K Determination In Extract	Flame Emission Spectroscopy	Flame Emission Spectroscopy	Flame Emission Spectroscopy	Flame Emission Spectroscopy	Atomic Absorption

TABLE 4 (continued)

Parameter	Method				
	Mehlich No. 1	1N NH4OAc pH 7.0	Mehlich No. 3	Water	AB-DTPA[1]
Range in Soil K conc without Dilution, kg/ha	50-400	50-1000	50-1000	50-500	5-750
Sensitivity, kg/ha	5	5	5	5	1
Primary Reference	Mehlich (23)	Schollenberger & Simon (26)	Mehlich (31)	Doll & Lucas (26)	Soltanpour & Schwab (32)

[1] The AB-DTPA method is not suited for the determination of Ca or Mg.

TABLE 5 SOIL CONDITIONS AND CROPS WHERE MICRONUTRIENT DEFICIENCES MOST OFTEN OCCUR

Micronutrient	Sensitive Crops	Soil Conditions for Deficiency
Boron (B)	alfalfa, clover cotton, peanuts, sugar beet, cabbage	Acid sandy soils low in organic matter, overlimed soils, organic soils
Copper (CU	corn, onions, small grains, watermelon	Organic soils, mineral soils high in pH and organic matter content
Iron (Fe)	citrus, clover, pecan, sorghum, soybean	Leached sandy soils low in organic matter, alkaline soils, soils high in phosphate
Manganese (Mn)	alfalfa, small grains, soybean sugar beet	Leached acid soils, neutral to alkaline soils high in organic matter
Molybdenum (Mo)	alfalfa, cauliflower, soybean	Highly weathered acid soils
Zinc (Zn)	corn, field beans, pecan, sorghum	Leached acid sandy soils low in organic matter, neutral to alkaline soils and/or high in phosphate

326

Excellent references on soil test methods are the book by Jackson
(43) and the manual by Chapman and Pratt (44).

Although there is still need to improve some soil testing
procedures, those methods that have been described will adequately
identify the soil's fertility status in terms of crop production
potential, meeting the criteria as set forth by Fitts and Nelson
(13) for a soil test. The weakest link in the entire soil testing
procedure up to this point is sampling, the ability to obtain a
representative sample for laboratory analysis. The most significant advance has occurred with the analytical procedures used to
determine the elemental content in the extractant from the traditional colorimetric techniques (45) to automated continuous-flow
procedures (46), flame and atomic absorption spectrometry (47) and
up to the latest and most significant, ICP emission spectrometry
(33).

3.4 Interpretation

The interpretation of the soil test result is a subject of vast
interest and considerable controversy (48), but a subject beyond
the scope of this discussion. Those wishing to investigate further
would find the ASA Special Publication No. 29 (49), the book by
Davidescu and Davidescu (7), and the articles by Cope and Rouse
(50), and Hanway (51) of interest.

4. TISSUE TESTING

Tissue testing usually refers to on-site determination of elemental presence and/or concentration in selected fresh plant tissue,
usually petioles, leaf midribs, stems or stalks, using special
testing kits. Krantz, Nelson and Burkhart (52) give instructions
for field testing of corn, cotton and soybean plants using sap
pressed from fresh tissue for the semiquantitative determination of
nitrate (NO_3), phosphate (PO_4) and potassium (K) employing test
papers, vials, reagents and color charts. Wickstrom (5) has also
described the techniques required to conduct tissue tests for
diagnosis of various field crops using similar test procedures.
Syltie, Melsted and Walker (53) provide considerable detail on
their tissue test method for the determination of NO_3, PO_4 and K,
plus magnesium (Mg) and manganese (Mn) in corn and soybean tissues,
providing instructions for reagent preparation and elemental concentration data for interpretative use.

Tissue tests as described above were in frequent use in the
1950s and 60s, but in more recent years have been replaced by plant
analyses. However, there seems to be renewed interest in field
testing, particularly for NO_3. There is an abundance of data on
the critical level of NO_3 in plant tissues for a number of crops,
such as the vegetables (54,55,56,57) and ornamentals (58,59).
Determination of the NO_3 level in cotton leaf petioles (60) and
wheat stems (61) are recent examples of renewed interest in this

test for those crops which have very specific N requirements for
maximum yield achievement.

The NO_3 content of plant tissue can be determined in the field
on sap extracted from fresh plant material, or tissue may be taken
into the laboratory for determination by extraction from either
fresh (62) or dried (80C) and ground plant material (63). In the
field, NO_3 content in plant sap can be determined using either the
Bray's Nitrate Powder (53), or Mercko-Quant Test Strips, a procedure
recently devised by Scaife and Stevens (64) which gives a quantitative determination.

Iron (Fe) is an element that can be determined by means of a
tissue test, a method developed by Bar-Akiva, Maynard and English
(65) for field use. Peroxidase activity is measured by floating
leaf discs in a reactive reagent with the development of a blue
color indicating adequate Fe in the plant tissue.

Tissue tests conducted in the field are usually qualitative in
reaction and semiquantitative in determination. The "reading" of
the test result comes in gained experience from their use. Some
see the advantage for the tissue test in terms of time and savings
as compared to the laboratory conducted plant analysis. The limitations, however, are their general lack of adequate quantification,
reliable interpretative data and fewness of elements determinable.

5. PLANT ANALYSIS

The objective of a plant analysis (sometimes referred to as
leaf analysis) is to determine the essential[1] element status of a
plant at some critical stage of growth. Krantz, Nelson and Burkhart
(52) have given four principle objectives for a plant analysis:
1) to aid in determining the nutrient-supplying power of the
   soil;
2) to aid in determining the effect of treatment on the nutrient
   supply in the plant;
3) to study relationship between the nutrient status of the
   plant and crop performance as an aid in predicting fertilizer
   requirements; and
4) to help lay the foundation for approaching new problems or
   for surveying unknown regions to determine where critical
   plant nutritional experimentation should be conducted.
None of these early stated objectives fits the primary use for plant
analyses today--as a means of diagnosing suspected essential element
deficiencies. Objective number 3 as stated above fits the second
most used objective, primarily for tree fruits and nuts. Munson and
Nelson (66) describe the use of plant analyses for verification of
deficiency symptoms, analysis of normal and abnormal plants, and
crop logging, each one an important application for a plant analysis.

---

1. It is common practice in the literature to refer to an essential element as a "nutrient" or "nutrient element." In this
   volume, the preferred method of expression is essential element.

The successful application of a plant analysis result is based
in the concept of "critical values" which Ulrich (62) defines as
"that range of concentrations at which the growth of the plant is
restricted in comparison to that of plants at a higher nutrient
level." In a later review article, Ulrich (67) discussed the
"physiological bases for assessing the nutrient requirements for
plants," the foundation on which the interpretation of a plant
analysis has been built.

The literature on plant analysis is considerable covering a
long history dating back into the 1800s. There has been considerably more written about plant analysis than soil testing even in
more recent times. Plant analysis has had greater application to
those crops considered horticultural, such as the fruits, nuts,
vegetables and ornamental plants, as compared to field crops, such
as corn, soybeans, alfalfa, wheat, rice, cotton, etc.

A plant analysis is carried out in a series of steps: sampling,
sample preparation, laboratory analysis and interpretation, each
equally important to the success of the technique.

5.1 Sampling

Careful sampling is required in order to obtain the proper
plant tissue for analysis since plant part and time samples will
affect the interpretation of the laboratory analysis result. A
list of typical sampling instructions for several representative
crops is given in Table 6 (68). Those requirements noted for soil
necessary to obtain a representative sample, have equal application
for plant tissue sampling. Plants mechanically or insect damaged,
disease infested, heavily contaminated by dust, or having sustained
a long period of stress, should not be selected for sampling. A
sufficient number of samples must be taken to adequately represent
the area under test. Sets of plant tissue differing in visual
appearance provide a means of comparative evaluation. This is
particularly helpful in the interpretation when dealing with a suspected essential element insufficiency.

Fresh plant tissue should never be placed in air-tight bags
unless kept refrigerated. If more than 24 hours will be required
to deliver the tissue samples to the laboratory, they should be
partially air-dried and placed into clean paper bags.

5.2 Sample Preparation

Decontamination to remove foreign substances may be necessary,
particularly if the tissue is covered with dust or spray materials
that contain elements to be included in the analysis. Washing in
a dilute (2%) detergent solution of the fresh tissue is essential
if iron (Fe) is an element of particular interest. Not washing
tissue that appears to be relatively free of dust and other contam-
inates will not affect the elemental analysis. Oven drying at 80°C
and sample size reduction (grinding) prepares the sample for
laboratory analysis. A thorough discussion of these preparation
procedures has been given by Jones and Steyn (69).

TABLE 6   SAMPLING PROCEDURES FOR COLLECTING PLANT TISSUE FOR ANALYSIS
Source: Jones, Large, Pfleiderer and Klosky (68)

Stage of Growth	Plant Part to Sample	Number of Plants to Sample
FIELD CROPS		
CORN		
(1) Seedling stage (less than 12")	All the above ground portion	20-30
or		
(2) Prior to tasselling	The entire leaf fully developed below the whorl	15-25
or		
(3) From tasselling and shooting to silking	The entire leaf at the ear node (or immediately above or below it)	15-25
Sampling after silking occurs is not recommended.		
SOYBEANS OR OTHER BEANS		
(1) Seedling stage (less than 12")	All the above ground portion	20-30
or		
(2) Prior to or during initial flowering	Two or three fully developed leaves at the top of the plant	20-30
Sampling after pods begin to set not recommended.		
SMALL GRAIN (INCLUDING RICE)		
(1) Seeding stage (less than 12")	All the above ground portion	50-100
or		
(2) Prior to heading	The 4 uppermost leaves	
Sampling after heading not recommended.		

329

TABLE 6 (continued)

Stage of Growth	Plant Part to Sample	Number of Plants to Sample
	FIELD CROPS	
	SUGARCANE	
Up to 4 months old	Third or fourth fully developed leaf from top	15-25
	PEANUTS	
Prior to or at bloom stage	Mature leaves from both the main stem and either cotyledon lateral branch	40-50
	COTTON	
Prior to or at first bloom or when first squares appear	Youngest fully mature leaves on main stem	30-40
	VEGETABLE CROPS	
	POTATO	
Prior to or during early bloom	Third to sixth leaf from growing tip	20-30
	HEAD CROPS (CABBAGE, ETC.)	
Prior to heading	First mature leaves from center of whorl	10-20
	TOMATO (FIELD)	
Prior to or during early bloom stage	Third or fourth leaf from growing tip	20-25

TABLE 6 (continued)

Stage of Growth	Plant Part to Sample	Number of Plants to Sample
	VEGETABLE CROPS	
	TOMATO (GREENHOUSE)	
Prior to or during fruit set	(1) Young plants: leaves adjacent to 2nd and 3rd clusters.	20-25
	(2) Older plants: leaves from 4th to 6th clusters	20-25
	BEANS	
(1) Seedling stage (less than 12")	All the above ground portion	20-30
(2) Prior to or during initial flowering	Two or three fully developed leaves at the top of the plant	20-30
	ROOT CROPS (CARROTS, ONIONS, BEETS, ETC.)	
Prior to root or bulb enlargement	Center mature leaves	20-30
	MELONS (WATER, CUCUMBER, MUSKMELON)	
Early stages of growth prior to fruit set	Mature leaves near the base portion of plant on main stem	20-30
	FRUITS AND NUTS	
	APPLE, APRICOT, ALMOND, PRUNE, PEACH, PEAR, CHERRY	
Mid season	Leaves near base or current year's growth or from spurs	50-100

TABLE 6 (continued)

Stage of Growth	Plant Part to Sample	Number of Plants to Sample
	FRUITS AND NUTS	
	PECAN	
6 to 8 weeks after bloom	Leaves from terminal shoots, taking the pairs from the middle of the leaf	30-45
	WALNUT	
6 to 8 weeks after bloom	Middle leaflet pairs from mature shoots	30-35
	LEMON, LIME	
Mid season	Mature leaves from last flush of growth or non-fruiting terminals	20-30
	ORANGE	
Mid season	Spring cycle leaves, 4 to 7 months old from fruit-bearing terminals.	20-30
	GRAPES	
End of bloom period	Petioles from leaves adjacent to fruit clusters	60-100

5.3 Laboratory Analysis

The organic matter in the prepared green, dried and ground tissue must be destroyed by either wet oxidation or dry ashing, procedures that are well described in the books by Gorsuch (70) and Block (71). However, there is considerable controversy as to which procedure is best in terms of elemental recovery and consistent results. Munter, Halverson and Anderson (72) have studied this problem and recommended a dry ashing procedure followed by acid digestion of the ash. When dry ashing, the temperature should not exceed 500°C, with 8 to 12 hours required for complete oxidation.

The use of various acid mixtures for doing a wet oxidation is discussed by Tolg (73). Wet digestion of plant tissue using block digestors is described by Zasoski and Burau (74), and Halvin and Soltanpour (75), procedures that simplify the method. It should be remembered that boron (B) is lost in wet digestion and sulfur (S) when dry ashing.

The traditional wet chemistry methods (43,45) and even atomic absorption spectroscopy (47) have not been the procedures normally used for the elemental determination in plant tissue digests. It is by emission spectroscopy that the elemental content has been determined using a progression of excitation sources from AC and DC arcs (76), to AC spark (77), and just recently by inductively coupled (33,78,79) or DC (80) plasmas, using either single, scanning or multielement fixed slit spectrometers. High speed multielement capacity, and computer or microprocessor control, are the analytical requirements being sought today by most analysts, and plasma emission spectroscopy meets those requirements for doing plant analyses.

X-ray emission spectroscopy (81) is another technique applicable for the elemental assay of plant tissues. Although the technique of analysis is non-destructive, matrix effects have seriously hampered its acceptance and wide use.

Most of the elements of interest can be determined by one or more of the analytical procedures mentioned above except for N, and also S, unless one is using a vacuum plasma emission spectrometer or X-ray emission spectrometer. Kjeldahl digestion is the usual procedure for N determination, while a number of analytical techniques can be used for determining S in plant tissue.

The Kjeldahl digestion procedure dates back to the 1800s, the first paper published in 1883 (82). Hundreds of papers have been written about the procedure as numerous modifications have been proposed to speed the analysis, and improve accuracy and precision (83). The method is in two steps; first, high temperature (330-350°C) digestion in concentrated sulphuric acid (H_2SO_4) in the presence of a catalyst (Cu, Hg or Se) converting organic N to inorganic ammonium (NH_4); and then, determination of the formed NH_4 by either alkaline steam distillation (84), spectrophotometry (85), or specific-ion electrode (86).

Near infrared reflectant (NIR) spectroscopy looks promising as a non-destructive method of N determination in dried ground plant tissue (87). A comparison of Kjeldahl N with N determined by NIR using a Technicon InfraAlyzer 400 found N concentrations in corn tissues to be within 0.1% of each other for 90% of the samples and within 0.2% for all of the samples (88).

Another recently proposed procedure for N determination in plant tissue is by direct distillation (89), a technique that requires precise control of the distillation process in terms of alkali concentration, time and temperature. The N content is determined indirectly based on the release of amide-N times a factor determined by correlation with total Kjeldahl N.

The S content in plant tissue can be determined following wet oxidation which converts organic S to sulfate (SO_4), the concentration of formed SO_4 determined by either spectrophotometric or turbidimetric procedures (90). The turbidity method can be automated using an Auto-Analyzer as described by Wall, Gehrke and Suzuki (91). Sulfur can be also determined by combustion in a steam of oxygen (92), a procedure described by Jones and Isaac (93) using a LECO Sulfur Analyzer equipped with either a titrator or sulfur dioxide (SO_2) detector.

5.4 Interpretation

The interpretation of a plant analysis is based on the use of either standard values as suggested by Kenworthy (94), or critical values and sufficiency ranges as described by Smith (95) and Ulrich (67). These values or ranges are based on the concept that there is a significant relationship between the elemental content of the plant (or one of its parts) and its growth or yield. This relationship is shown graphically in Figure 1 (95), although for some elements, such as the micronutrients, the left hand side of the response curve is considerably steeper as has been shown by Ulrich and Hills (96).

The sufficiency ranges have been very useful indices for interpreting a plant analysis result. There are a number of excellent sources for these values (9,97,98,99,100,101,102), the Chapman book (98) being the most complete listing.

The other concept for plant analysis interpretation is the Diagnosis and Recommendation Integrated System, frequently referred to as DRIS which was developed by Beaufils (103). An excellent review on the technique and its application has been published by Summer (104), as well as a computer program for calculating DRIS indices (105).

Although plant analysis has had its most significant use as a diagnostic tool (confirming the existence of possible essential element insufficiencies), its most useful role for "tracking" the essential element status of plants has yet to be fully utilized by farmers. The usefulness of this application is discussed by Clements (106), and Jones, Pallas and Stansell (107).

FIGURE 1

GENERAL RELATIONSHIP BETWEEN PLANT GROWTH OR YIELD AND ELEMENTAL
CONTENT OF THE PLANT

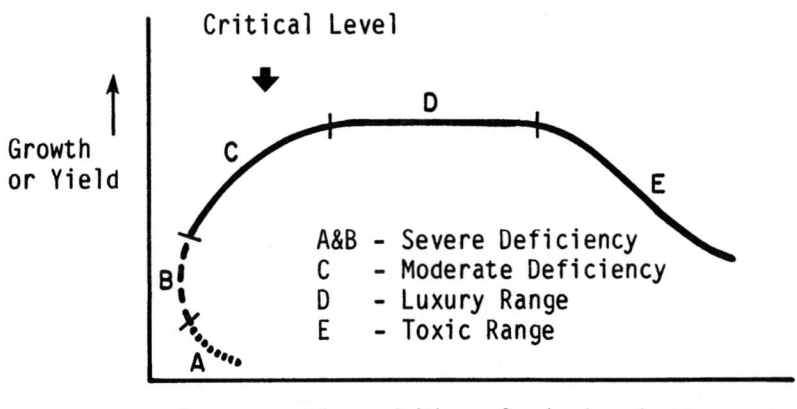

Source: Smith (95)

6. USE OF SOIL TESTS AND PLANT ANALYSES

Soil tests and plant analyses should be used as a matter of routine by farmers to evaluate their soil/crop essential element status. Unfortunately, both procedures are considerably under utilized in much of the developed world and used very little in much of the under developed world. The former is a matter of education, the latter due primarily to a lack of testing facilities and laboratories. A new turnkey laboratory, AGROLAB (108), has been proposed to solve the lack of facility problem, while a different approach for the use of soil tests and plant analyses as proposed by Jones and Budzynski(109) with more details given by Jones (110), is needed to fully utilize the test results effectively. Their approach to the incorporation of soil test and plant analysis data into the crop production plan is illustrated in Figures 2 and 3. The objectives for soil testing and plant analyses, as described by Jones and Budzynski (109), would be to bring the soil-plant system into proper balance and then maintain it. Both test procedures then serve as the means of monitoring the soil-plant system, quite a different concept than what was described in a recent issue of the magazine, BETTER CROPS (111).

The under utilization of both soil tests and plant analyses is of considerable concern as the productivity of many of the world's cropland soils is decreasing. The demand today is to increase the per unit area production with the hope of stopping the ever expanding cultivation of marginal soils which results in the loss of forests, increased soil erosion and a change in the ecological balance. The intelligent and proper use of chemical fertilizers demands the requirement to determine the need-based chemical tests of the soil/crop system in order to avoid the hazards of under and over utilization.

The cost and time factors associated with chemical testing, and the requirement for a sophisticated laboratory have been factors of concern, particularly in the under developed world as well as the Western world. Some see field testing using portable soil testing and plant analysis kits, such as the HACH kits[2], is a partial resolution, although their range of application is quite limited as was discussed earlier. Presently, there are no other techniques that can replace the traditional soil testing and plant analysis methods as they have been described in this paper. The use of electrical and electrochemical procedures to monitor ion movement in the soil or plant have not been investigated for field application, although these techniques have useful application for research when dealing with single plants, plant parts or cells as was discussed in considerable detail at this conference. It is doubtful that any of these research techniques will find their way into the field in the near future. Therefore, we are left with those methods of soil testing and plant analysis that have been proven to be effective if properly utilized, and solutions to their under utilization have been offered.

---

2. HACH Chemical Company, P.O. Box 907, Ames, IA 10050.

FIGURE 2

PRIMARY STEPS FOR ACHIEVING ESSENTIAL ELEMENT SUFFICIENCY

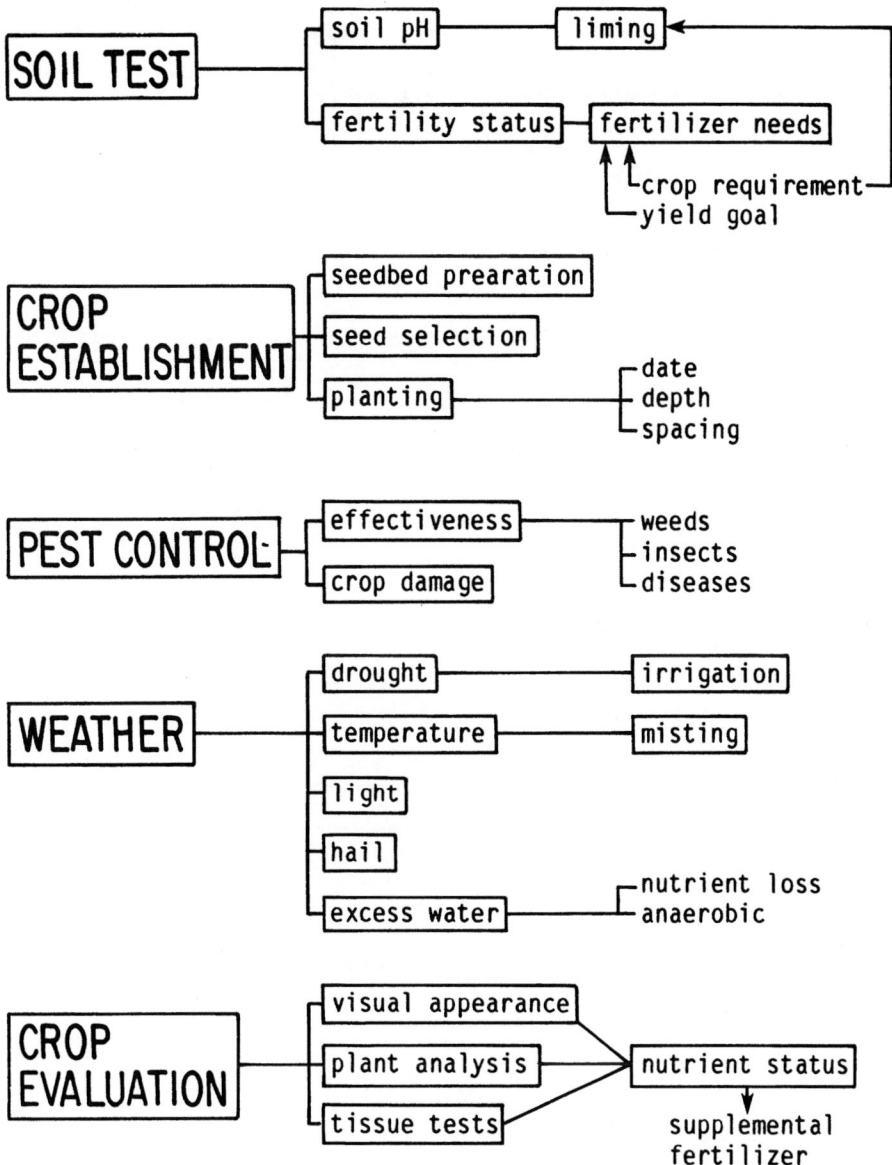

338

FIGURE 3

EVALUATION PROCEDURE FOR ACHIEVING ESSENTIAL ELEMENT SUFFICIENCY

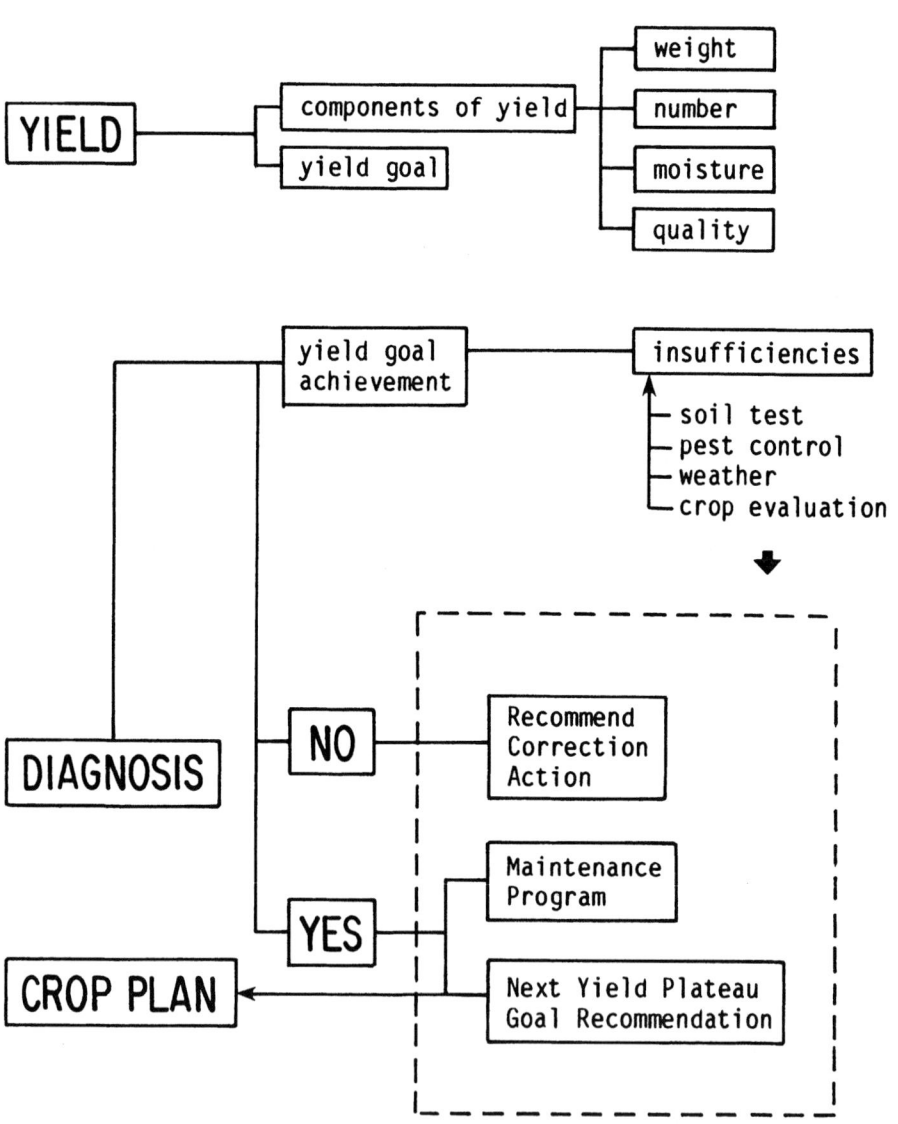

REFERENCES

1. The Global 2000 Report to the President: Entering the Twenty-First Century, Vol. 1. The U.S. Government Printing Office, Wshington, D.C. A/NO41-011-003-8, 1980.

2. Dudal, R. Inventory of the major soils of the World Special Reference to Mineral Stress Hazards. In: Plant Adaptation to Mineral Stress in Problem Soils (M.J. Wright, ed.) Cornell University Press, Ithaca, NY, pp. 3-14, 1976.

3. Fertilizer Manual. International Fertilizer Development Center, Muscle Shoals, AL, 1979.

4. Melsted, S.W. and Peck, T. R. The principles of soil testing. In: Soil Testing and Plant Analysis (L.M. Walsh and J.D. Beaton, eds.) Soil Science Society of America, Madison, WI, pp. 13-22, 1978.

5. Wickstom, G.A. Use of tissue testing in field diagnosis. In: Soil Testing and Plant Analysis: Plant Analysis, Part II (G.W. Hardy, ed.) Special Publication No. 2, Soil Science Society of America, Madison, WI, pp. 109-112, 1967.

6. Munson, R.D. and Nelson, W.L. Principles and practices in plant analysis. In: Soil Testing and Plant Analysis (L.M. Walsh and J.D. Beaton, eds.) Soil Science Society of America, Madison, WI, pp. 223-248, 1973.

7. Davidescu, D. and Davidescu, V. Evaluation of Fertility by Plant and Soil Analysis. Editura Academiei, Bucuresti, Romania, 1982.

8. Sprague, H.E., ed. Hunger Signs in Crops. David McKay Company, New York, NY, 1949.

9. Wallace, T. The Diagnosis of Mineral Deficiencies in Plants by Visual Symptoms. Chemical Publishing Company, NY, 1961.

10. Robinson, J.B.D., ed. Diagnosis of Mineral Disorders in Plants, Volume 1, Principles. Chemical Publishing Company, NY, 1984.

11. Robinson, J.B.D., ed. Diagnosis of Mineral Disorders in Plants, Vegetables, Volume 2. Chemical Publishing Company, NY, 1984.

12. Nelson, W.L., Fitts, J.W., Kardos, L.T., McGeorge, W.T., Parks, R.Q. and Fielding Reed, J. Soil Testing in the United States. National Soil & Fertilizer Research Committee, U.S. Government Printing Office, Publication No. O-979953, Washington, D.C., 1951.

340

13. Fitts, J.W. and Nelson, W.L. The determination of lime and fertilizer requirements of soils through chemical tests. In: Advances in Agronomy, Vol. 8 (A.G. Norman, ed.) Academic Press Inc., New York, NY, pp. 242-282, 1956.

14. Peck, T.R. and Melsted, S.W. Field sampling for soil testing. In: Soil Testing and Plant Analysis (W.L. Walsh and J.D. Beaton, eds.) Soil Science Society of America, Madison, WI, pp. 67-76, 1973.

15. Wolf, B. An improved universal extracting solution and its use for diagnosing soil fertility. Communi. in Soil Testing and Plant Analysis 13:1005-1022, 1982.

16. McLean, E.O. Soil pH and lime requirement. In: Methods of Soil Analysis, Part 2, Chemical and Microbiological Properties, 2nd Edition (A.L. Page, ed.) American Society of Agronomy, Madison, WI, pp. 199-224, 1982.

17. Jones, J.B., Jr., ed. Handbook on Reference Methods for Soil Testing. Council on Soil Testing and Plant Analysis, Athens, GA, 1980.

18. McLean, E.O., Trierweiler, J.E. and Eckert, D.J. Improved SMP buffer method for determining lime requirement of acid soils. Communi. in Soil Sci. Plant Analysis 8:667-675, 1977.

19. Adams, F. and Evans, C.E. A rapid method for measuring lime requirement of red-yellow podzolic soils. Soil Science Society American Proceedings 26:355-357, 1962.

20. Thomas, G.W. and Peaslee, D.E. Testing soils for phosphorus. In: Soil Testing and Plant Analysis (L.M. Walsh & J.D. Beaton, eds.) Soil Science Society of America, Madison, WI, pp. 115-132, 1973.

21. Olsen, S.R. and Sommers, L.E. Phosphorus. In: Methods of Soil Analysis, Part 2, Chemical and Microbiological Properties, 2nd Edition (A.L. Page, ed.) American Society of Agronomy, Madison, WI, pp. 403-430, 1982.

22. Bray, R.H. and Kurtz, L.T. Determination of total, organic and available forms of phosphorus in soils. Soil Science 59:39-45, 1945.

23. Mehlich, A. Determination of P, Ca, Mg, K, Na and NH Mineo. North Carolina Soil Testing Division, Raleigh, NC, 1953.

24. Olsen, S.R., Cole, C.V., Watanabe, F.S. and Dean, L.A. Estimation of Available Phosphorus in Soils by Extraction with Sodium Bicarbonate. USDA Circular No. 939.

25. Schollenberger, C.J. and Simon, R.H. Determination of exchange capacity and exchangeable bases in soil-ammonium acetate method. Soil Science 59:13-24, 1945.

26. Doll, E.C. and Lucas, R.E. Testing soils for potassium, calcium and magnesium. In: Soil Testing and Plant Analysis (L.M. Walsh and J.D. Beaton, eds.) Soil Science Society of America, Madison, WI, pp. 133-152, 1976.

27. Berger, K.C. and Troug, E. Boron tests and determination for soils and plants. Soil Science 57:25-36, 1944.

28. Grigg, J.L. Determination of available soil molybdenum. New Zealand News No. 3:37-40, 1953.

29. Lindsay, W.L. and Norwell, W.A. Equilibrium relationships of Zn, Fe, Ca, and H with EDTA and DTPA in soils. Soil Science Society American Proceedings 33:62-68, 1969.

30. Cox, F.R. and Kamprath, E.J. Micronutrient soil tests. In: Micronutrients in Agriculture (J.J. Mortvedt, P.M. Giordano and W.L. Lindsay, eds.) Soil Science Society of America, Madison, WI, pp. 289-318, 1972.

31. Mehlich, A. Comprehensive methods in soil testing. In: Proceedings Soil and Plant Analysis Seminar, Council on Soil Testing and Plant Analysis, Athens, GA, 1982.

32. Soltanpour, P.N. and Schwab, A.P. A new soil test for simultaneous extraction of macro- and micro-elements in alkaline soils. Communi. in Soil Sci. and Plant Analysis 8:195-207, 1977.

33. Soltanpour, P.N., Jones, J.B., Jr. and Workman, S.M. Optical emission spectrometry. In: Methods of Soil Analysis, Part 2, Chemical and Microbiological Properties, 2nd Edition (A.L. Page, ed.) American Society of Agronomy, Madison, WI, pp. 29-66, 1982.

34. Baker, D.E. A new approach to soil testing. Soil Science 112:381-391, 1971.

35. Baker, D.E. and Amacher, M.C. The development and interpretation of a diagnostic soil testing program. Bulletin 826, Pennsylvania Agriculture Experiment Station, College Station, PA, 1981.

36. McLean, E.O. Chemical equilibrations with soil buffer systems as bases for future soil testing procedures. Communi. in Soil Sci. and Plant Analysis 13:411-433, 1982.

37. Dahnke, W.C. and Vasey, E.H. Testing soils for nitrogen. In: Soil Testing and Plant Analysis (L.M. Walsh and J.D. Beaton, eds.) Soil Science Society of America, Madison, WI, pp. 97-114, 1973.

38. Tabatabai, M.A. Sulfur. In: Methods of Soil Analysis, Part 2, Chemical and Microbiological Properties, 2nd Edition (A.L. Page, ed.) American Society of Agronomy, Madison, WI, pp. 501-538, 1982.

39. Rhoades, J.D. Soluble salts. In: Methods of Soil Analysis, Part 2, Chemical and Microbiological Properties, 2nd Edition (A.L. Page, ed.) American Society of Agronomy, Madison, WI, pp. 167-180, 1982.

40. Dahnke, W.C., ed. Recommended Chemical Soil Test Procedures for the North Central Region, Bulletin No. 499, revised, North Dakota Agricultural Experiment Station, Fargo, ND, 1980.

41. Issac, R.A., ed. Reference Soil Test Methods for the Southern Region of the United States, Southern Cooperative Series Bulletin 289, Georgia Agricultural Experiment Station, Athens, GA, 1983.

42. Quick, J., ed. California Soil Testing Procedures. California Fertilizer Association, Sacramento, CA, 1980.

43. Jackson, M.L. Soil Chemical Analysis. Prentice-Hall, Englewood, NJ, 1958.

44. Chapman, H.D. and Pratt, P.F. Methods of Analysis for Soils, Plants, and Waters, Publication No. 4034, Agricultural Sciences Publications, Berkeley, CA, 1982.

45. Piper, C.S. Soil and Plant Analysis. Hassell Press, Adelaide, Australia, 1942.

46. Flannery, R.L. and Markus, D.K. Determination of phosphorus, potassium, calcium, and magnesium simultaneously in northern Carolina, ammonium acetate, and Bray P1 soil extracts by autoanalyzer. In: Instrumental Methods for Analysis of Soils and Plant Tissue (L.M. Walsh, ed.) Soil Science Society of America, Madison, WI, pp. 97-112, 1971.

47. Issac, R.A. and Kerber, J.C. Atomic absorption and flame photometry: techniques and uses in soil, plant, and water analysis. In: Instrumental Methods for Analysis of Soils and Plant Tissue (L.M. Walsh, ed.) Soil Science Society of America, Madison, WI, pp. 17-38, 1971.

48. Jones, J.B., Jr. Soil test works when used right. _Solutions_ _27_:25-37, 1983.

49. Peck, T.R., ed. _Soil Testing: Correlating and Interpreting the Analytical Results_. ASA Special Publication No. 39 American Society of Agronomy, Madison, WI, 1977.

50. Cope, J.T., Jr. and Rouse, R.D. Interpretation of soil test results. In: _Soil Testing and Plant Analysis_ (L.M. Walsh and J.D. Beaton, eds.) Soil Science Society of America, Madison, WI, pp. 35-54, 1973.

51. Hanway, J.J. Experimental methods for correlating and calibrating soil tests. In: _Soil Testing and Plant Analysis_ (L.M. Walsh and J.D. Beaton, eds.) Soil Science Society of America, Madison, WI, pp. 55-66, 1973.

52. Krantz, B.A., Nelson, W.L. and Burkhart, L.F. Plant-tissue tests as a tool in agronomic research. In: _Diagnostic Techniques for Soils and Crops_ (H.B. Kitchen, ed.) The American Potash Institute, Washington, D.C., pp. 137-156, 1948.

53. Syltie, P.W., Melsted, S.W. and Walker, W.H. Rapid tissue tests as indicators of yield, plant composition, and soil fertility for corn and soybean. _Communi. in Soil Sci. and Plant Analysis_ _3_:37-49, 1972.

54. Lorenz, O.A. and Bartz, J.F. Fertilization for high yields and quality of vegetable crops. In: _Changing Patterns in Fertilizer Use_ (L.B. Nelson, ed.) Soil Science Society of America, Madison, WI, pp. 327-352, 1968.

55. El-Sheikh, A.M., El-Hakam, A. and Ulrich, A. Critical nitrate levels of sqsash, cucumber and melon plants. _Communi. in Soil Sci. and Plant Analysis_ _1_:63-74, 1970.

56. Maynard, D.N., Barker, A.V., Minotti, P.L. and Pack, N.H. Nitrate accumulation in vegetables. In: _Advances in Agronomy_, Vol. 28 (N.C. Brady, ed.) Academic Press, Inc., NY, pp. 71-118, 1976.

57. Lorenz, O.A. and Tyler, K.B. Plant tissue analysis of vegetable crops. In: _Soil and Plant Tissue Testing in California_ (H.M. Reisnenaur, ed.) California Cooperative Extension Bulletin 1879, University of California, Berkeley, CA, 1978).

58. Woodson, W.R. and Boodley, J.W. Petiole nitrate concentration as an indicator of geranium nitrogen. _Communi. in Soil Sci.and Plant Analysis_ _14_:363-372, 1983.

59. Prasad, M. and Spiers, T.M. Evaluation of a simple sap nitrate test for some ornamental crops. In: Plant Nutrition 1982, Proceedings of the 9th International Plant Nutrition Colloquium (A. Scaife, ed.) Commonwealth Agricultural Bureaux, Slough, England, pp. 474-479, 1982.

60. Sabbe, W.E. and MacKenzie, A.J. Plant analysis as an aid to cotton fertilization. In: Soil Testing and Plant Analysis (L.M. Walsh and J.D. Beaton, eds.) Soil Science Society of America, Madison, WI, pp. 289-314, 1972.

61. Papastylianou, I., Graham, R.D. and Puckridge, D.W. The diagnosis of nitrogen deficiency in wheat by means of a critical nitrate concentration in stem bases. Communi. in Soil Sci. and Plant Analysis 13:473-485, 1982.

62. Ulrich, A. Plant analysis. In: Diagnostic Techniques for Soils and Crops (H.B. Kitchen, ed.) American Potash Institute, Wshington, D.C., pp. 157-198, 1948.

63. Baker, A.S. and Smith, R. Extracting solution for potentiometric determination of nitrate in plant tissue. J. Agri. and Food Chem. 17:1284-1287, 1969.

64. Scaife, A. and Stevens, K.L. Monitoring sap nitrate in vegetable crops: Comparison of test strips with electrode methods and effects of time of day and leaf position. Communi. in Soil Sci. Plant Analysis 14:761-771, 1983.

65. Bar-Kiva, A., Maynard, D.N. and English, J.E. A rapid tissue test for diagnosing iron deficiencies in vegetable crops. Hortscience 13:284-285, 1978.

66. Munson, R.D. and Nelson, W.L. Principles and practices in plant analysis. In: Soil Testing and Plant Analysis (L.M. Walsh and J.D. Beaton, eds.) Soil Science Society of America, Madison, WI, pp. 223-248, 1973.

67. Ulrich, A. A physiological base for assessing the nutritional requirements of plants. Ann. Rev. of Plant Physiol. 3:207-228, 1952.

68. Jones, J.B., Jr., Large, R.L., Pfleiderer, D.R. and Klosky, H.S. How to properly sample for a plant analysis. Crops & Soils 28: 114-120, 1971.

69. Jones, J.D., Jr. and Steyn, W.J.A. Sampling, handling, and analyzing plant tissue samples. In: Soil Testing and Plant Analysis (L.M. Walsh and J.D. Beaton, eds.) Soil Science Society of America, Madison, WI, pp. 249-270, 1973.

70. Gorsuch, T.T. Destruction of organic matter. International Series of Monographs in Analytical Chemistry, Vol. 39, Pergamon Press, New York, NY, 1970.

71. Block, R.A. Handbook of Decomposition Methods in Analytical Chemistry. John Wiley & Sons, New York, NY, 1979.

72. Munter, R.C., Halverson, J.A. and Anderson, L.M. Quality assurance for plant analysis by ICP-AES. Communi. in Soil Sci. Plant Analysis 15, (in press).

73. Tolg, G. The basis of trace elements. In: Methodocium Chimicum, Vol. 1, Analytical Method, Part B: Micromethods, Biological Methods, Quality Control, Automation (F. Korte, ed.) Academic Press, Inc., New York, NY, 1974.

74. Zasoski, R.J. and Burau, R.G. A rapid nitric-perchloric acid digestion method for multi-element tissue analysis. Communi. Soil Sci. Plant Analysis 8:425-436, 1977.

75. Halvin, J.L. and Soltanpour, P.N. A nitric acid plant tissue digest method for use with inductively-coupled plasma spectrometry. Communi. Soil Sci. Plant Analysis 11:969-980, 1980.

76. Mitchell, R.L. The Spectrographic Analysis of Soils, Plants, and Related Materials. Technical Communication 44, Commonwealth Bureau of Soils, Harpenden, Herts, England, 1956.

77. Jones, J.B., Jr. Elemental analyses of biological substances by direct-reading spark emission spectroscopy. Amer. Lab. 8: 15-20, 1976.

78. Dahlquist, R.L. and Knoll, J.W. Inductively-coupled plasma atomic emission spectroscopy: Analysis of biological materials and soil for major, trace, and ultra-trace elements. Applied Spectroscopy 32:1-30, 1978.

79. Munter, R.C. and Grande, R.A. Plant tissue and soil extract analysis by ICP-atomic emission spectrometry. In: Developments in Atomic Plasma Spectro-Chemical Analysis (R.M. Barnes, ed.) Heyden & Son Ltd., London, England, pp. 653-672, 1981.

80. DeBolt, D.C. Multi-element emission spectroscopic analysis of plant tissue using DC agron plasma source. J. Assoc. Official Analytical Chemists 63:802-805, 1982.

81. Kubota, J. and Lazar, V.A. X-Ray emission spectrograph: Techniques and uses for plant and soil studies. In: Instrumental Methods of Analysis of Soils and Plant Tissue (L.M. Walsh, ed.) Soil Science Society of America, Madison, WI, pp. 67-82, 1971.

82. Morries, P. A century of Kjeldahl (1883-1983). J. Assoc. Public Analysts 21:53-58, 1983.

83. Nelson, D.W. and Sommers, L.E. Total nitrogen analysis of soil and plant tissues. J. Assoc. Official Analytical Chemists 63:770-778, 1980.

84. Horwitz, W., ed. Official Methods of Analysis of the Association of Official Analytical Chemists, Section 2.057. Assoc. Official Analytical Chemists, Arlington, VA, 1980.

85. Isaac, R.A. and Johnson, W.C. Determination of total nitrogen in plant tissue. J. Assoc. Official Analytical Chemists 59:98-100, 1976.

86. Gallaher, R.N., Weldon, C.O. and Bowell, F.C. A semi-automated procedure for total nitrogen in plant and soil samples. Soil Sci. Soc. Amer. Proc. 40:887-889, 1976.

87. Dorsheimer, W.T. and Isaac, R.A. Application of NIR analysis. Amer. Lab. 14:58-63, 1982.

88. Issac, R.A. and Johnson, W.C. Determination of protein nitrogen in plant tissue using near infrared spectroscopy. J. Assoc. Official Analytical Chemists 66:506-509, 1983.

89. Anonymous. Rapid determination of protein content in grains by using the Kjeltec auto system DD. Tecator Application Note AN 33/81, Tecator, Inc., Hernodon, VA, 1983.

90. Beaton, J.D., Burns, G.R. and Platou, J. Determination of sulphur in soils and plant material. Technical Bulletin 14, The Sulphur Institute, Washington, D.C., 1968.

91. Wall, L.L., Gehrke, C.W. and Suzuki, J. Automated turbidimetric determination of sulfate sulfur in soils and fertilizers and total sulfur in plant tissues. J. Assoc. Official Analytical Chemists 63:845-853, 1980.

92. Bremner, J.M. and Tabatabai, M.A. Use of automated combustion techniques for total carbon, total nitrogen, and total sulfur analysis of soils. In: Instrumental Methods for Analysis of Soils and Plant Tissue (L.M. Walsh, ed.) Soil Science Society of America, Madison, WI, pp. 1-16, 1971.

93. Jones, J.B. and Isaac, R.A. Determination of sulfur in plant material using a LECO sulfur analyzer. J. Agri. Food Chem. 20:1292-1294, 1972.

94. Kenworthy, A.L. Interpreting the balance of nutrient-elements in leaves of fruit trees. In: Plant Analysis and Fertilizer Problems, Publication No. 8 (W. Reuther, ed.) American Institute of Biological Science, Washington, D.C., pp. 28-43, 1960.

95. Smith, P.F. Mineral analysis of plant tissues. Ann. Rev. Plant Physiol. 13:81-108, 1962.

96. Ulrich, A. and Hills, F.J. Plant analysis as an aid in fertilizing sugar crops. In: Soil Testing and Plant Analysis (L.M. Walsh and J.D. Beaton, eds.) Soil Science Society of America, Madison, WI, pp. 271-288, 1973.

97. Goodall, D.W. and Gregory, F.G. Chemical composition of plants as an index to their nutritional status. Technical Publication 17, Imperial Bureau of Horticulture and Plantation Crops, Penglais, Aberyswyth, Wales, 1947.

98. Chapman, H.D. Plant analysis values suggestive of nutrient status of selective crops. In: Soil Testing and Plant Analysis, Part II, Plant Analysis (G.W. Hardy, ed.) Special Publications Series 2, Soil Sci. Soc. of Amer., Madison, WI, pp.77-92, 1976.

99. Neubert, P., Wrazidlo, W., Vielemeyer, H.P., Hundt, I., Gullmick, F. and Bergmann, W. Tabellen zur pflanzenernalzre-erste orientierende ubersicht. Institut fur Pflanzerenahrung Jerna, Berlin, Germany, 1969.

100. Jones, J.B., Jr. Plant Analysis Handbook for Georgia. Georgia Cooperative Extension Bulletin 735, University of Georgia, Athens, GA, 1974.

101. Childers, N.F., ed. Temperate to Tropic Fruit Nutrition. Horticultural Publications, Rutgers, New Brunswick, NY, 1966.

102. Reisenauer, H.M., ed. Soil and Plant Tissue Testing in California. California Cooperative Extension Bulletin 1879, University of California, Berkeley, CA, 1978.

103. Beaufills, E.R. Diagnosis and Recommendation Integrated System (DRIS). Soil Science Bulletin 1, University of Natal, South Africa, 1973.

104. Sumner, M.E. Use of the DRIS system in foliar diagnosis of crops at high yields. Communi. Soil Sci. Plant Analysis 8:251-268, 1977.

105. Letzch, W.S. and Sumner, M.E. Computer program for calculating DRIS indices. Communi. Soil Sci. Plant Analysis 14:811-815, 1983.

106. Clements, H.F. Crop logging of sugar cane in Hawaii. In: Plant Analysis and Fertilizer Problems (W. Reuther, ed.) Publication No. 8, American Institute of Biological Sciences, Washington, D.C., pp. 131-147, 1960.

107. Jones, J.B., Jr., Pallas, J.E., Jr. and Stansell, J.R. Tracking the elemental content of leaves and other plant parts of the peanut under irrigated culture in the sandy soils of south Georgia. Communi. Soil Sci. Plant Analysis 11:81-92, 1980.

108. Jones, J.B., Jr. A Turnkey laboratory concept for agricultural testing. Amer. Lab. 16:64-72, 1984.

109. Jones, J.B., Jr. and Budzynski, W.W. A Professional Seminar on Agronomics, Benton Laboratories, Inc., Athens, GA, 1980.

110. Jones, J.B., Jr. Soil test works when used right, Part 2. Solution 27:61-70, 1983.

111. Armstrong, D., ed. The Diagnostic Approach. Better Crops 68: 1-39, 1984.

THE DEVELOPMENT OF ELECTROCHEMICAL SENSORS

W. John Albery, Barry G.D. Haggett, L. Robert Svanberg

Imperial College
London, England

1. INTRODUCTION

Our text is taken from Experiment 7.01 of the Physical Chemistry Course in the University of Oxford: "The currents measured will therefore be distinctly fluctuating and by no means exactly reproducible." In this experiment, the wretched undergraduate was required to immerse two copper rods in a solution of copper sulphate, apply an electrode potential and measure the current as a function of time. It was an extremely successful experiment, since it had been designed by spectroscopists to convince generations of Oxford students that nothing good could be expected from electrochemistry. The only result obtained from the experiment was the general Albery equation

$$i = f(N_W, L, \partial JD/\partial t) \tag{1}$$

where i is the current, N_W is the number of windows open in the Laboratory, L the number of lorries (trucks) passing by outside and $\partial JD/\partial t$ is a complicated periodic function describing the supervision of a typical junior demonstrator. Equation (1) shows that the current at an electrode is not well defined unless we control not only the electrode potential but also the supply of electrode reactant to the electrode surface.

2. POTENTIOMETRIC AND AMPEROMETRIC SENSORS

Figure 1 shows the shape of a typical current-voltage curve. Near the point of zero current the variation of the current with potential is steep; typically a change of only 60 mV can alter the current by a factor of 10.

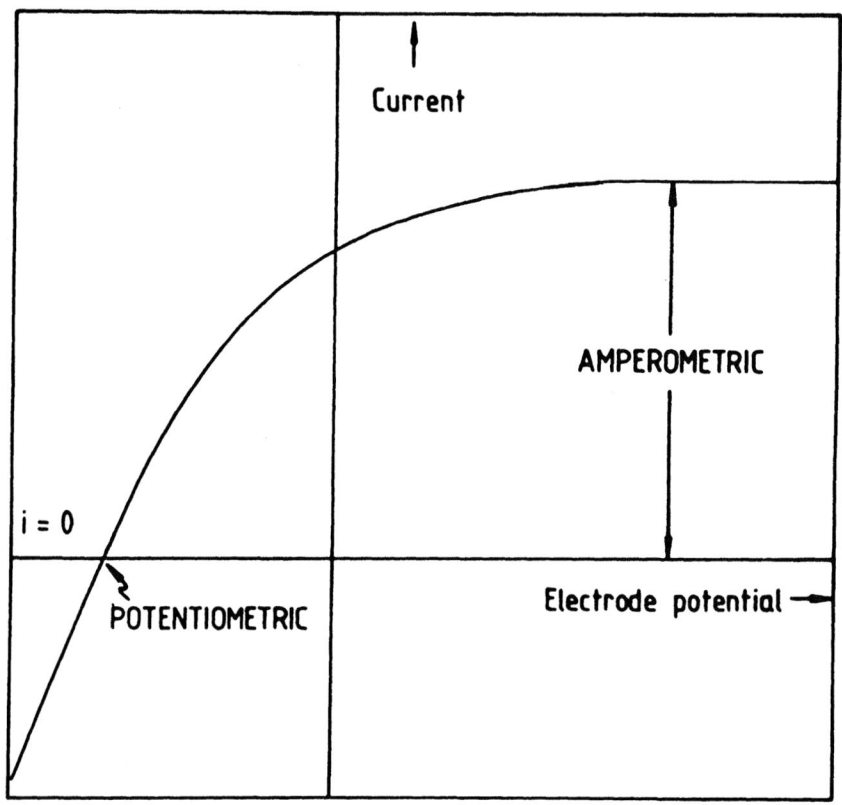

Figure 1.   A typical current voltage curve showing potentiometric
            measurement at the zero current point and amperometric
            measurement of the transport limited current.

An important feature of electrochemistry is that the equilibrium
and rate constants of electrode processes are so sensitivie to the
electrode potential. For a one electron reaction, a change of one
volt will alter the equilibrium constant by 10^{17} and the rate constant by 10^8. Hence we can use the electrode potential to select
the electrochemical reaction of our choice.

     Returning to Figure 1, as more potential is applied, the cur-
rent voltage curve flattens out and the "transport limited current"
is reached on the plateau. Here the electrode is so active that
every molecule of reactant which reaches the electrode surface is
destroyed; the current is then determined by the transport of
reactant to the electrode.

     Electrochemical sensors can operate on two parts of the current
voltage curve. In the first type of sensor, the "potentiometric"
sensor, the electrode potential is measured at the point of zero
current. This has the advantage that there is no net consumption of
reactant. Hence the.mass transport is unimportant. However this
type of sensor has two great disadvantages. First, sensible

information will only be obtained if there is local thermodynamic equilibrium at the electrode interface. This requires that the electrode kinetics be rapid with a standard electrochemical rate constant greater than 10^{-2} cm s^{-1}. This is a severe requirement. Compared to the limited number of systems that meet this requirement, there are many more electrochemical systems which are kinetically controlled. The second disadvantage of a potentiometric sensor arises from the exponential dependence of the current on the voltage which gives rise to the steep part of the current voltage curve in Figure 1. With a relation for the concentration, c, of the type

$$\ln(c) = \text{constant} + nEF/RT \tag{2}$$

we give in Table 1 the error in c that will be caused by an error, ΔE, in E. It can be seen that quite small errors cause significant errors in the measurement of concentration. The logarithmic relation in equation (2) does, however, have the advantage of giving a good dynamic range to the sensor.

TABLE 1

Errors in Concentration for Potentiometric Sensors [a]

n = 1		n = 2	
ΔE/mV	% Error	ΔE/mV	% Error
2	3	2	7
5	9	5	19
10	19	10	40

[a] Calculated at 25°C from equation (2).

In the second type of sensor, the "amperometric" sensor, the current is measured on the pleateau of the current voltage curve in Figure 1. Here the current is insensitive to the electrode potential, and so the control of the electrode potential can be relatively crude. On the other hand, since the current is determined by the transport of the reactant from the bulk of the solution, that transport has to be defined and controlled. We shall describe various methods of achieving the desired control. For these sensors the transport limited current is directly proportional to the concentration of the target species. Hence there is no equivalent problem to that for potentiometric sensors, where the exponential relation in equation (2) magnifies the errors. In Table 2 we summarize the advantages and disadvantages of the two types of sensor. As a rule of thumb, both types of sensor have much the same sensitivity and can measure concentrations down to about 10^{-6} mol dm^{-3}.

TABLE 2

Comparison of Potentiometric and Amperometric Sensors

	Potentiometric	Amperometric
Method of operation	Measure potential at $i = 0$	Measure transport limited current
Electrode kinetics	Must be fast	Electrode potential can drive reaction
Response	c is exponential function of EF/RT giving good dynamic range but making c sensitive to errors in measurement of E	c is linear function of current giving normal dynamic range and normal response to errors in measurement of i
Mass transport	Unimportant	Must be controlled
Sensitivity	$\approx 10^{-6}$ mol dm^{-3}	$\approx 10^{-6}$ mol dm^{-3}

3. MEMBRANE ELECTRODES

The first type of electrochemical sensor we wish to discuss is the membrane electrode invented by Clark (2) for the determination of O_2. This type of sensor, illustrated in Figure 2, is an amperometric sensor. The electrochemistry all takes place behind a thin plastic membrane. The membrane may be made of such materials as teflon, silicone rubber or polypropylene. (In our early work very satisfactory membranes were supplied by the London Rubber Company.) Small neutral gas molecules are sufficiently soluble in the membrane to diffuse through it and into the thin film of electrolyte. The electrode is set at a potential, such that every molecule of the target gas reaching the electrode is destroyed. The mass transport of the gas to the electrode is controlled by its diffusion through the membrane. A great advantage of this type of electrode is that the membrane separates the dirty external world from the pristine interior where the electrochemistry takes place. This prevents poisoning of the electrode by macromolecules when the whole assembly is exposed to biological fluids such as blood or sewage.

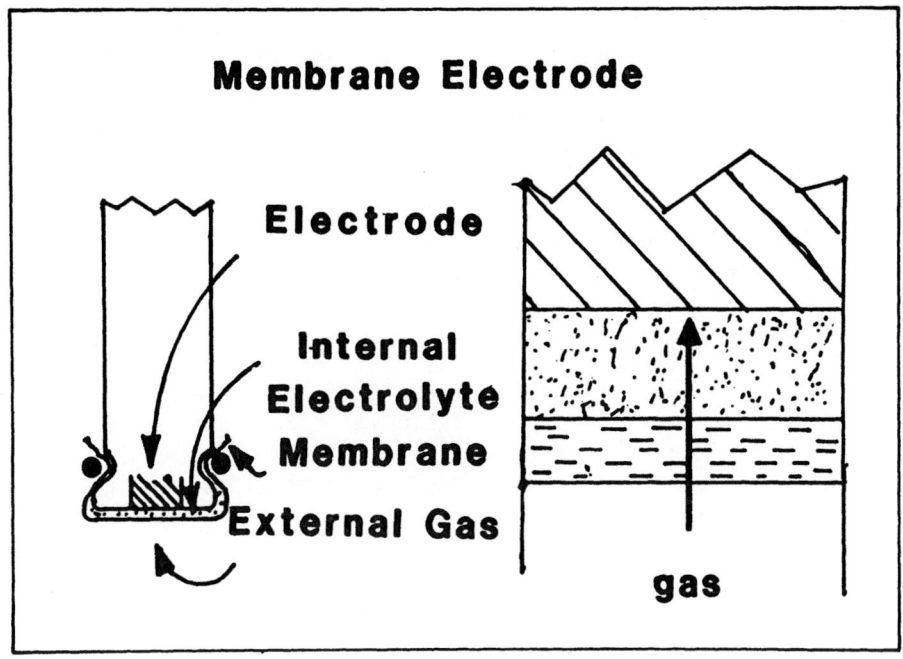

Figure 2. A Clark membrane electrode for gas determination.

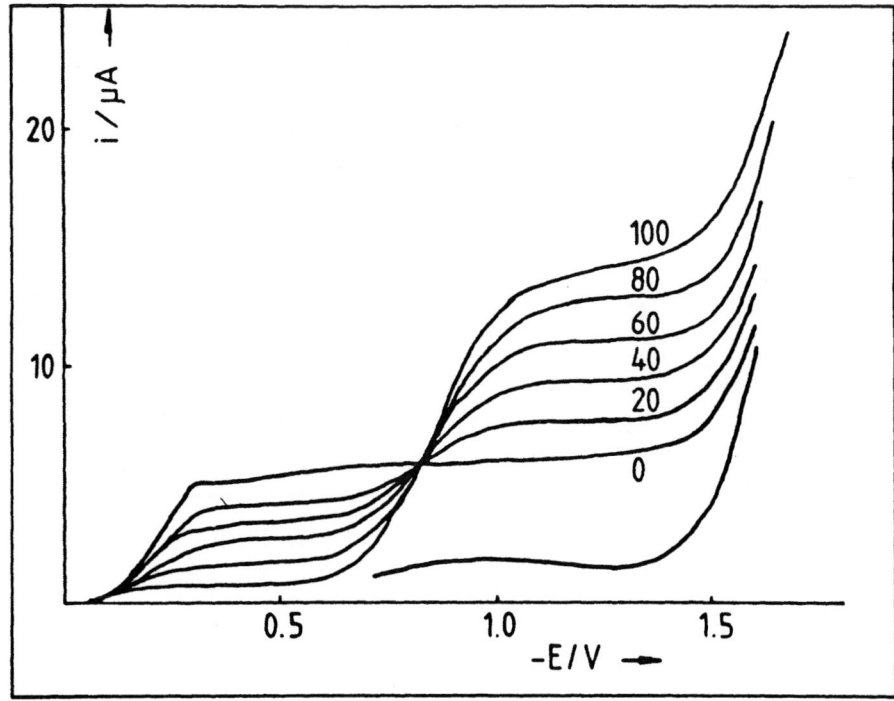

Figure 3. Determination of mixtures of O_2 and N_2O on a silver cathode. Each current voltage curve is labeled with the % of N_2O. The first wave is caused by the O_2 and the second wave by the N_2O.

Much work has been done on developing oxygen membrane electrodes. This work has been very well reviewed in a monograph by
Hitchman (3). Our work has concentrated on extending this technique
to other gases besides O_2 and in particular to the development of a
range of sensors for gases of clinical interest. This work has been
carried out in collaboration with Dr. Hahn of the Radcliffe Infirm-
ary in Oxford (4-8). We have, for instance, shown that N_2O can be
determined quantitatively using a silver cathode (4,5). Typical
results are shown in Figure 3. In agricultural practice N_2O is used
as a tracer gas to measure circulation in cow sheds.

Another gas of interest is CO_2. We have recently developed a
CO_2 electrode for breath by breath analysis (9). This electrode uses
a non-aqueous solvent. Typical results are shown in Figures 4 and
5. While the sensitivity is satisfactory for determining CO_2 in
expired breath, lower levels of CO_2 (up to 1000 ppm) are used in
glasshouses, and we are currently working on developing a low level
CO_2 sensor for this purpose.

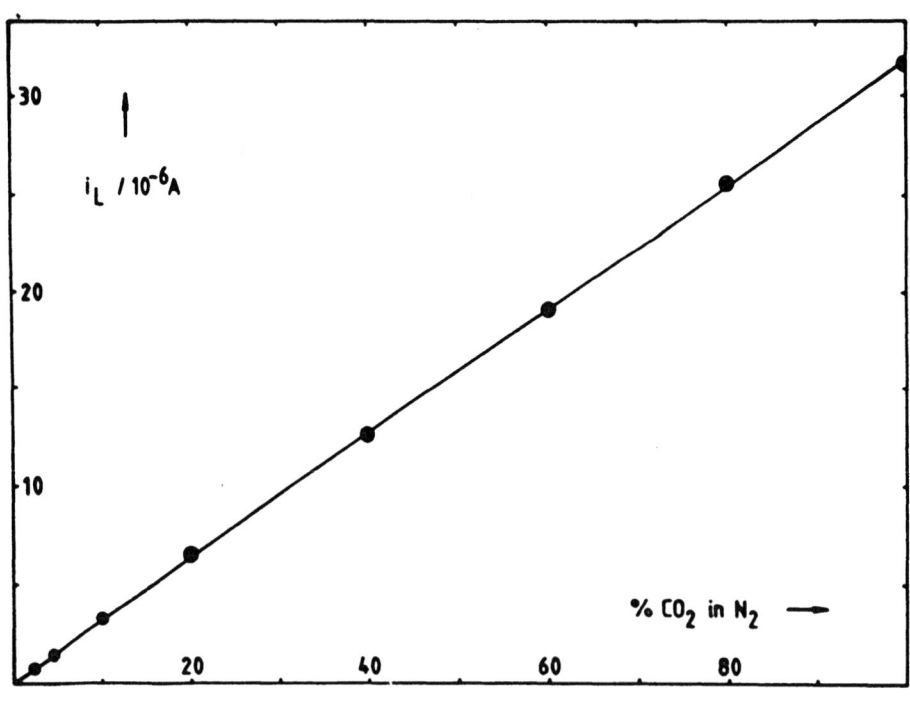

Figure 4. Variation of the limiting current for the reduction of
CO_2 with the concentration of CO_2.

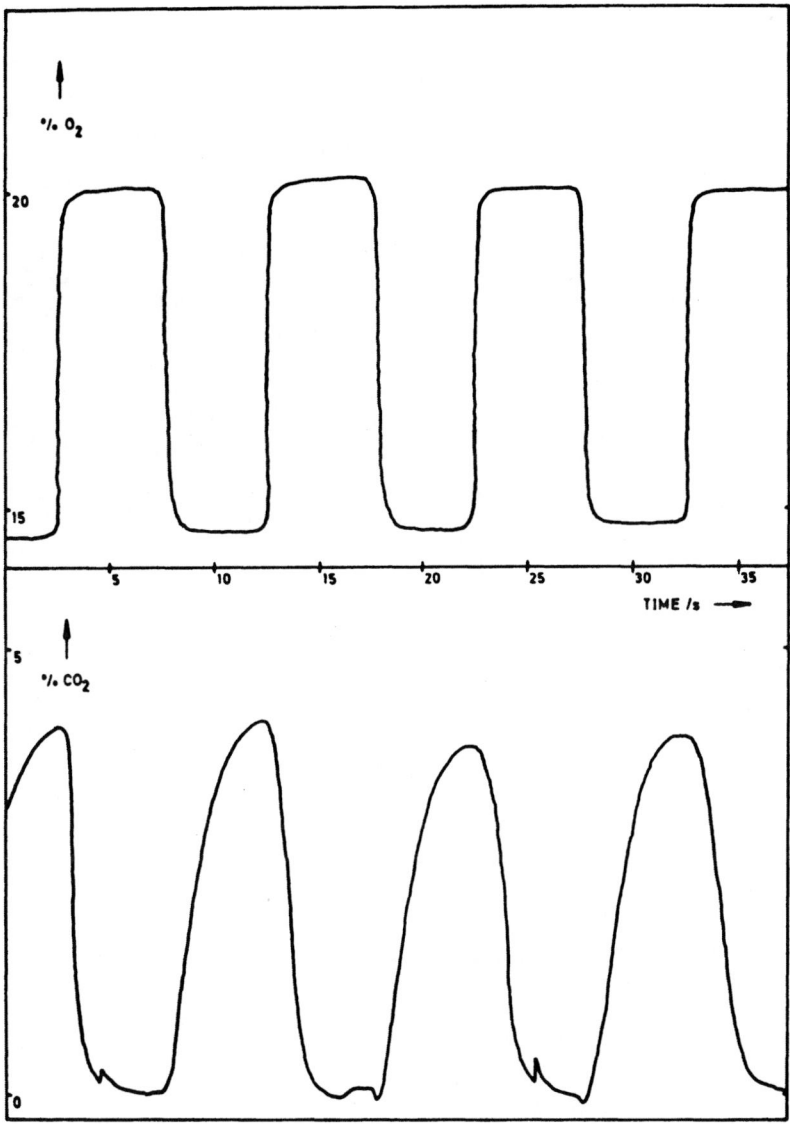

Figure 5. Breath by breath determination of O_2 (upper trace) and CO_2 (lower trace).

4. ION SELECTIVE ELECTRODES

Next we turn to ion selective electrodes (10-14). Ions such as K^+, Na^+ and Ca^{2+} are electrochemically inert under most conditions. However their concentrations can be determined by ion selective electrodes. The classifical form of this electrode is illustrated in Figure 6.

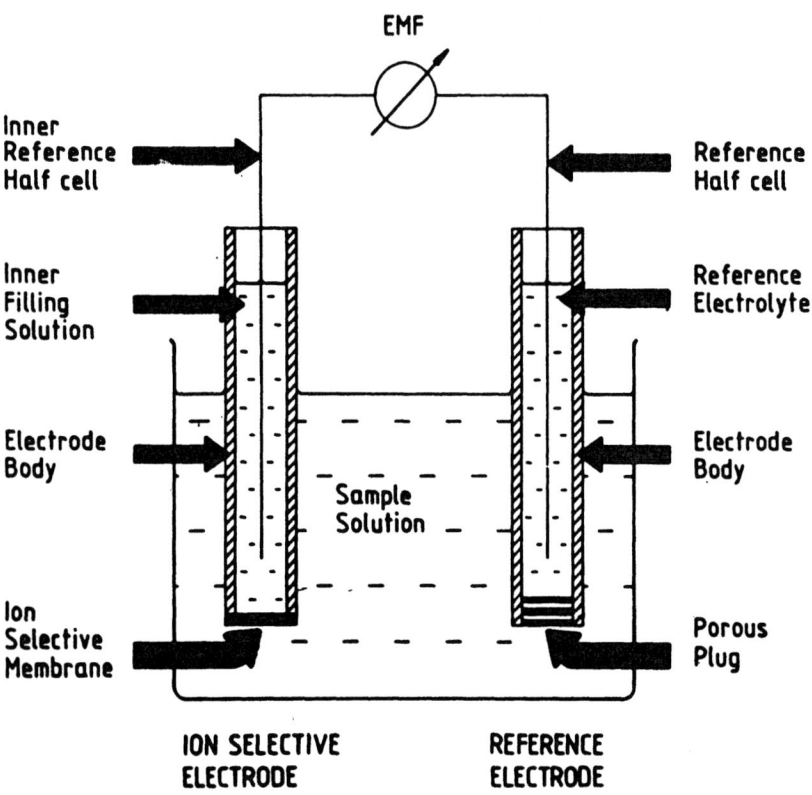

Figure 6. A typical ion selective electrode.

The electrodes in the system are reference electrodes, and their function is to provide as stable a voltage platform as possible, from which to measure the all important potential difference across the membrane or thin film of glass separating the test solution from the reference solution. To develop this potential, the membrane or glass must select the target ion and reject the other ions in the test solution. In the case of the membrane type of electrode, the membrane contains a carrier which binds the target ion. For instance, valinomycin is often used for K^+. A perfectly selective electrode would give a Nernstian response as given in equation (2), and these potentiometric sensors suffer the disadvantages discussed above and illustrated in Table 1. We shall discuss below in more detail the use of this type of electrode to measure the composition of nutrient film solutions in hydroponics.

In recent years ion selective electrodes are being realized as CHEMFETS. In this development, as shown in Figure 7, the inner test solution, and inner reference electrode are abolished, and the membrane and first stage amplification are realized on a single chip.

357

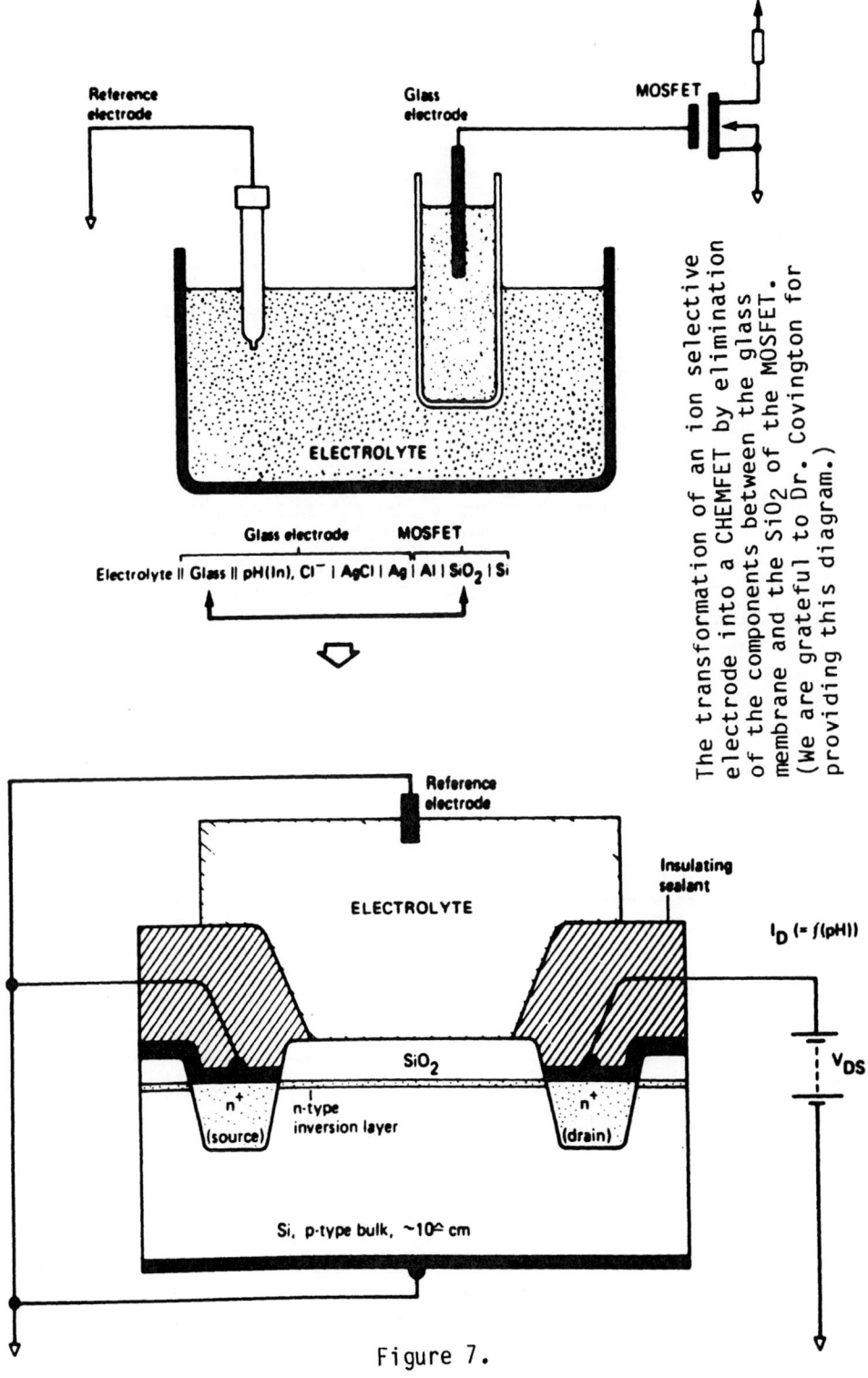

The transformation of an ion selective electrode into a CHEMFET by elimination of the components between the glass membrane and the SiO2 of the MOSFET. (We are grateful to Dr. Covington for providing this diagram.)

Glass electrode   MOSFET

Electrolyte ‖ Glass ‖ pH(In), Cl^- | AgCl | Ag | Al | SiO_2 | Si

I_D (= f(pH))

V_{DS}

Reference electrode

Insulating sealant

ELECTROLYTE

SiO_2

n⁺ (source)

n-type inversion layer

n⁺ (drain)

Si, p-type bulk, ~10Ω cm

Figure 7.

In the UK, Dr. Covington's group has been particularly active in developing these devices (16-20). They are, for instance, using a K^+ CHEMFET for clinical tests on blood. The advantages of these new devices are that, first, they can be mass produced using the techniques of the electronics industry, and secondly, they can be miniaturized. Covington has made a chip containing four sensors that can pass through the eye of a needle. There are, however, problems with the sealing of the membrane on to the chip and preventing the electrolyte corroding the gate. While integration of function may be desirable for the reasons given above, it is our view that these advantages can be overemphasized. As we shall see below, there are great advantages in shortening the distance between the electrode and the first amplifier, but it may be better to keep the different functions separate, choosing the best ion selective membrane and the best amplifier.

5. ELECTRODES WITH CONTROLLED HYDRODYNAMICS

Although it is desirable to use a membrane to protect the vulnerable electrochemistry from the dirty hazards of the external world, there are many analytes which will not diffuse through a membrane with sufficient rapidity to provide a measurable current. Amperometric electrodes for these systems use controlled hydrodynamic regimes to provide a known flux of reactant from the bulk of the solution to the electrode. The classical system is the dropping mercury electrode. For academic investigations, the rotating disc electrode is being increasingly applied (21,22). However, these electrodes are normally used in a conventional cell. For on-line applications a more useful regime is the "wall-jet" electrode (23-25). In this system, as shown in Figure 8, solution is forced through a small jet (diameter ≈ 0.5 mm) and then impinges on the center of a disc electrode. The flow pattern and mass transport to the electrode can be calculated. The advantages of this system for on-line analysis are that, first, there is a small dead space, secondly, a fast response time, and thirdly, a reasonably high percentage of the reactant ($\approx 7\%$) reaches the electrode surface.

In our own work (26-28), we have further developed this electrode system by surrounding the central disc electrode with a second concentric ring electrode to make a ring-disc electrode (29,30). This double electrode system can be used as a bromine microtitrator. We have found that a typical protein molecule, P, will react rapidly with several hundred molecules of bromine:

Upstream
Disc Electrode $2Br^- \longrightarrow Br_2 + 2e$

Solution $P + nBr_2 \longrightarrow PBr_{2n}$

Downstream $Br_2 + 2e \longrightarrow 2Br^-$
Ring Electrode

359

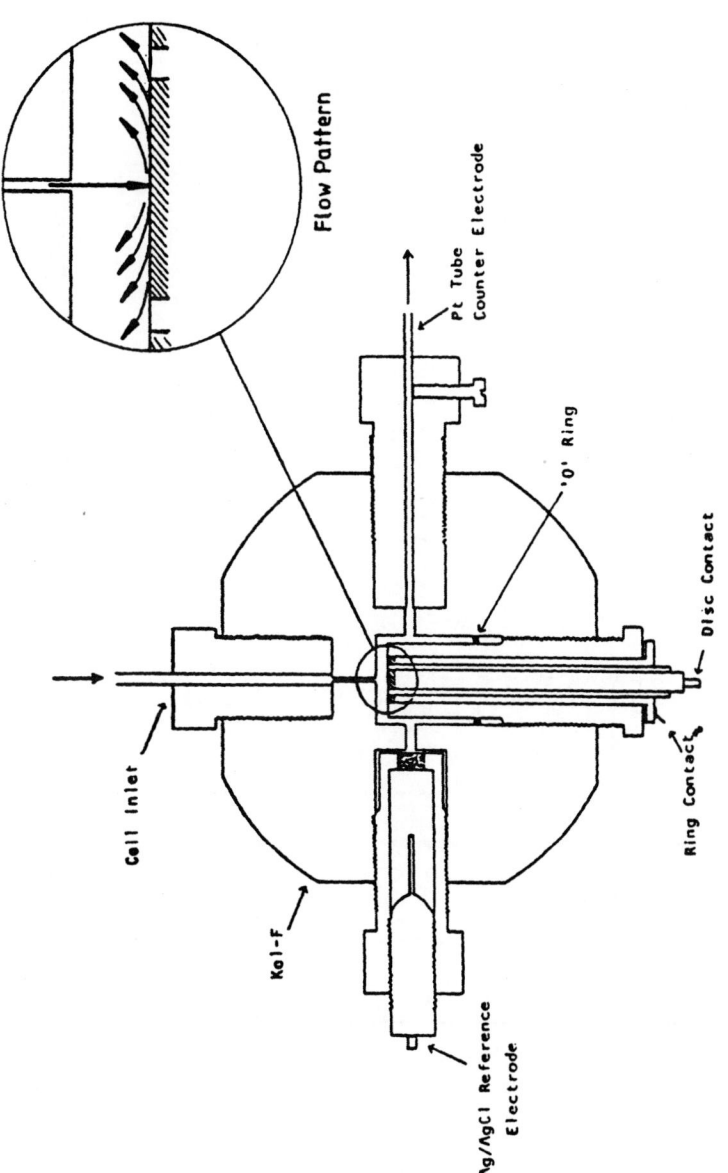

Figure 8. A wall-jet ring-disc electrode. The inset shows the flow pattern.

The unreacted bromine is measured on the ring electorde. Figure 9
shows the response of the ring current to increasing bromine generation on the disc. The insets show the concentration patterns in the
vicinity of the ring-disc electrode. The displacement of the titration curve to higher disc currents is proportional to the concentration of protein in the solution.

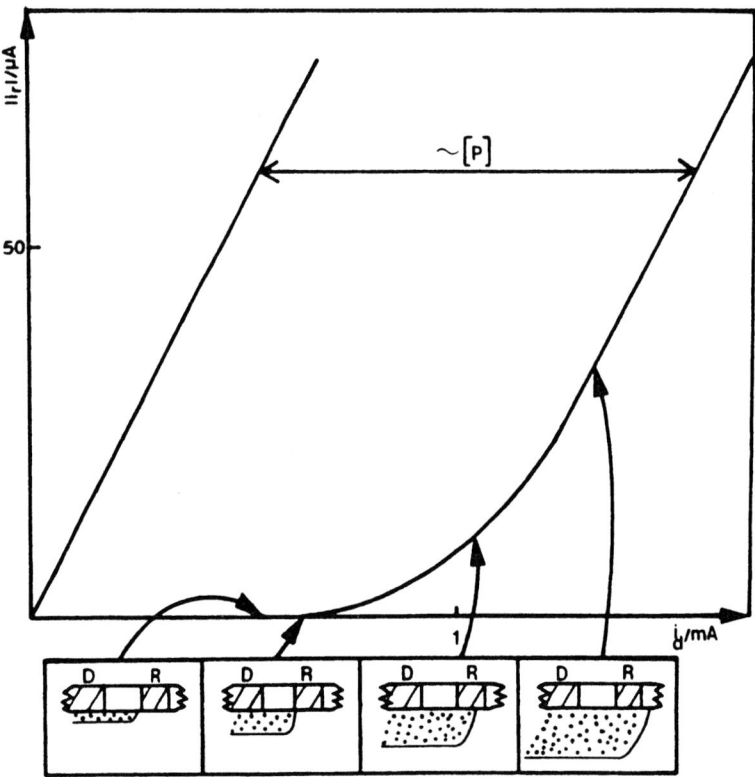

Figure 9. A ring-disc titration curve showing the displacement of
the ring current by the titration of protein, P. The
insets show the spreading out of the Br_2 region as the
disc current increases. The displacement is proportional
to [P].

Typical results for cytochrome c are shown in Figure 10. We have
shown that many proteins proteins can bve determined in this way
(28). Furthermore, by carrying out the titration at different pH
one can obtain information that is characteristic of the particular
protein. This electrode system can be used as a detector on the end
of an HPLC column. We discuss below how this method can be adapted
for use as an indicator of root death in hydroponics.
A variant of the wall-jet electrode which we are finding particularly useful is the "packed bed wall-jet electrode" (31). In
this electrode, shown in Figure 11, a packed bed is inserted upstream
of the jet. The packed bed can be used to precondition the solution

before it reaches the electrode. This arrangement is very suitable for enzyme electrodes (see below) where the enzyme can be immobilized on the packed bed. The bed can also be an electrode itself.

A problem with solid electrodes is that they gradually poison. Pletcher and Poorabedi (32) have described how NO_3^- can be reduced on a copper electrode:

$$NO_3^- + 9H^+ + 8e \ ------> \ NH_3 + 3H_2O$$

We have been adapting this method as an amperometric sensor for NO_3^- determinations in agricultural applications. Unfortunately, we found that the current decays over a period of half an hour because of electrode poisoning. This problem can be overcome with a packed bed electrode made of copper filings. Before each determination, the packed bed electrode is made positive and Cu^{2+} dissolves. The platinum disc electrode is held at a negative potential and a fresh film of copper is deposited. The determination is made and afterwards, the copper film is stripped off the platinum by taking the disc electrode to a positive potential. In this way we can overcome the problems of the poisoning of solid electrodes and confer on them the advantage of a freshly renewed surface, so long enjoyed by the mercury drop electrode (31).

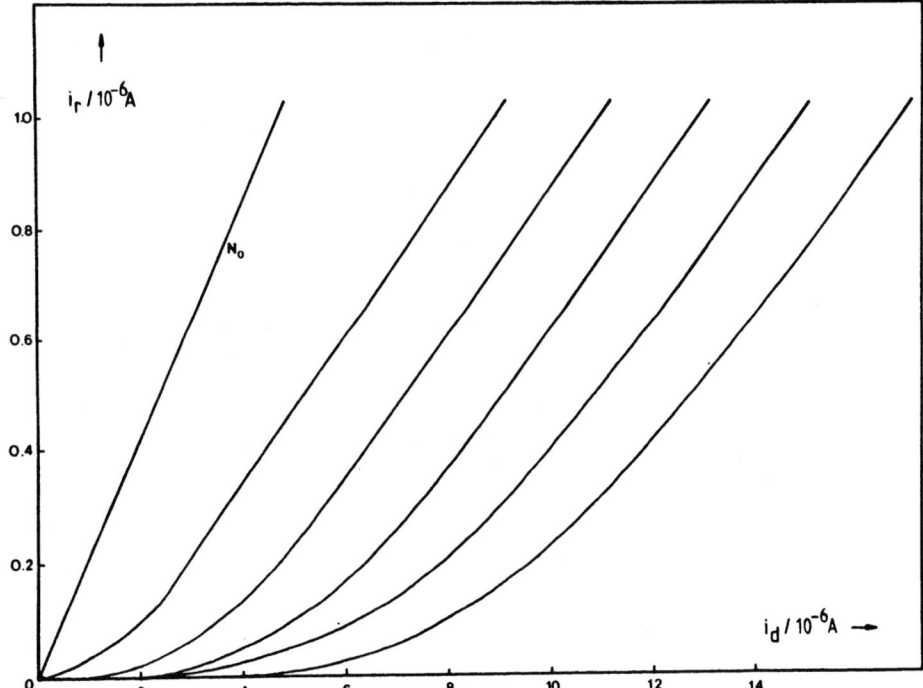

Figure 10. Typical titration curves for cytochrome c at pH 9.2, showing the effect of adding successive aliquots increasing the concentration each time by 1.9 µg cm^{-3}.

362

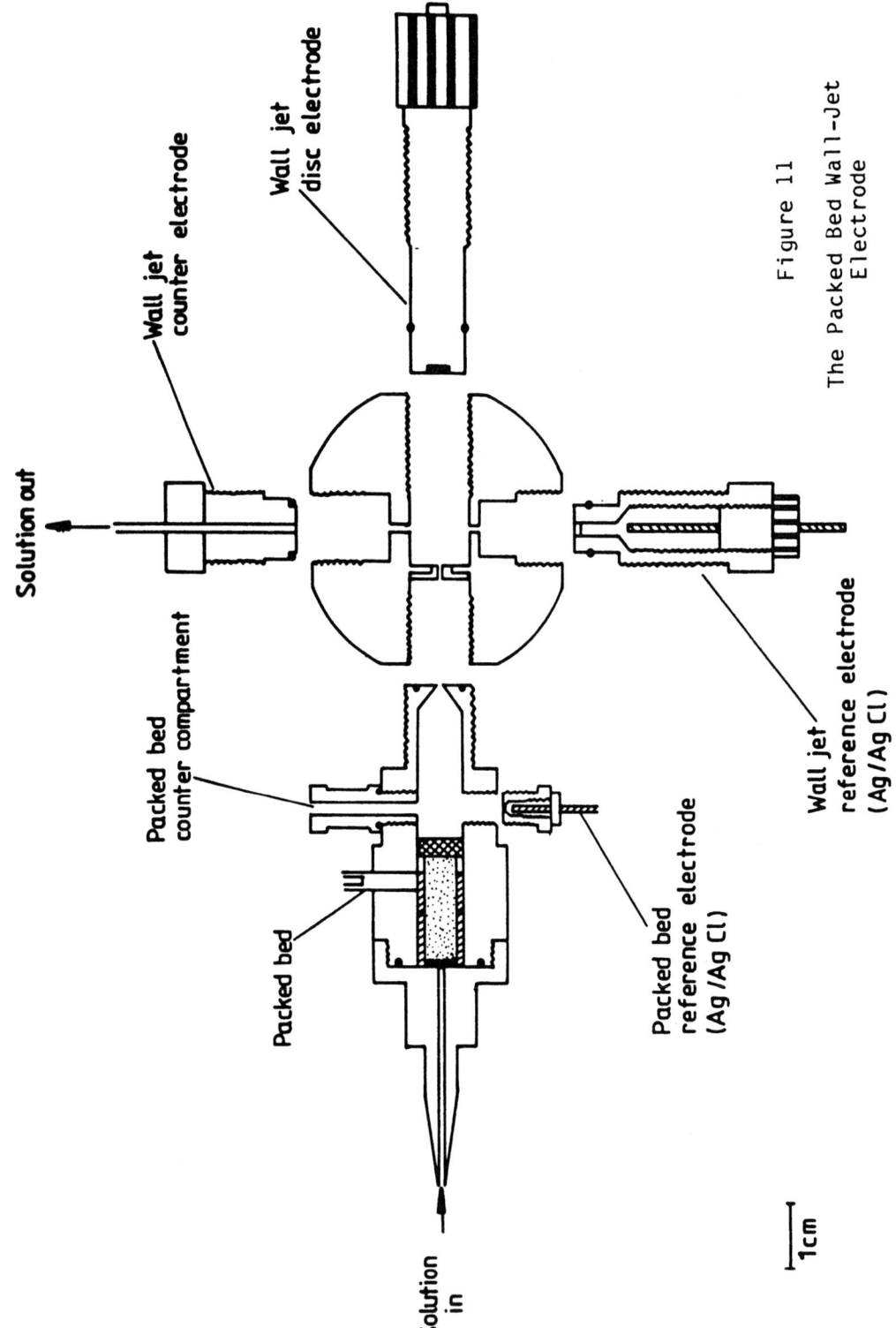

Wall jet counter electrode

Wall jet disc electrode

Solution out

Packed bed counter compartment

Packed bed

Solution in

Packed bed reference electrode (Ag/Ag Cl)

Wall jet reference electrode (Ag/Ag Cl)

1cm

Figure 11

The Packed Bed Wall-Jet Electrode

6. ENZYME ELECTRODES

Greater selectivity for compounds of biochemical interest can be obtained by using an enzyme to recognize the target species. An enzyme electrode is usually an amperometric sensor where the rate of an enzyme catalyzed reaction is measured electrochemically (33-35). Many enzymes involved in oxidation and reduction reactions contain redox groups such as iron, copper, or quinone centers. However these centers are surrounded by a coat of protein, and this coat prevents efficient electron transfer to ordinary electrodes. For this reason the first generation of enzyme electrodes used membranes electrochemically to detect the product of the natural enzyme reaction. The classic example is the glucose sensor using glucose oxidase. The reaction scheme is as follows:

Solution   Glucose + FAD -------> Gluconolactone + $FADH_2$
              $GADH_2 + O_2$ -------> $FAD + H_2O_2$

Electrode H_2O_2 -------> $O_2 + 2H^+ + 2e$

The device is illustrated in Figure 12. It is a fairly complicated device with two membranes, one to keep the enzyme in place and one to protect the electrode from being poisoned by the enzyme.

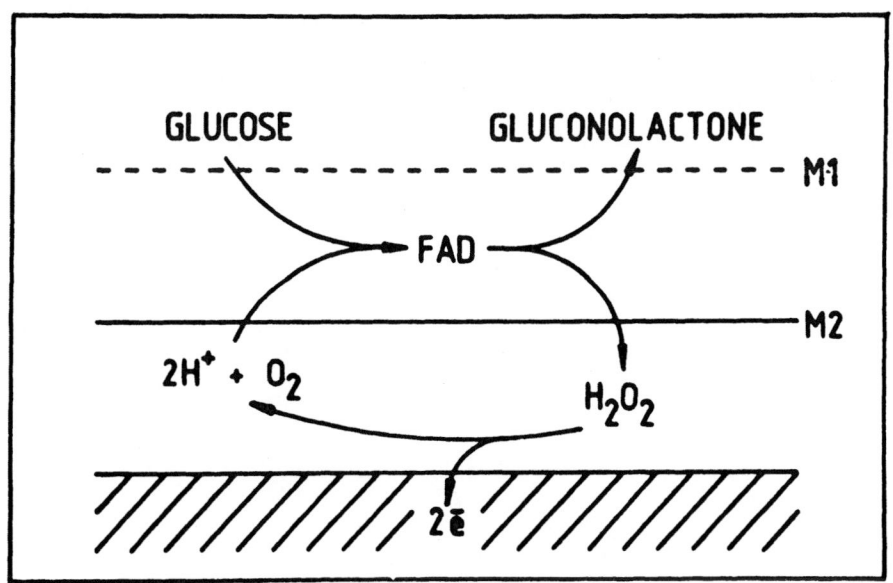

Figure 12. The reaction scheme for the classical glucose electrode. The dialysis membrane, M1, retains the enzyme but is permeable to the substrate and the product. The membrane M2 is only permeable to small neutral molecules such as H_2O_2 and O_2.

364

Since O_2 is involved in the reaction scheme, the response of the system is sensitive to the ambient O_2 concentration. These disadvantages can be overcome by eliminating the O_2 reaction and using instead an electron transfer mediator.

These second generation enzyme electrodes have been developed for instance by Hill and Higgins (36-38). In their glucose oxidase electrode, they use ferrocene/ferrocinium($Fe(Cp)_2/Fe(Cp)_2^+$) as a mediator:

Solution   Glucose + FAD -------> Gluconolactone + $FADH_2$
                $FADH_2 + 2Fe(Cp)_2^+$ -------> $FAD + 2Fe(Cp)_2 + 2H^+$

Electrode $2Fe(Cp)_2$ -------> $2Fe(Cp)_2^+ + 2e$

This electrode does not need a second membrane, and the ferrocene mediator can be directly immobilized on the electrode surface. The electrode has been developed for testing diabetic blood samples. Typical results on whole blood are shown in Figure 13.

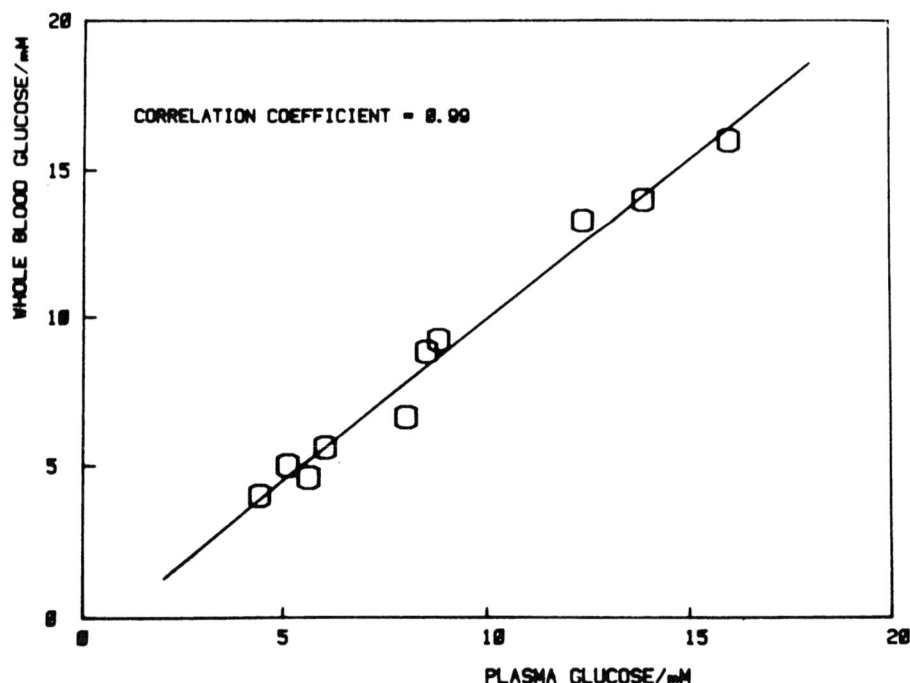

FERROCENE ENZYME ELECTRODE ASSAY OF DIABETIC SAMPLES

Figure 13. Typical results for the use of the ferrocene glucose electrode on whole blood samples. We are grateful to Dr. Allen Hill for supplying us with these results.

Our own work has been concerned with the development of third generation devices in which the enzyme reacts directly on the electrode itself. We have found that conducting organic salts like NMP^+TCNQ^- are particularly good electrode materials for the direct

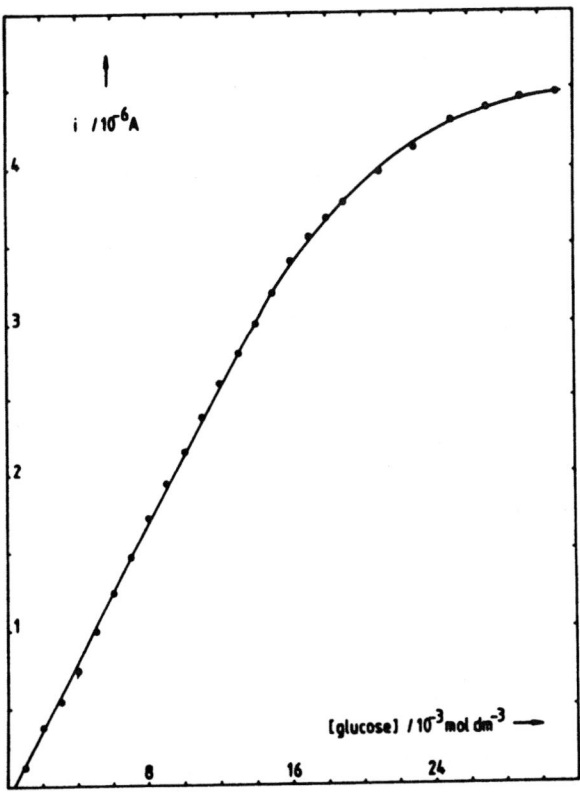

transfer of electrons to and from enzymes. Similar conclusions have been reached by Kulys and his school (39,40). The reaetion scheme is now as simple as it can be:

Solution   Glucose + FAD -------> Gluconolactone + $FADH_2$

NMP^+TCNQ^-
Electrode $FADH_2$ -------> FAD + $2H^+$ + 2e

Typical results obtained by Bartlett and Craston in my group for the response of this electrode to different concentrations of glucose are shown in Figure 14 (41).

Figure 14. Typical results for a glucose electrode make of NMP^+TCNQ^-.

We have also shown that these materials make good electrodes for the oxidation of NADH (42). Over 250 enzymes use NAD^+ as a cofactor, and hence we can design sensors using a wide variety of different dehydrogenases for a typical substrate SH_2:

$$\text{Solution} \qquad SH_2 + NAD_+ \xrightarrow{\text{Dehydrogenase}} S + NADH + H^+$$

$$\begin{array}{l} NMP^+TCNQ^- \\ \text{Electrode} \qquad NADH \xrightarrow{\hspace{1cm}} NAD^+ + H^+ + 2e \end{array}$$

In Table 3 we collect together some examples of the possible application of this type of sensor.

TABLE 3

NAD/NADH Enzyme Systems

ANALYTE	ENZYME	APPLICATION
Alcohol	Alcohol Dehydrogenase	Fermentation
Lactate	Lactate Dehydrogenase	Dairy Industry
Malate	Malate Dehydrogenase	Fermentation
Glutamate	Glutamate Dehydrogenase	Fermentation, Food Industry
Glucose	Glucose Dehydrogenase	Fermentation Clinical
Clycerol	Clycerol Dehydrogenase	Fermentation
Bile Acids	11$_B$ Hydroxysteroid Dehydrogenase	Clinical
Nitrate	Nitrate Reductase	Agriculture, Water Industry
Oestradiol	Oestradiol 17$_B$ Dehydrogenase	Agriculture, Food Industry, Clinical
Amino Acids	Amino Acid Dehydrogenase	Fermentation, Clinical, Food Industry

Enzyme electrodes are already finding use in clinical medicine. Looking to the future we have no doubt that their use in agricultural science will be equally important and will enable the concentrations of specific metabolites to be traced and measured.

7.   SENSORS FOR HYDROPONICS

Having surveyed different types of electrochemical sensors, we now turn to the use of these sensors to monitor the concentrations of species in nutrient films used in hydroponics. Hydroponics is the science of growing plants without soil. The nutrients for the plant are supplied to the roots in a solution which can either be static or flowing. The word was first defined in 1930 by Gericke (43). In 1860 Sachs and Knop (44) were the first to grow plants successfully using a nutrient solution and no soil. These early investigations showed that plants need certain elements in large quantities. These "macronutrients" or "macroelements" are nitrogen, oxygen, phosphorous, potassium, sulphur, calcium and magnesium. Other nutrients are required in much smaller concentrations and are known as "micronutrients" or "trace elements." These are iron, manganese, zinc, molybdenum, boron and chlorine. In the early 1930's Gericke, the Father of modern hydroponics, had developed the technique, and greenhouses using flowing nutrient solutions were established in California producing vegetables such as beet, potatoes, lettuce and tomatoes. Hydroponics is usually carried out in a glasshouse since it is desirable to control the total environment of the growing crop and gives the following advantages: more efficient nutrient regulation and use of water and fertilizers, higher planting density and better yields. In Table 4 we compare normal and hydroponic yields. There are several useful reviews of hydroponic techniques (45-50).

TABLE 4

Comparative Yield per Acre in Soil and Hydroponic Culture

Crop	Soil	Hydroponics
Soya	600 lb	1550 lb
Beans	5 tons	21 tons
Peas	1 ton	9 tons
Wheat	600 lb	4100 lb
Rice	1000 lb	5000 lb
Oats	1000 lb	2500 lb
Beet	4 tons	12 tons
Potatoes	8 tons	70 tons
Cabbage	13,000 lb	18,000 lb
Lettuce	9000 lb	21,000 lb
Tomatoes	5-10 tons	60-300 tons
Cucumbers	7000 lb	28,000 lb

8. THE NUTRIENT FILM TECHNIQUE

In classical hydroponics with a flowing nutrient solution, each plant grows in a container, which is filled with an inert material such as pumice stone, wood chips, glass wool or sand. The nutrient solution permeates this material. The plant roots also penetrate the material, and the plant stands upright supported on its root system. While with this system the mass transfer of the nutrients from the fresh incoming solution to the extended root system is more efficient than a similar system growing in soil, nevertheless the

transport by convection and diffusion still leaves much to be desired desired. The solution around the roots becomes depleted of
the vital nutrients including, in particular, oxygen. Mass transport
to the roots can be improved by doing away with the container and
its filler material.

In the 1970s, Cooper (51) at the Glasshouse Crops Research
Institute (GCRI) in Littlehampton, invented the nutrient film technique (NFT). In this technique, as shown in Figure 15, the roots
are bathed in a thin film of flowing nutrient solution.

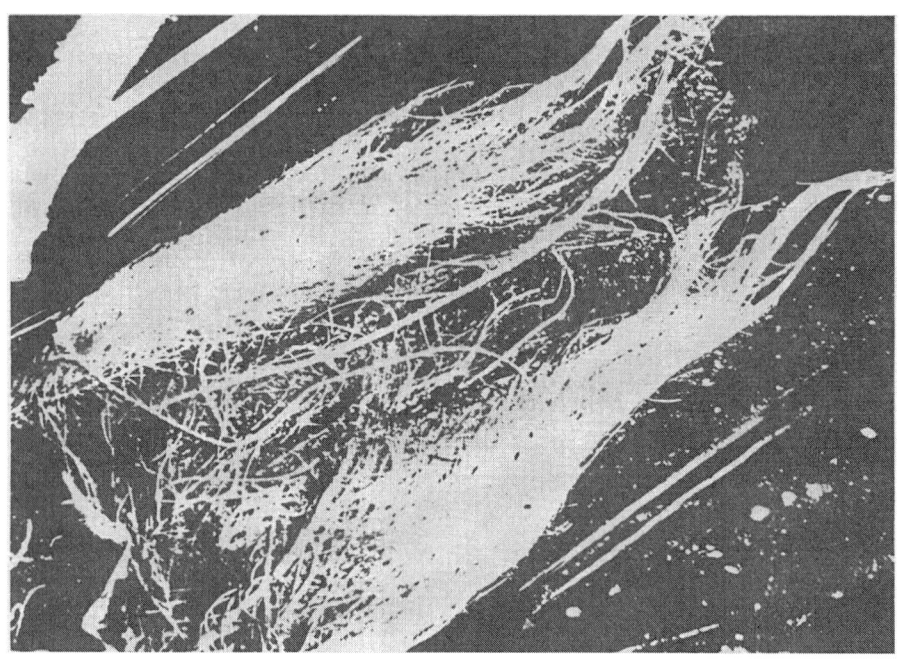

Figure 15. Root mat growing in the nutrient film at NIAE.

A complete layout is illustrated in Figure 16. The plants grow in
gently sloping gullies. A pump feeds nutrient solution to the high
end of the gully, from where it flows by gravity over the roots of
the plants down to a catchment trough. This trough is also sloping
in a direction which allows flow to the storage tank. It is this
tank that supplies the pump with solution, thus closing the loop. A
float valve regulates the liquid level in the storage tank, and the
correct level of nutrients is maintained by a control system.

The advantages of nutrient film technique compared to classical
hydroponics have been discussed in an excellent review by Graves
(52). First, as discussed above, the chemical environment of all
the roots is more closely controlled; secondly, nutrient supply is
uniform and can be matched to the plant's needs as it grows (53,
54); thirdly, instead of heating the whole greenhouse, good results
have been obtained by simply heating the nutrient solution (55,56);

fourthly, it is easy to disperse chemicals required at low concentration for crop protection throughout the whole system (57); fifthly, there is a quicker turn around between crops because growing medium does not have to be sterilized. Finally, greater crop densities can be achieved. Since the roots do not penetrate any solid material, these plants have to be supported from above. This can be used to good advantage to train the plants in a horizontal direction. The nutrient film technique is so efficient that, for instance, tomato plants will grow to a height of 20 feet producing fruit and giving a cropping season of 10 months. Horizontal growing is essential to accommodate the monster 20' tomato plant in the average glasshouse. High cropping densities can thereby be achieved giving an increased output per cropped area (58,59).

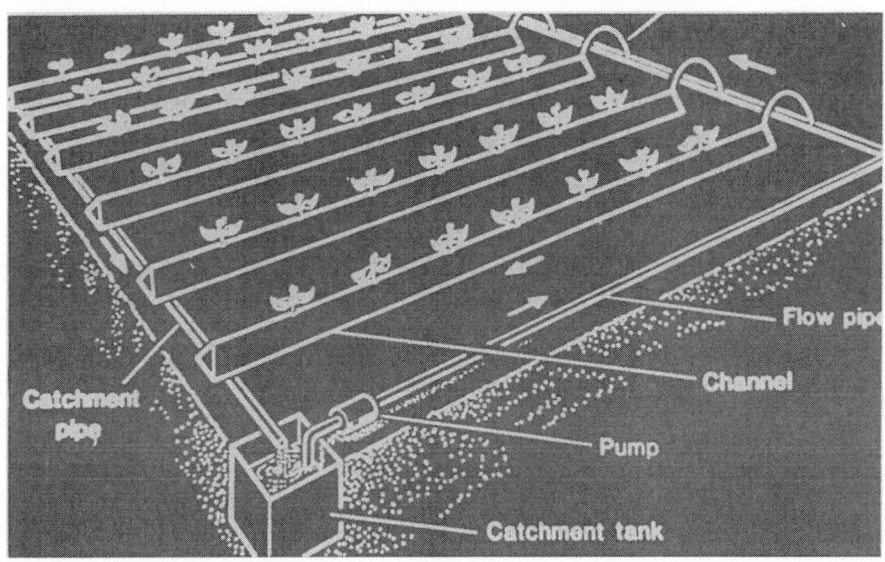

Figure 16. Schematic layout of a nutrient film system.

9. THE NUTRIENT SOLUTION

In Table 5 we give the composition of typical nutrient solutions for the growing of tomatoes (52). The concentrations have been given in two sets of units. The ppm scale is the scale used in practice in the greenhouse; the molarity scale is more familiar to the scientist. Whereas the first 12 components of the mixture from NO_3^- to Mo are essential for the growth of the plant, the last two components Na^+ and Cl^- can be toxic to the plant and must be kept below the prescribed levels. In its transpiration, the plant consumes considerable quantities of water and this water has to be supplied from the mains. It inevitably contains ions such as Na^+,

Ca^{2+}, Mg^{2+} and Cl^-. Hence during the growth of the crop, the concentration of Na^+ and Cl will increase and care has to be taken that the toxicity levels are not reached.

TABLE 5

Nutrient Concentrations and Chemicals for Tomatoes in NFT

Element	Desirable concentrations (ppm)	Chemicals
nitrate nitrogen	150-200	KNO_3, NH_4NO^e, $Ca(NO_3)_2$
ammonium nitrogen	0-20	NH_4NO_3, $(NH_4)_2 SO_4$
potassium	300-500	KNO_3, K_2SO_4, KH_2PO_4
phosphorus	50	KH_2PO_4, NaH_2PO_4, $CaHPO_4$
calcium	150-300	$Ca(NO_3)_2$, $CaSO_4$, $CaHPO_4$
magnesium	50	$MgSO_4$, $Mg(NO_3)_2$
iron	3*	FeEDTA, FeEDDHA
manganese	1	$MnSO_4$
copper	0.1	$CuSO_4$
zinc	0.1	$ZnSO_4$
boron	0.3-0.5	H_3BO_3
molybdenum	0.5	$(NH_4)_6 MO_7O_{24}$
sodium	maximum 250	--
chlorine	maximum 200	--

* 5-10 ppm is preferable during the early stages of growth before the start of fruit picking.

10. PRESENT CONTROL AND SENSING SYSTEMS

The automatic nutrient control system used by most growers today consists of two controllers, one for the Ph and another for the conductivity. Both controllers work on a set-point basis, generally on an hourly basis, i.e. if the sensor input is above or below a set point the controller reacts.

Conductivity depends on all ions in solution so it does not yield any information about the concentration of individual nutrients. The mains water that is used is to make up the nutrient solution has a background conductivity, mainly from Ca^{2+}, Na^+ and Cl^-. A typical solution volume for a 1 ha greenhouse is 50,000 1 and on a hot summer's day an equivalent amount of water may be lost by transpiration and evaporation. This leads to a fast build up of undesired chemicals in the system, and the controller does not provide nutrients when the plants need them most. Furthermore, simple conductivity measurements do not monitor the build up of toxic ions like Cl^-. While pH and conductivity are measured continuously, analysis for the other elements is performed (if at all) in a batch mode by taking samples once a week. Such a procedure is unsatisfactory from the point of view of supplying the nutrients at the desired level and preventing crop damage. Many research workers have recognized the inadequacy of the present systems (60,61).

A particularly bad feature of the present system is the use of conductivity to control the supply of nutrients because the technique is not a specific sensor for important ions like NO_3^- and K^+. This means that the levels of nutrient in Table 1 have to be higher than necessary in order to make sure that the crop has sufficient NO_3^-, and that the conductivity meter is not just measuring the hardness of the mains water.

It would clearly be more desirable to monitor continuously as many of the constituents of the nutrient film mixture as possible. The purpose of this is twofold. First, when plants are grown hydroponically, the grower has control over the amount of nutrients fed to the plant. Hence one can investigate in a systematic fashion what factors affect the crop yield. This should mean that the nutrient levels in Table 5 can be set more precisely. It is quite probable that the desired levels will be significantly smaller than those used today. This will save the grower the cost of his chemicals and perhaps more important there will be less pollution of the environment from the disposal of waste solutions of NO_3^-. Running the greenhouse at lower levels of nutrients will require better sensing and control. So the second purpose in developing a set of sensors for hydroponics is to provide the grower with a cheap automated system that cannot only analyze the nutrient film solution but also replenish the nutrients as they are used. At present the level of nutrients is controlled to certain set points. However, as our knowledge increases, we will develop more complex and efficient programs for varying nutrient levels with time as the crop grows. For instance, young plants need a good supply of NO_3^- to encourage the development of vegetative growth. Early fruit can then be developed by putting the plants under "nutrient stress" (starving them). The regime for this practice should be able to be optimized and controlled. McDonald and Parsby (62) have shown that different proportions of carbon and nitrogen will be incorporated into the different parts of the plant depending on the nutrient supply. The effect of these variations on the taste and quality of the fruit is perhaps more a matter of taste rather than science but it is still important.

11. AUTOMATED NUTRIENT FILM ANALYZER (ANFA)

To solve these problems in collaboration with the National Institute of Agricultural Engineering, we have developed the Automated Nutrient Film Analyzer (ANFA). The apparatus is illustrated in Figure 17. The whole apparatus is controled by a microcomputer. This microprocessor selects which solutions to be measured, selects which sensor is to be interrogated, acquires the data, processes the data and plots the data out. A special electrochemical interface has been developed connecting the microcomputer to the sensors. This interface may be divided into two parts, the digital interface and the analogue interface. The sensors and their flow systems are all housed in a Faradaic cage in order to reduce noise. These different elements of the system will now be described, starting with the sensors.

Figure 17. The automated nutrient film analyzer (ANFA) in the Venlo glasshouse at NIAE.

12. THE SENSORS

While our long term aim is to develop a complete range of sensors for hydroponics, at present the ANFA has 6 ion selective electrodes for the following ions in Table 5, H^+, K^+, Na^+, Ca^{2+}, Cl^- and NO_3^-. Ion selective electrodes are the method of choice for these ions, and electrodes for these ions are commercially available. However, the use of such electrodes in the greenhouse environment is not straightforward. Earlier work by Weaving (63) had shown that while reasonable results could be obtained in the laboratory, the electrodes were not sufficiently reliable to be left in charge of a greenhouse.

The task before us has, therefore, been to take the commercial electrodes and develop the system together with its microprocessor to provide a reliable instrument. In our philosophy, one cannot divorce the sensor from its instrumentation and we, therefore, developed the sensors and their instrumentation in an integrated fashion.

13. SELECTIVITY

The emf, E, of an ion selective electrode for ion A is given by the Nicolsky Eisenman equation:

$$E = E_A^\theta + \frac{RT}{n_AF} \ln \left\{ a_A + \sum_B k_{A,B} a_B^{n_A/n_B} \right\} \tag{3}$$

where E_A^θ is the standard electrode potential.

n_A is the charge on ion A

a_A is the activity of ion A

$k_{A,B}$ is the selectivity coefficient for the interference of ion B in the determination of ion A

a_B is the activity of ion B

n_B is the charge on ion B

The Σ in eqn (3) contains a term for each of the ions of the same charge as ion A; normally cations do not interfere with anions and vice versa. When $n_A = n_B$, the interference will start to be significant when $a_A/a_B \approx k_{A,B}$. Hence a small value of $k_{A,B}$ allows a_A to be measured in the presence of a larger activity of B. This behavior is illustrated in Fig. 18. We have determined the selectivity coefficients, $k_{A,B}$, for the principal ions in the nutrient film mixture. Typical results for the cations are collected in Table 6.

In nearly all cases, the selectivity coefficients are small enough for the interferences to be negligible. This can be illustrated in another way.

TABLE 6

Values of Selectivity Coefficients $k_{A,B}$ for Cations

Interfering Ion B	Target Ion A Na^+	K^+	Ca^{2+}
Na^+	(1.0)	3.2×10^{-4}	2.2×10^{-3}
K^+	4.0×10^{-3}	(1.0)	7.9×10^{-4}
Ca^{2+}	1.7×10^{-7}	1.5×10^{-8}	(1.0)
Ng^{2+}	1.7×10^{-7}	1.2×10^{-8}	1.3×10^{-2}
NH_4^+	4.4×10^{-4}	1.2×10^{-2}	1.8×10^{-3}
H^+	32*	--*	--*

* Negligible interference in pH range 5.5 to 6.5.

Figure 19 shows the response of the different ion selective electrodes to their target ion activity. For H^+, K^+, Ca^{2+}, NO_3^-, the two vertical lines in each diagram show the required range of composition of the nutrient solution. For Na^+ and Cl^-, the single vertical line shows the toxicity level above which the activity must not be allowed to rise. In all cases, the electrodes respond well to the target ion activity.

Equation (3) predicts that the slopes of the response curves should ideally be 59 mV (at 298 K) per decadic change in activity for singly charged ions and 29.5 mV per decadic change for Ca^{2+}. The results in Figure 19 were obtained with new electrodes and the slopes are fairly close to the theoretical values. However, over a period of months, the slopes of the membrane types ISEs (K^+, Ca^{2+}, and NO_3^-) decrease typically to two thirds of their initial values. Our approach to ISEs is not to expect perfect ideality but to be thankful that the electrode is showing sufficient response for a concentration to be measured.

A great avantage of microcomputer control is that the electrodes can be calibrated as often as the operator desires. In our system, we use two calibration solutions. The composition of each calibration solution is chosen so that the concentration of the target ions in the nutrient solution fall between their values in the two calibrating solutions. This ensures that each measurement is an interpolated value lying betwen the calibrating values. This procedure extends the useful life of the electrode from a matter of weeks to three months or more. The difference between the readings for the high and low calibrating solutions allows a calculation of the electrode response. For the singly charged ions, the electrodes are replaced where the response has fallen to 40 mV per decadic change in concentration. (For Ca^{2+} the corresponding figure is 20 mV per decadic change in concentration.)

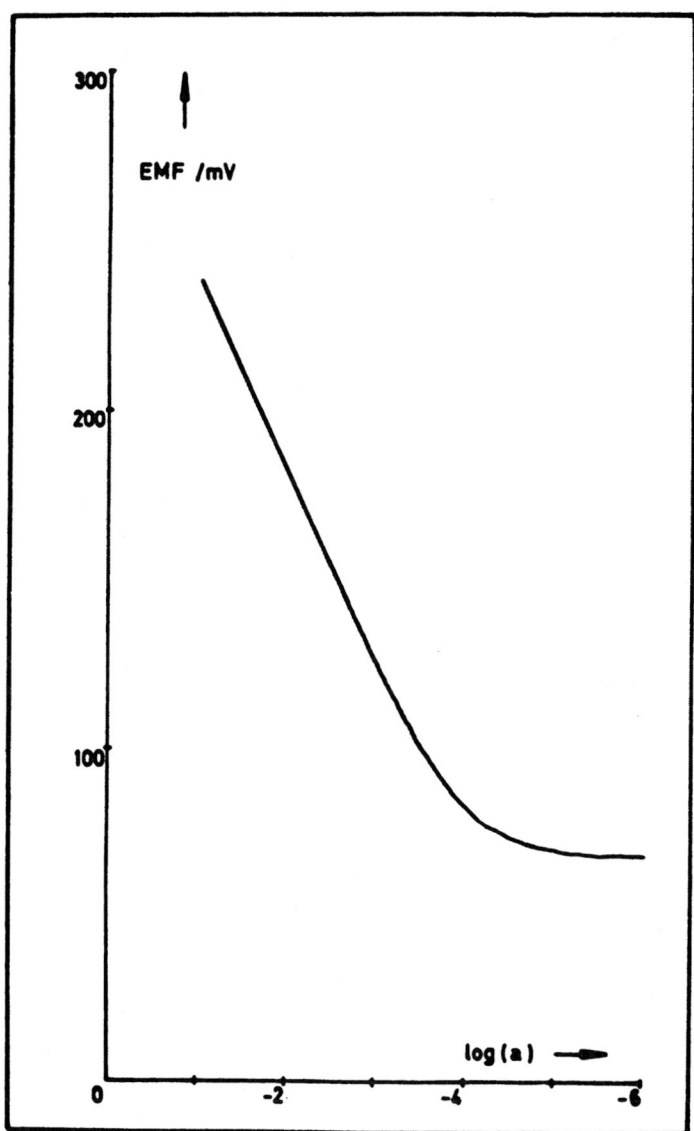

Figure 18. Typical response of an ion selective electrode. At low concentrations of the target ion, the interference from other ions in the solution reduces the gradient of the EMF against log (a), where a is the target ion activity, to zero.

376

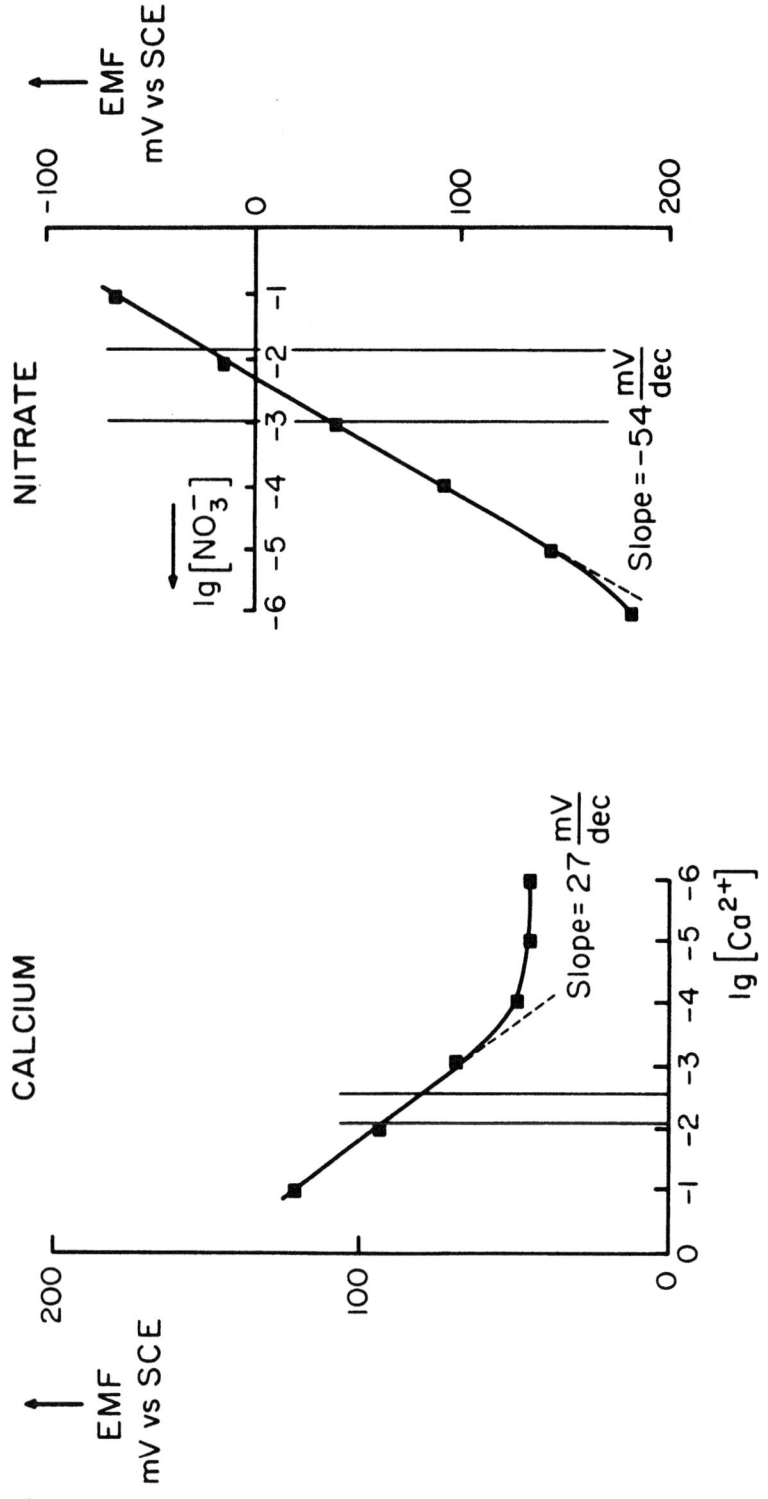

Figure 19. Response of ion selective electrodes to target ion activity ion in the nutrient solution. For Ca^{2+}, NO_3^-, and H^+ the two vertical lines show the working range for the target ion. For Na^+ the single vertical line shows the permitted upper limit of Na^+ activity.

377

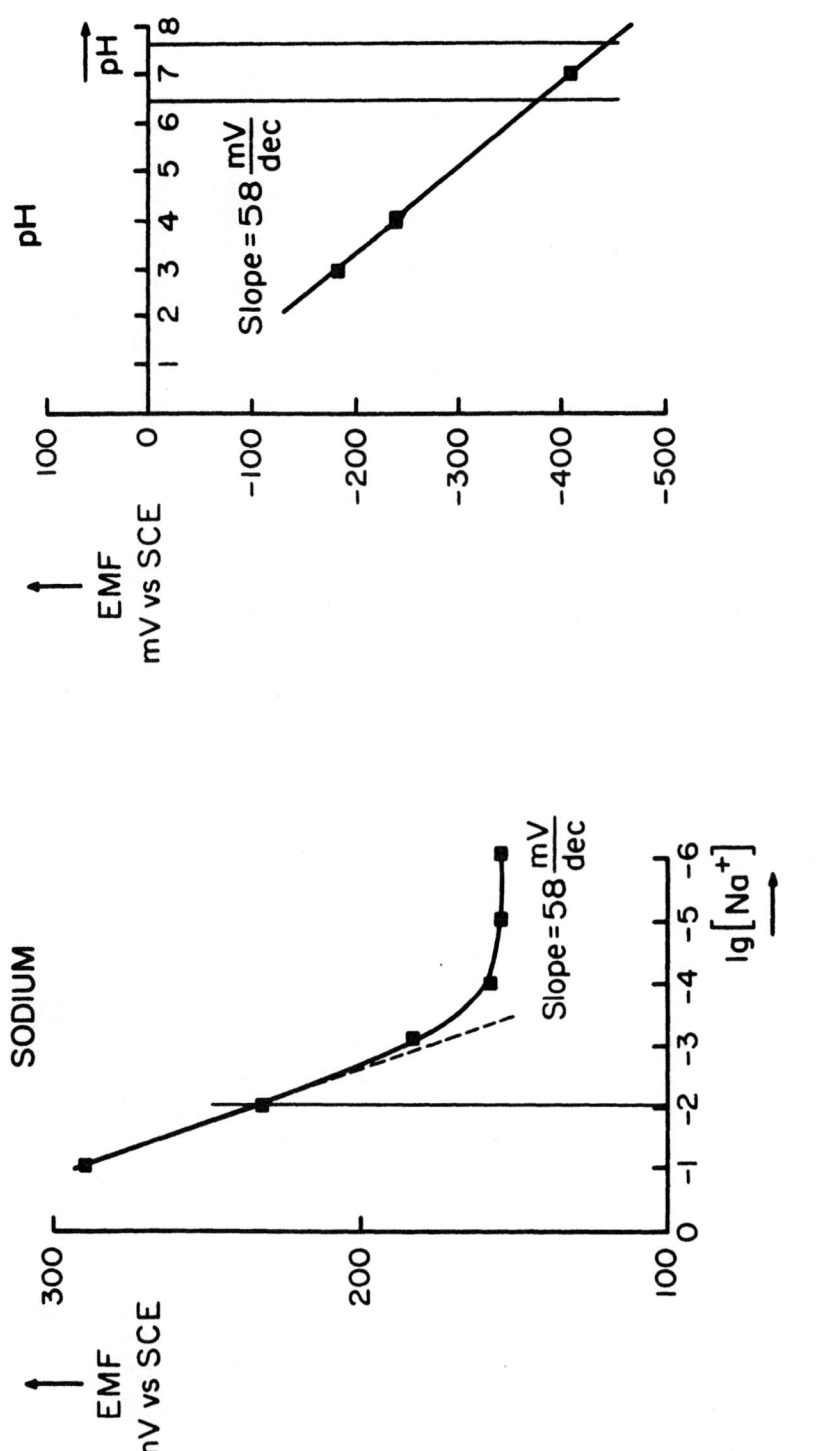

Figure 19. Continued

378

14. THE FLOW SYSTEM

To present the nutrient solution to the array of sensors, we
have constructed and developed a flow system containing modular compartments for each electrode. The flow cell for the electrodes is
illustrated in Figure 20. Each of the cell compartments is made of
perspex. Problems were encountered with bubbles being trapped under
the electrodes, and the present design includes a bubble trap at the
start of the sequence of electrodes. The solution is pumped with a
peristaltic pump controlled by the microcomputer. During measurement,
the pump is switched off to prevent interference from the inevitable
pulsations produced by the peristaltic pump. The sequence includes
a temperature sensor, and the microcomputer includes the measured
temperature in its conversion of measured e.m.f. to concentration.
It was found that noise could be minimized using the actual electrolyte solution as the common ground for the whole system, and this is
achieved by inserting a plain platinum wire electrode in the measuring sequence. To further reduce noise, all the tubing that carried
the electrolyte solution is surrounded by a cylinder of earthed
copper braid giving us coaxial tubing. The switching between the
two calibrating solutions and the nutrient solution is carried out
by 3 solenoid valves controlled by the microcomputer.

Figure 20. The ion selective electrodes and the modular flow through
        cells.

15. IMPEDANCE CONVERSION

The ion selective electrodes have high impedances ($\approx 10^9$ Ω).
This means that the generation of 1 nA of current in the lead to the
electrode will generate 1 V of noise. The discussion above and the
results in Table 1 show that to obtain reliable results, noise must
be kept below a few mV. An important and essential development in
our work was to locate an impedance converter on the top of each ion
selective electrode. The impedance converter is a voltage follower
(Burr Brown 3527).

Comparative tests were made of a number of commercially available ion selective electrodes. Little difference was found in their
performance. In some cases this is not surprising since companies
buy their sensing elements from the same manufacturer. However, the
ISEs supplied by Russell pH of Auchtermuchty have the great advantage of a coaxial screw top on which the impedance converter can be
easily mounted. These electrodes are, therefore, now the electrodes
of choice for this work.

16. ION SELECTIVE ELECTRODE INTERFACE UNIT

The signal from each ion selective electrode is passed to the
Ion Selective Electrode Interface Unit (ISEIU) shown in Figure 21.
This unit was specially designed and constructed. The first stage
amplifies the ISE signal of typically 100 mV tenfold to give signals
to the 1 to 10 volt range. This was found to be necessary because
the noise in the digital interface to the microcomputer is $\approx$ 20 mV,
and this noise produced serious errors when we tried to measure un-
amplified voltages from the ISEs. The signal then passes to the
Multiplexer (AMUX) which is the microcomputer controlled switchboard
that selects the different sensors. Here the signal becomes somewhat
corrupted with digital noise, and so it is sent back to the ISEIU to
be cleaned up by a differential amplifier and a low pass filter.
Finally, the signal returns to the digital interface to be converted
(ADC) to a binary number. An important finding in our development
work was the necessity to separate the analogue operations in one
box (ISEIU) from the digital operations in the digital interface.
We believe that this may be a useful general practice.

17. THE MICROPROCESSOR SYSTEM AND DIGITAL INTERFACE

The system is based on an Oxford Research Machine 380Z microprocessor, and is illustrated in Figure 17. It was designed by Dr.
Nicholas Goddard of the Imperial College Department of Chemistry
Microprocessor Unit. The central core of this system is the 380Z
and its associated disc drive, which is the computer's memory, where
programs and instructions are stored. The 380Z is controlled through
inputs from the keyboard. Outputs are either seen on the VDU or
hard copies can be obtained on the Digiplot. Nearly all the work
in developing the system has been in the interface.

380

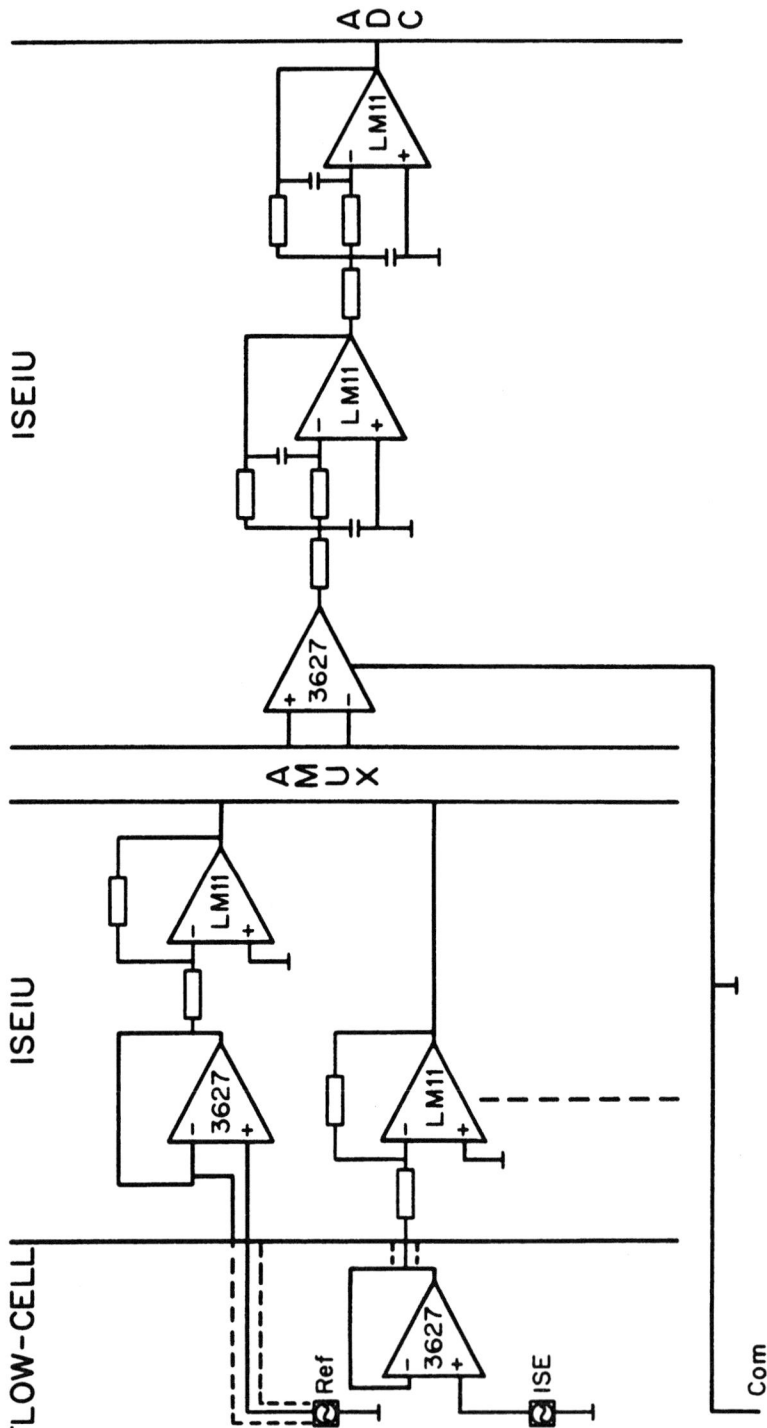

Figure 21. The ion selective electrode interface unit (ISEIU).

The interface is of modular construction so that up to 16 different subunits can be selected. Each technique has its own set of modules but when we have several sensors using the same technique, in this case a set of different ion selective electrodes, then the sensors are connected to a multiplexer. This allows each sensor to be called individually to be interfaced to the computer through the same set of modules. In one direction the microporcessor controls the electrochemical sensors and in the other, data is collected and analyzed according to programs lodged in the disc drive.

Considerable time has had to be spent developing the software for this versatile electrochemical interface. The whole system is "user friendly" at two different levels. First, for the scientist, the programs he writes are written in Basic rather than in a machine language. This more transparent language allows easier development of the software. The disadvantage of using a high level language like Basic is that it can be rather slow to execute the operator's commands. We have, therefore, developed for each module subroutines written in Assembly Code.

The transparent "slow" program in Basic, therefore, calls up these "opaque" subroutines which then rapidly carry out their functions. This dual approach combines the advantages of transparency and speed.

The second level at which the whole system is user friendly is at the technician or operator level. The programs we have developed display on the screen numerous prompts, helpful questions and instructions to the user of the system. A typical screen display is shown in Figure 22.

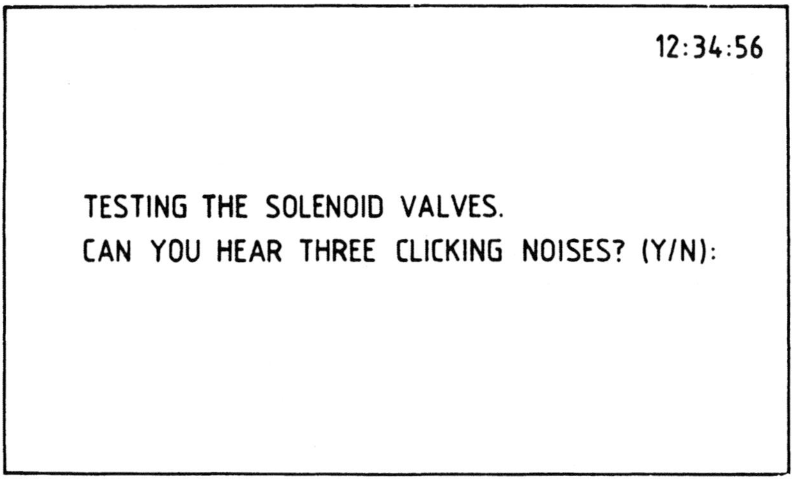

```
                                                    12:34:56




        TESTING THE SOLENOID VALVES.
        CAN YOU HEAR THREE CLICKING NOISES? (Y/N):




```

Figure 22.   Typical prompt on the ANFA screen during setting up. The figures in the top right hand corner show the time in h, m and s.

This type of prompting allows the expertise of the scientist to be used on the spot by the operator. As we acquire more experience with the system, we are continually updating this feature of our software.

18. PERFORMANCE OF THE ANFA

After extensive testing at Imperial College, the ANFA has been installed at NIAE. The first periods of testing the unit at NIAE took place in the summer of 1983 when the unit was run for several weeks. During this time it was found that the unit could not be placed in the greenhouse itself because, firstly, the dust and dirt gummed up the computer disc drives and secondly, the heat and humidity (especially in summer) caused the computer to crash from time to time. Problems were also encountered with the stability of the mains supply and a special transformer and mains conditioner had to be inserted. However, useful results were obtained and the unit ran continuously with little maintenance over periods of a week. Typical results are shown in Figure 23.

On February 13, 1984, a tomato crop was planted in the greenhouse at NIAE. With the exceptions of demonstrations at the Royal Institution and at the NATO Advanced Studies Institute on Advanced Agricultural Instrumentation, the unit has been monitoring the nutrient film continuously since that date. At present we can, therefore, claim that ANFA has produced reliable results for a period of several months. The unit has been serviced by a technician who has had no previous acquaintance with computer controlled instrumentation. Typical results for a week are shown in Figure 24. It is still too early to draw definite conclusions about the patterns of nutrient uptake or of the optimum conditions for crop yield, but collaborative work with NIAE and with the Glasshouse Crops Research Institute is in progress.

19. DEVELOPMENT OF NEW ELECTROCHEMICAL SENSORS

Besides the construction and application of the ANFA, we have also been designing and developing new electrochemical sensors for use in hydroponics. As these sensors are developed and tested in the laboratory, we will add them to the ANFA. We have described above work on CO_2 and amperometric NO_3^- sensors. Here we will now describe recent work on a sensor for total iron and the bromine microtitration technique as an early indicator of root death.

20. TOTAL IRON SENSOR

Iron is added to the nutrient solution as an EDTA complex. However, the soluble iron can exist in both the reduced (Fe(II)) and oxidized state (Fe(III)). It is desirable to know the total iron content (Fe(II) + (Fe(III)). To determine the total iron, we use the packed bed wall-jet electrode described above. In this case the packed bed consists of a porous carbon bed (vitrecarb from Fluorocarbon).

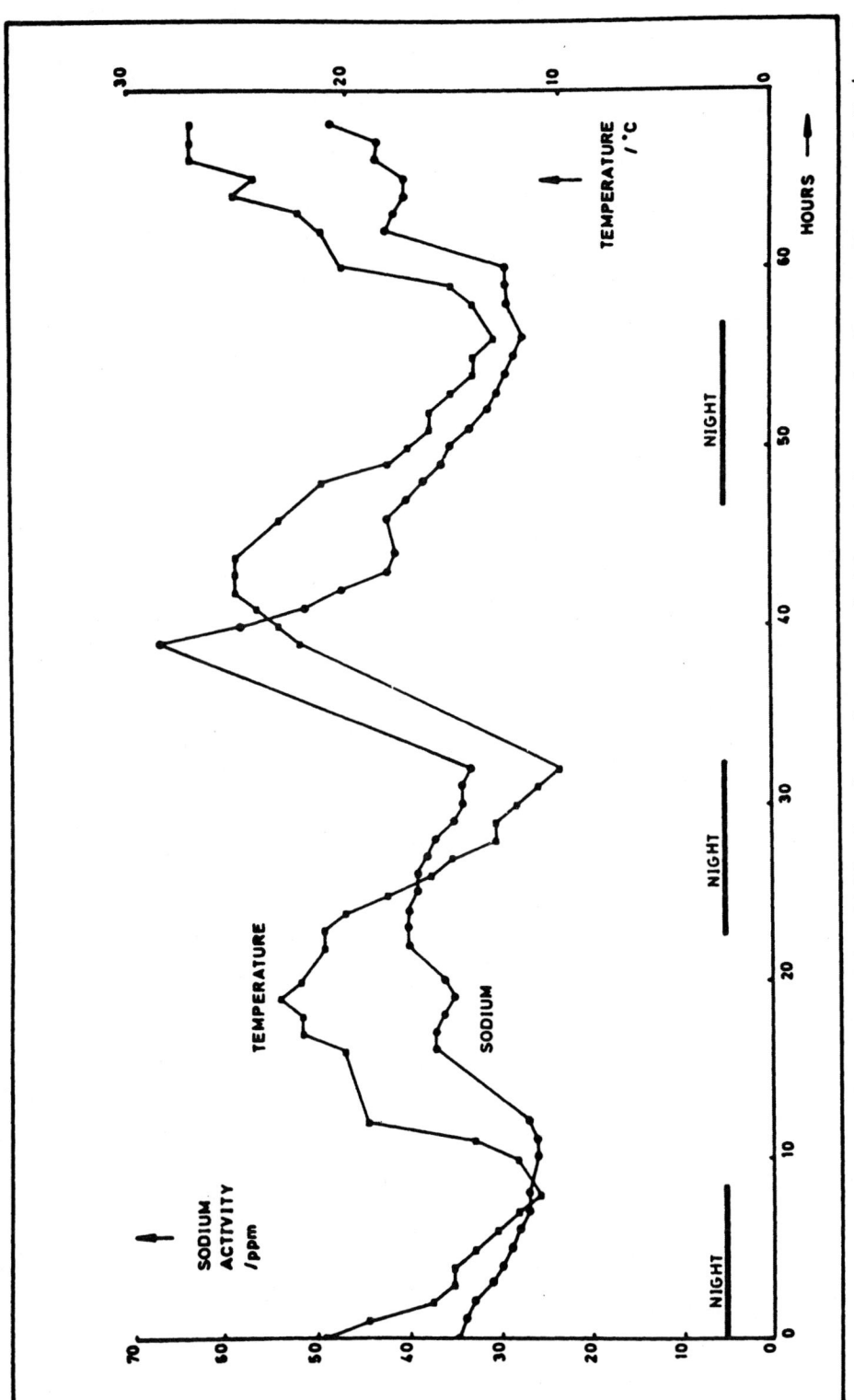

Figure 23. Typical results from the ANFA showing the correlation between air temperature and Na$^+$ activity in the nutrient film. The correlation arises because higher temperature causes increased transpiration and hence increased injection of Na$^+$ from the mains water supply.

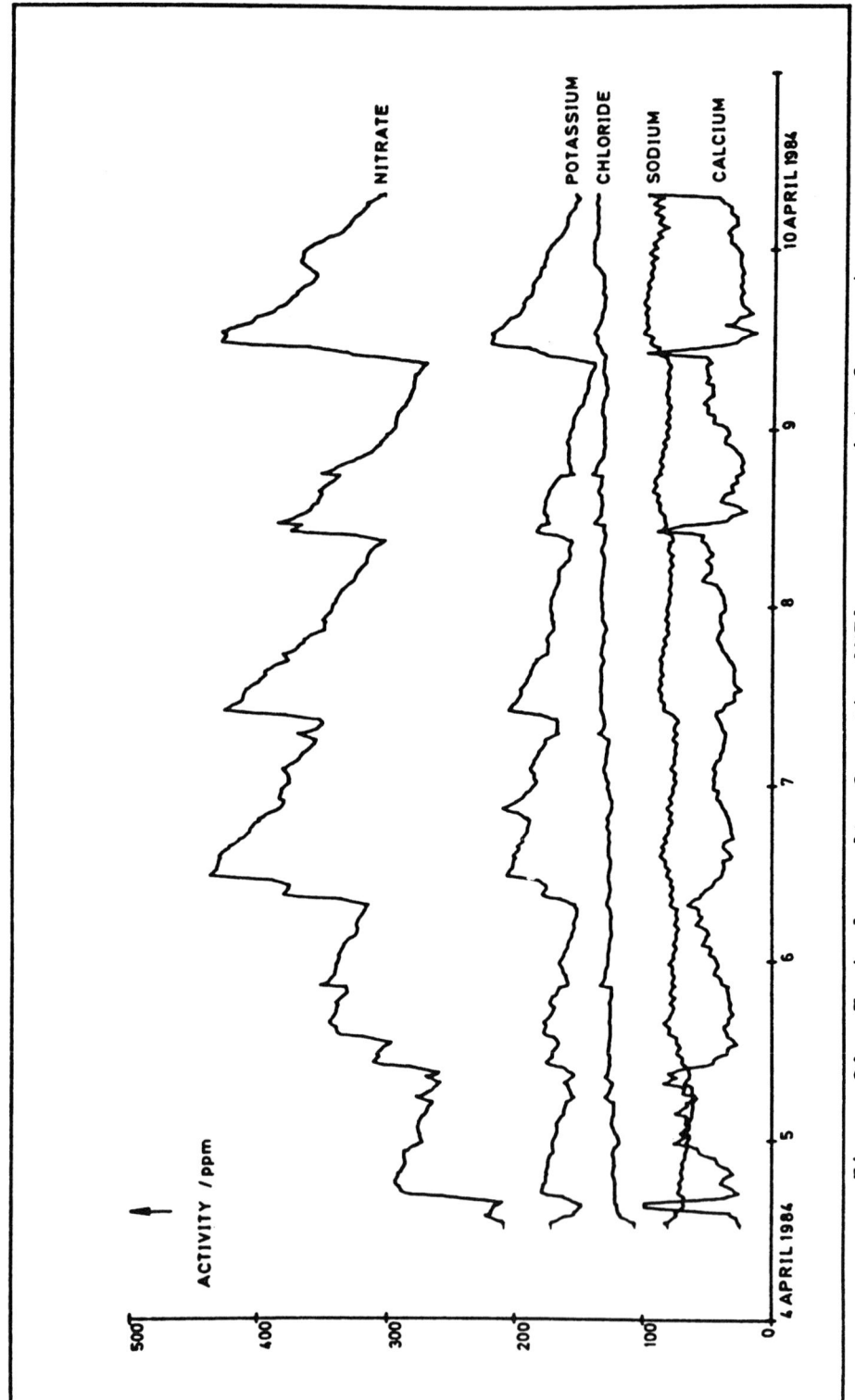

Figure 24. Typical results from the ANFA over a period of a week.

The bed is used to reduce Fe(III) to Fe(II) and also to remove O_2:

Packed Bed $e + Fe(III) \longrightarrow Fe(II)$

Electrode $4e + O_2 + 4H^+ \longrightarrow 2H_2O$

After the packed bed all the soluble iron is present as Fe(II) and can now be determined by oxidation on the wall-jet electrode:

Wall-jet Electrode $Fe(II) \longrightarrow Fe(III) + e$

Figure 25 shows typical results obtained in the laboratory with this system. Work is in hand on adding this electrode to the ANFA.

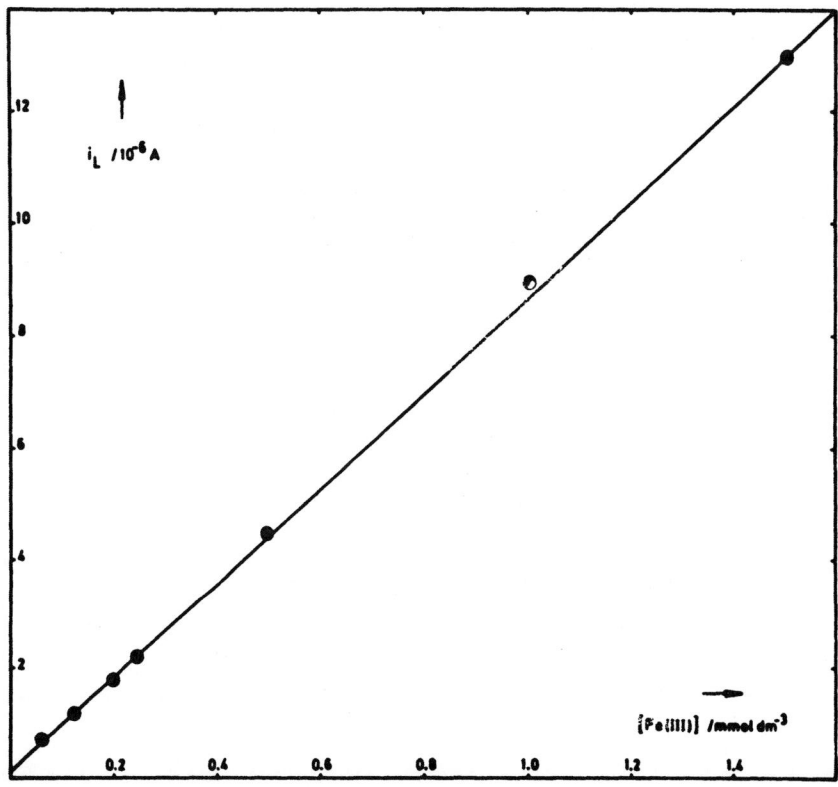

Figure 25. Typical results for the determination of Fe using the packed bed wall-jet electrode.

21. ROOT DEATH SENSOR

When root death occurs we expect that organics and proteins will be released into the nutrient film. As discussed above, these molecules have many groups that react rapidly with bromine in solution. Hence we use the bromine microtitration technique to monitor the number of brominatable groups in the nutrient solution. A rise in this number should be an indication of trouble. We have tested this concept in the laboratory.

Nutrient solution provided from the Glasshouse Crops Research Institute (GCRI) was used to grow Impatiens (Petersiana) plants with reasonable roots. A sample nutrient solution was taken after adding background electrolyte and buffer, a titration curve was recorded (Fig. 26 - curve a). Only a small amount of organic material was shown to be present, since the titration curve obtained was very close to the collection efficiency curve working only with the background electrolyte. In order to kill the plant it was kept in the dark for a few days.

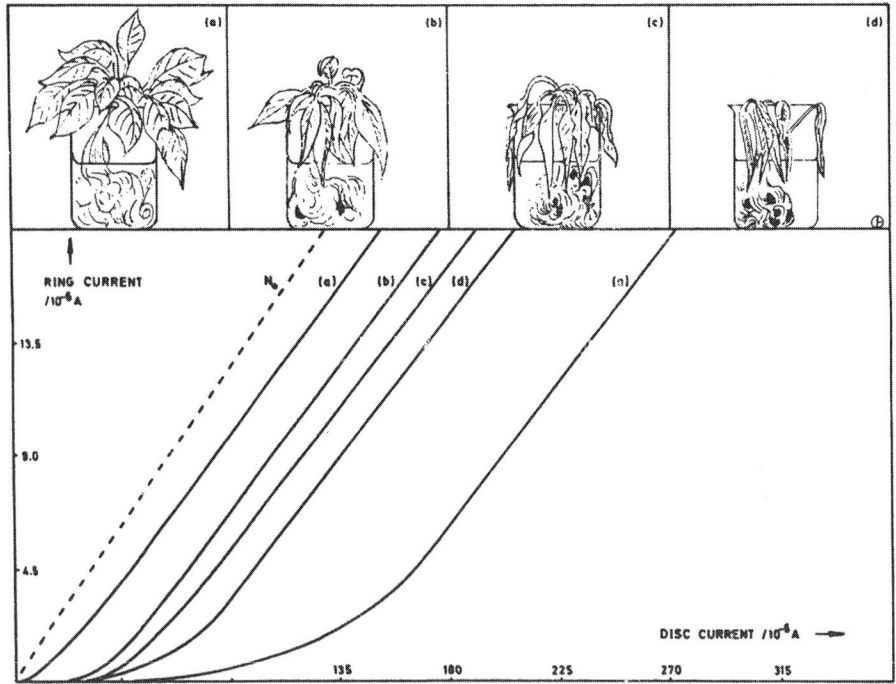

Figure 26. The murder of Busy Lizzy. The successive displacements of the ring-disc titration curves show the increasing amounts of brominatable organic material in the nutrient solution attending her demise.

Figure 26 shows different stages of the plant's demise: a) healthy plant; b) the leaves begin to wilt; c) advanced wilting; d) dead plant. The corresponding titration curves show the increase of organic materials. The correlation between the health of the plant and the "bromine titration concentration" is clear. The great advantage of the microprocessor controlled ANFA is that these new sensors can be added as they become available. We expect that this type of instrumentation will lead to greater understanding of plant physiology and to better control for the practicing commercial horticulturalist.

ACKNOWLEDGEMENTS

We would like to thank the AFRC for financial support. We are grateful to our colleagues, Dr. Brian Legg, Dr. Bernard Bailey, George Weaving and Alan Hunter of NIAE and to Dr. Peter Adams and Dr. Christopher Graves of GCRI. We thank the following of our co-workers at Imperial College for their contributions to this program: Dr. Nicholas Goddard and Humphrey Drummond of the Microprocessor Unit, Dr. Philip Bartlett and Derek Craston for their work on enzyme electrodes, Mrs. Nani Mendes-Neto for her work on the total iron and root death sensors, Dr. Guogang Pu for measuring the ISE selectivity coefficients and Christopher Jones for his work on the packed bed wall-jet electrode. Finally, we are grateful to NATO and to Professor William Gensler for organizing the NATO Workshop.

REFERENCES

1. Albery, W.J. Electrode Kinetics, p. 48, Oxford: Clarendon Press, 1975.

2. Clark, L.C., Jr. Monitor and control of blood and tissue oxygen tensions. Trans. Am. Soc. Artificial Internal Organs 2:41-49, 1965.

3. Hitchman, M.L. Measurement of Dissolved Oxygen, New York: Wiley-Interscience, 1978.

4. Albery, W.J., Brooks, W.N., Gibson, S.P. and Hahn, C.E.W. An electrode for PN_0 and PO_2 analysis in blood and gas. J. Applied Physiology 45:637-642, 1978.

5. Albery, W.J., Brooks, W.N., Gibson, S.P., Heslop, M.W. and Hahn, C.E.W. An electroanalytical method for the determination of N_2O. Electrochimica Acta 24:107-108, 1979.

6. Hahn, C.E., Brooks, W.N., Albery, W.J. and Rolfe, P. O_2 and N_2O analysis with a single intravascular catheter electrode. Anaesthesia 34:263-266, 1979.

7. Brooks, W.N., Hahn, C.E.W., Foex, P., Maynard, P. and Albery, W.J. The simultaneous measurement of PO_2 and PN_2O in vivo with a single catheter intravascular electrode. Br. J. Anaesth. 50:1082-1085, 1978.

388

8.   Albery, W.J., Hahn, C.E.W. and Brooks, W.N. The polarographic measurement of halothane. Br. J. Anaesth. 53:447-453, 1981.

9.   Albery, W.J. and Barron, P. A membrane electrode for the determination of CO_2 and O_2. J. Electroanal. Chem. 138:79-87, 1982.

10. Bailey, P.L. Analysis With Ion Selective Electrodes, London: Heyden, 1978.

11. Midgley, D. and Torrence, K. Potentiometric Water Analysis, London: Wiley, 1978.

12. Koryta, J. Ion Selective Electrodes, Cambridge: Cambridge University Press, 1975.

13. Freiser, H. Ion-selective Electrodes in Analytical Chemistry, Vol. 1, New York: Plenum Press, 1987.

14. Koryta, J. Medical and Biological Applications of Electro-chemical Devices, London: Wiley, 1980.

15. Arnold, M.A. and Meyerhoff, M.E. Ion-selective electrodes. Anal. Chem. Reviews 20R-48R, 1984.

16. Covington, A.K., Harbinson, T.R. and Sibbald, A. A field effect transistor sensitive to phenobarbitol anion. Anal. Lett 15:1423-1429, 1982.

17. Sibbald, A., Covington, A.K, Cooper, E.A. and Carter, R.F. On-line measurement of potassium in blood by chemical sensitive field-effect transistors: a preliminary report. Clin. Chem. 29:405-406, 1983.

18. Sibbald, A., Covington, A.K. and Carter, R.F. Simultaneous on-line measurement of blood potassium, calcium, sodium and pH with a four-function ChemFET integrated circuit sensor. Clin. Chem. 30:135-137, 1984.

19. Sibbald, A., Whalley, P.D. and Covington, A.K. A miniature flow-through cell with a four-function ChemFET integrated circuit for simultaneous measurements of potassium, hydrogen, calcium and sodium ions. Anal. Chim. Acta. 159:47-62, 1984.

20. Sibbald, A. and Covington, A.K. Encapsulated chemoresponsive microelectronic device arrays. Eur.Pat.Appl. EP63, 455, 1982.

21. Albery, W.J. Electrode Kinetics, pp. 49-71, Oxfgord, Clarendon Press, 1975.

22. Bard, A.J. and Faulkner, L.R. Electrochemical Methods, pp. 282-311, New York, Wiley, 1980.

23. Yamada, J. and Matsuda, H. Limiting diffusion currents in hydrodynamic voltammetry. III. Wall-jet electrodes. J. Electroanal. Chem. 162:189-198, 1984.

24. Fleet, B. and Little, C.J. Design and evaluation of electro-chemical detectors for HPLC (high-pressure liquid chromatography). J. Chromatogr. Sci. 12:747-752, 1974.

25. Varadi, M. and Pungor, E. Turbulent hydrohynamic voltammetry. III. Analytical investigations with a turbulent voltammetric cell and applications to amino acid analysis. Anal. Chim. Acta 94:351-356, 1977.

26. Albert, W.J. and Brett, C.M.A. The wall-jet ring-disc electrode. I. Theory. J. Electroanal. Chem. 148:201-210, 1983.

27. Albert, W.J. and Brett, C.M.A. The wall-jet ring-disc electrode. II. Collection efficiency, titration curves and anodic stripping voltammetry. J. Electroanal.Chem. 148:211-220, 1983.

28. Albery, W.J., Svanberg, L.R and Wood, P. The estimation and identification of proteins by ring-disc titration. I. Titration and identification. J. Electroanal.Chem. 162:29-43,1984.

29. Albery, W.J., Svanberg, L.R. and Wood, P. The estimation and identification of proteins by ring-disc titration. II. Application to liquid chromatography. J. Electroanal. Chem. 162:45-53, 1984.

30. Albery, W.J. and Hitchman, M.L. Ring-disc Electrodes, Oxford, Clarendon Press, 1971.

31. Albery, W.J., Haggett, B.G.D., Jones, C.P. and Svanberg, L.R. Packed bed wall-jet electrode and the determination of nitrate. J. Electroanal. Chem., submitted, 1984.

32. Pletcher, D. and Poorabedi, Z. The reduction of nitrate at a copper cathode in aqueous solution. Electrochimica Acta 24:1253-1256, 1979.

33. Guilbault, G.G. and Sadar, M.H. Preparation and analytical uses of immobilized enzymes. Accounts of Chem. Res. 12:344-350, 1979.

34. Guilbault, G.G. Enzyme electrode probes. Enzyme Microb. Technol. 2:258-264, 1980.

35. Skogberg, D. and Richardson, T. Enzyme electrodes for the food industry. J. Food Protection 42:808-819, 1980.

36. Cass, A.E.G., Davis, G., Francis, G.D., Hill, H.A.O., Aston, W.J., Higgins, I.J., Plotkin, E.V. and Scott, L.D.L. Ferrocene-mediated enzyme electrode for amperometric determination of glucose. Anal. Chem. 56:667-671, 1984.

37. Cass, A.E.G., Davis, G., Hill, H.A.O., Higgins, I.J., Plotkin, E.V., Turner, A.P.F. and Aston, W.J. Amperometric enzyme electrode for blucose determination. In: Charge and Field Effects in Biosystems (M.J. Allen and P.N.R. Usherwood, eds.) pp. 475-482, U.K., Abacus Press, 1984.

38. Cass, A.E.G., Davis, G., Hill, H.A.O. and Nancarrow, D.J. The reaction of flavocytochrome b_2 with cytochrome c and ferricinium carboxylate. Comparative kinetics by cyclic voltammetry and chronoamperometry. Biochim. Biophys. Acta, in press, 1984.

39. Kulys, J.J., Samalius, A.S. and Svirmickas, G.J.S. Electron exchange between the enzyme active center and organic metal. FEBS Lett. 114:7-10, 1980.

40. Kulys, J.J. and Samalius, A.S. Kinetics of biocatalytic current generation. Bioelectrochem. Bioenerg. 10:385-393, 1983.

41. Albery, W.J., Bartlett, P.N. and Craston, D. Amperometric enzyme electrodes. J. Electroanal. Chem., submitted, 1984.

42. Albery, W.J. and Bartlett, P.N. An organic conductor electrode for the oxidation of NADH. J.Chem. Soc. Chem. Comm. 234, 1984.

43. Gericke, W.F. The Complete Guide to Soiless Gardening. London, Putnam, 1940.

44. Ticquet, C.E. Successful Gardening Without Soil. London, Pearson, 1952.

45. Hollis, H.F. Profitable Growing Without Soil. London, The English University Press, 1964.

46. Schwarz, M. Guide to Commercial Hydroponics. Jerusalem, Israel University Press, 1968.

47. Douglas, J.S. Hydroponics: The Bengal System. Oxford, Oxford University Press, 1976.

48. Douglas, J.S. Advanced Guide to Hydroponics. London, Pelham Books, 1976.

49. Cooper, A. Nutrient Film Technique of Growing Crops. London, Grower Books, 1976.

50. Resh, H.M. Hydroponic Food Production. Santa Barbara, Wood-bridge Press, 1978.

51. Symposium on research on recirculating water culture. Acta Horticulturae 98, 1980.

52. Graves, C.J. The nutrient film technique. Horticultural Reviews 5, 1983.

53. Richardson, S.J. Crop nutrition in nutrient film culture. Fert. Soc. London, December 1981.

54. Winsor, G.W. Some nutritional aspects of nutrient film culture (NFT). Proc. 9th Int. Pl. Nutr. Col., England, 722-727, 1982.

55. Moorby, J. Effects of manipulating root and air temperatures on tomato growth and the efficient use of energy. In: Opportunities of Increasing Crop Yields (R.G. Hurd, P.V. Biscoe and C. Denis, eds.) pp. 183-194, London, Pitman, 1982.

56. Orchard, B. Solution heating for the tomato crop. Acta Hort. 98:19-28, 1980.

57. Price, D. and Dickinson, A. Fungicides and the nutrient film technique. Acta Hort. 98:277-282, 1980.

58. Morgan, J.V. and Tan, A. Production of greenhouse lettuce at high densities in hydroponics. Proc. 21st Inst. Hort. Congr., Hamburg Abstract Vol. 1, 1620.

59. Giacomelli, G.A., Roberts, W.J., Mears, D.R. and Janes, H.W. An alternative concept in greenhouse tomato production. Annual Meeting North Atlantic Region of Amer. Soc. Agri. Eng., University of Vermont, Burlington, 1982.

60. Graves, C.J. The nutrient film technique. Hort. Rev. 5:22-23, 1983.

61. Attenburrow, D.C. and Waller, P.L. Sodium chloride: its effect on nutrient uptake and crop yields with tomatoes in NFT. Acta Hort. 98:229-236, 1980.

62. McDonald, J. and Parsby, J. Implications of small plant (laboratory) studies to the design of field instrumentation and measurement systems. Paper given at NATO Workshop on Advanced Agricultural Instrumentation, Il Ciocco, Pisa, Italy, 1984.

63. Weaving, G.S. and Hunter, A. An assessment of ion selective electrodes for monitoring nutrient culture systems, National Institute of Agricultural Engineering, Divisional Note 1143, December 1982.

PRESENTATION AND IMPLICATIONS OF AN EXPERIMENTAL GROWTH SYSTEM

A.J.S. McDonald and J. Parsby

Swedish University of Agricultural Sciences
Uppsala, Sweden

The intention of this paper is to present, in summary form, a system for the experimental regulation of plant growth rate. The use of this system has led to results which might hopefully promote discussion concerning a priority of experimentation with regard to understanding crop growth, and what this might entail in terms of experimental design and measurement systems. This is a quite different emphasis from that pertaining to the direct use of instrumentation in agricultural management (7). Nor is the aim primarily one of predicting harvest yield, although that information may be forthcoming. If yield prediction on a more extensive geographical basis is the sole intention, then measurement or calculation of total absorbed radiation during the growth season is likely to be highly correlated with above-ground dry matter production (15,18).

On the other hand, neither of the above emphases gives direct information on what growth processes have been affected and how they contribute to differences in production. This more functional information has, of course, been the subject of much ecophysiological experimentation. At the outset it might be stated that more functional or mechanistic information is not necessarily of immediate benefit to crop management. It might be emphasized that the aim of ecophysiological experimentation is to understand the biology associated with production; the genetic adaptation which is associated with a certain acclimation to stress. In the long-term, such information may or may not lead to a more intelligent management of crop production.

Although a more mechanistic or functional approach may neither be essential nor necessarily the most practical for crop yield prediction, it might still be argued that process information is only of potential use to crop management when set in an ecophysiological context of whole plant or crop growth. This has not always been the case. Only a few, for example, of the many studies pertaining to photosynthesis and respiration in forest tree species have been related to growth (13,14). This is perhaps not so surprising from a number of points of view. Firstly, the amount of feedback in carbon gain and water loss at the single leaf/shoot level renders the study of either water relations or carbon gain processes apart from one another arguably meaningless with respect to plant ecophysiology (19). Secondly, it has become apparent that the often a priori assumption that photosynthetic studies should be given priority in understanding growth is extremely dubious. Much evidence points to the limiting importance of subsequent carbon allocation after photosynthesis in determining growth (2,5,6), which raises the intriguing possibility that net assimilation rate of crop leaves is rarely limiting to production on a seasonal (and perhaps even diurnal) basis. This does not necessarily point to a reduction of effort with respect to understanding photosynthesis, but it should at least suggest the inadequacy of studying net carbon gain to the exclusion of subsequent allocation processes. Thirdly, the intention of relating ecophysiological studies to plant or crop growth raises problems since, in most laboratory or field studies, growth rate is unknown. Ideally, the data pertaining to crop growth measurement should be independent of that pertaining to growth processes being studied.

To achieve a more functional understanding of crop growth, it would seem necessary to experiment with an appropriate range in those driving variables which are most likely to affect growth. In practice this has meant field experimentation with water and nutrient variables. However, although it would seem possible to investigate the affects of limited nutrient supply on growth at near-optimal soil water availability (16), it is extremely difficult to achieve limited water supply at optimal nutrition. In other words, when studying the effects of limited water supply on growth, nutrition may not always be independent of water supply. This means that the interpretation of field water stress experiments is complicated since the affect of the water variable may be confounded by simultaneous changes in the nutrient variable. This is true of both field and laboratory experiments.

The choice of nutrient variable in laboratory and field growth experiments has been the subject of much discussion (9). In the majority of studies a concentration variable has been used. This is true of laboratory studies where constant volumes of nutrient solutions of different concentrations are periodically added (e.g. to potted plants) or where nutrient solutions of different concentrations have been changed or replenished at regular intervals. In field experiments, nutrients are normally given as single or repeated dosages of stock fertilizer of various amounts. In all such instances, the concentration variable is inappropriate to maintaining a constant level of nutrient stress since it is not related to the growth of the plant. This is intuitively correct since, as plants grow larger, the uptake rate of nutrient must increase in proportion to dry matter gain if the rates of nutrient-dependent growth conversions in the plant are to be maintained. This phenomenon has been discussed more quantitatively by Ingestad et al. (11) and Linder and Rook (17).

The significance of nutrient uptake to plant growth rate has been demonstrated repeatedly with small plants from many species (3,4,8,10,12,20). The findings from a number of laboratory studies are summarized in Figure 1.

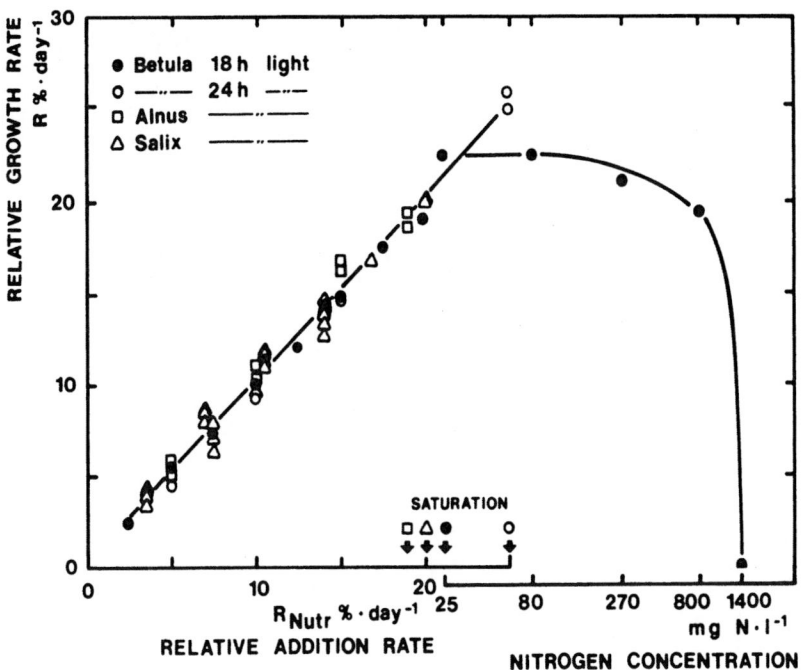

Figure 1. Relationships between relative growth rate (% d.w. increase per day) and the external nutrient factor during the exponential period of growth of some broad-leaf tree species. Within the range of sub-optimum and optimum nutrition (up to saturation), the relative nutrient addition rate is the driving variable, and within the supra-optimum range, the external concentration is relevant (11).

The data pertain to small plants (typically less than 5 g fresh weight) whose roots are bathed by a circulating nutrient solution (Figure 2). The important feature of the growth technique is the regulated delivery of nutrient amounts. This is achieved through the programmable opening of magnetic valves through which nutrients are added to a circulating volume of solution. This system is now commercially available through Biotronic, S-752 31 Sweden.

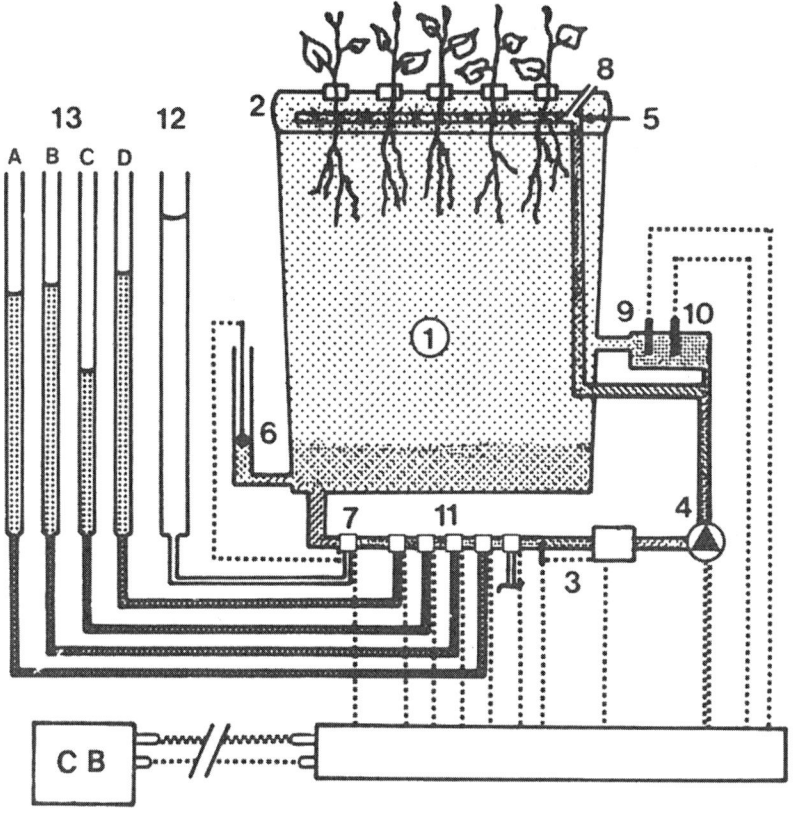

Figure 2. Schematic description of growth unit.
1. Growth box with nutrient solution mist.
2. Cover with holes for 120 seedlings.
3. Temperature sensor with cooling and heating unit.
4. Electrical pump with indirect power transmission for circulation of the nutrient solution.
5. Tubes with nozzles for spray of nutrient solution.
6. Level sensor.
7. Magnetic valve for addition of distilled water.
8. Aeration nozzle.
9. pH transmitter with measurement cell.
10. Conductivity transmitter with measurement cell.
11. Computer controlled magnetic valves.
12. Burette for distilled water.
13. Burettes for liquid fertilizers.

In Figure 1, exponentially increasing amounts of nutrient were added to the circulating solution at regular time intervals (e.g., 1 hr). Towards the end of each period, the conductivity of the solution is very low (typically less than 30 µS). This implies that, in the above experiments, the nutrient uptake rate is equal to the rate of nutrient supply. In Figure 1, the nutrient supply has been increased exponentially with different exponents, R_{Nutr} ($\% d^{-1}$). It is apparent that the relative growth rate, R ($\% d^{-1}$) is equal to the relative rate of nutrient addition. Thus, the exponent for dry matter gain is the same as that for nutrient uptake. This holds true for rates up to and including some maximum value of R which is dependent upon both genotype and growth climate (Fig. 1). The effect of increasing R_{Nutr} above maximum R is a time-dependent increase in concentration of nutrients in solution. The negative effect of increasing concentration on maximum R is shown on the concentration axis of Figure 1.

If plants are grown such that nutrient stress levels are maintained constant, an acclimation occurs involving characteristic values of structure and functional processes. This may be illustrated with respect to data pertaining to a shrub willow (<u>Salix dasyclados</u>). The data are from an experiment described by Waring, et al (20). In summary, replicate willow plants were grown in the laboratory at two light levels and with free access to nutrients. At the higher light level, a further treatment with constant nutrient stress (R_{Nutr} = 6 $\% d^{-1}$) was made. As might be expected from Figure 2, the relative growth rate of the nutrient stressed treatment (N_{low}) was approximately equal to the relative rate of nutrient supply. The free access treatments had maximum relative growth rates characteristic of the climate and genotype (Table 1). Partitioning of dry matter to roots was found to increase with increasing nutrient stress, but was only weakly dependent upon growth light level.

TABLE 1
Growth response data to nutrient and light stress in <u>Salix.</u>

Plants were grown at low light with optimum nutrient supply (C_{low}), high light with optimum nutrient supply (Opt) and high light with sub-optimal nutrient supply (N_{low}). Means followed by the same letters are not significantly different (LSD, 0.05 level). Number of plants per treatment = 9, except for gas-exchange and relative growth rate data, n=5. (Waring, 20). Nitrogen productivity and conductance data have been added.

Variable	C_{low}	Opt	N_{low}
Measured relative growth rate, R($\%d^{-1}$)	6.8[a]	16.1[b]	5.5[a]
Root dry weight: Total dry weight	.15[a]	.20[b]	.38[c]
Starch in leaves (% d.w.)	5.3[a]	5.1[a]	20.7[b]
Net photosynthesis (mg CO_2 dm^{-2} h^{-1})	1.9[a]	12.5[b]	13.1[b]
Nigrogen productivity of photosynthesis (mg CO_2 g leaf N^{-1} h^{-1})	135[a]	524[b]	863[c]
CO_2 conductance (mm s^{-1}) ... stomatal	2.6[a]	4.0[b]	4.2[b]
... mesophyll	0.1[a]	0.7[b]	0.7[b]

Although these data say something about growth response to nutrient stress (and something of the ecophysiology of Salix), they do not contribute much at the level of a more functional ecophysiology, in the sense of adaptive acclimation to stress. When measurement of carbon storage (starch) is made on the above Salix treatments, it is apparent (Table 1) that carbon uptake in the nutrient stressed treatment is in excess of the immediate carbon requirement for structural growth. The extent to which leaf carbon uptake is uncoupled from the immediate incorporation of carbon in new structural growth can be further investigated by measurements of leaf gas-exchange. Here, it was found that net carbon uptake per unit leaf area was independent of plant nutrient stress (although strongly dependent upon growth light level). In other words, the decrease in plant growth rate in response to nutrient stress is entirely attributable to the subsequent allocation of carbon within the plant (decreased leaf partitioning) and not the acclimation of carbon uptake in individual leaves. If the nitrogen productivity of photosynthesis is calculated (net photosynthetic rate per unit leaf nitrogen), it is found to increase with nutrient stress (Table 1). It would appear that nitrogen investment in photosynthetic processes has priority over that in subsequent growth conversions; less nitrogen is incorporated in enzymes of structural growth with increasing nutrient stress. Interestingly, stomatal and mesophyll conductances (and thus intercellular CO_2 concentration) were also independent of plant nutrient stress.

The ecophysiological interpretation of these data may not be striaghtfoward in terms of optimization of function. On the other hand, it may be concluded that the acclimation of leaf conductances to plant nutrient stress is such that net carbon uptake is unlikely to limit new structural growth and that transpiration loss is consistent with a cooling function common to single leaves from plants of different nutrient stress.

The type of result obtained with the above growth system may encourage some critical thought pertaining to the need for and/or the design of ecophysiological experiments in the field. At best, of course, the type of laboratory study described, may only give rise to hypotheses about how field plants function. However, this alone would seem to be a good motivation for such studies. In practice, it may be very difficult to test such hypotheses in the field.

It may, however, be appropriate to re-emphasize that the intended level of understanding has to do with the ecophysiology of crop plants in the field. This, then, is the dilemma: to carry out experiments in the laboratory which are somewhat compromised but under good experimental control or to attempt field experimentation with poor control over growth variables.

One of the most obvious differences between field and laboratory plants has to do with size. The type of laboratory experiment described above is with small plants which may be considered independent of one another as opposed to a crop stand with, for example, a large amount of plant shading. However, to investigate the eco-physiology of nutrient stress in larger (stand) plants, it is not immediately obvious that field experimentation is more appropriate than a laboratory approach. The laboratory system as presented is quite suitable for growth of much larger plants with self or stand shading at regulated rates of nutrient supply. This may have very real advantages over field experimentation where nutrient supply and uptake (as a function of root growth, mineralization and fertil-ization) may be extremely difficult to either regulate or quantify. It may well be that most progress in understanding crop response to nutrient stress will come from laboratory studies with larger plants and regulated nutrient supply. This approach might, for example, represent a very good intermediate step in translating the above results on leaf gas-exchange and plant growth with small plants, to leaf function in field crops.

There are many criticisms which might be leveled at the inter-pretation of field plant function from laboratory studies. For example, the field climate comprises a very variable pattern of weather in contrast to the typically more constant climate of laboratory studies. This may or may not be of great importance in determining acclimation to stress. For example, Chabot, et al (1) have shown that photosynthetic capacity is much more dependent on the integrated diurnal quanta absorption than on the absolute values of photon flux density. At least some of the criticisms pertaining to growth climate could be investigated by appropriate variation in laboratory climate.

In conclusion then, it is a great advantage in ecophysiological studies to have control over growth variables. This is possible for nutrient uptake in laboratory-grown plants, but not so readily in field experiments. On the other hand, many aspects of acclimation to light and nutrient stress can be investigated in the laboratory, likewise, the limitations of, for example, constancy in climate. This is true of growth structural and functional (e.g. leaf gas-exchange) response. In terms of understanding the acclimation to stress which results in decreased plant growth rates, it would appear that a much greater effort is required in the area of growth conversions other than net carbon uptake. This will involve at least as much effort in the choice of experimental design as in the development of existing measurement systems.

References

1. Chabot, B.F., Jurik, T.W. and Chabot, J.F. Influence of instantaneous and integrated light-flux density on leaf anatomy and photosynthesis. Amer. J. Bot. 66:940-945, 1979.

2. Elmore, C.D. The paradox of no correlation between leaf photo-synthetic rate and crop yields. In: Predicting Photosynthesis for Ecosystem Models Vol. 2 (J.D. Hesketh and J.W. Jones, eds.) pp. 155-167, 1980.

3. Ericsson, T. Effects of varied nitrogen stress on growth and nutrition in three _Salix_ clones. __Physiol. Plant.__ 51:423-429, 1981.

4. Ericsson, T. Growth and nutrition of three _Salix_ clones in low conductivity solutions. __Physiol. Plant.__ 52:239-244, 1981.

5. Evans, L.T. The physiological basis of crop yield. In: __Crop Physiology: Some Case Histories__ (L.T. Evans, ed.) pp. 327-355, 1975.

6. Gifford, R.M. and Evans, L.T. Photosynthesis, carbon partitioning, and yields. __Ann. Rev. Plant. Physiol.__ 32:485-509, 1981.

7. Gensler, W.G. Stem diameter and electrochemical measurements. See this volume.

8. Ingestad, T. Nutrition and growth of Birch and Grey Alder seedlings in low conductivity solutions and at varied relative rate of nutrition. __Physiol. Plant.__ 52:454-466, 1981.

9. Ingestad, T. Relative addition rate and external concentration: Driving variables used in plant nutrition research. __Plant, Cell and Environ.__ 443-453, 1982.

10. Ingestad, T. and Lund, A.B. Nitrogen stress in Birch seedlings. 1. Growth technique and growh. __Physiol. Plant__ 45:137-148, 1979.

11. Ingestad, T., Aronsson, A. and Agren, G.I. Nutrient flux density model of mineral nutrition in conifer ecosystems. __Stu. For. Suec.__ 160:61-71, 1981.

12. Jia, H. and Ingestad, T. Nutrient requirements and stress response of _Populus simonii_ and _Paulownua tomentosa_. __Physiol. Plant__ 62:117-124, 1984.

13. Linder, S. Photosynthesis and respiration in conifers: A classified reference list 1891-1977. __Stud. For. Suec.__ 149:71, 1979.

14. Linder, S. Photosynthesis and respiration in conifers: A classified reference list, Supplement 1. __Stud. For. Suec.__ 161:32, 1981.

15. Linder, S. Potential and actual production in Australian forest stands. In: __Research for Forest Management__ (J.J. Landsberg and W. Parsons, eds.) CSIRO, Melbourne, in press.

16. Linder, S. and Axelsson, B. Changes in carbon uptake and allocation patterns as a result of irrigation and fertilization in a young Pinus sylvestris stand. In: Carbon Uptake and Allocation in Subalpine Ecosystems as a Key to Management (R.H. Waring, ed.), Forest Research Laboratory, Oregon State University, Corvallis, OR, 1982.

17. Linder, S. and Rook, D.A. Effects of mineral nutrition on carbon dioxide exchange and partitioning of carbon in trees. In: Nutrition of Plantation Forests (G.D. Bowen and E.K.S. Nambian, eds.), Academic Press, pp. 211-236.

18. Monteith, J.L. Climate and efficiency of crop production in Britain. Phil. Trans. R. Soc. Lond. B. 281:277-294, 1977.

19. Norman, J.M. Instrumentation use in a comprehensive description of plant-environment interactions. See this volume.

20. Waring, R.H., McDonald, A.J.S., Larsson, S., Ericsson, T., Wiren, A., Arwidsson, E., Ericsson, A. and Lohammar, T. Differences in chemical composition in plants grown at constant relative growth rates with stable mineral nutrition. Oecologia (Berlin) 66:157-160, 1985.

ION TRANSPORT PROCESSES IN CORN ROOTS: AN APPROACH UTILIZING MICROELECTRODE TECHNIQUES

William J. Lucas and Leon V. Kochian

Department of Botany
University of California
Davis, California 95616

1. INTRODUCTION

There have been many studies conducted on ion or salt absorption by plant roots. This research has focused both on the basic mechanisms involved in transporting the ion into the plant and on the mineral nutrient status of the specific ion of interest (see, e.g. 7, 10, 17, 34). It is our intent here to present an outline of the previous work on K^+ transport, in order to provide the background from which the present studies arose.

1.1 Kinetics of Uptake

The early kinetic studies conducted by Epstein and his coworkers were particularly insightful. Epstein and Hagen (11), noting that ion uptake was selective and dependent on metabolism, hypothesized that ion transport was analogous to enzyme-substrate binding. Such a carrier system, locatd within the plasmalemma, could explain the kinetics obtained (rectangular hyperbola) when roots were exposed to low ionic concentrations (0-1.0 mM). Epstein and coworkers went on to show that at higher concentrations a second transport mechanism appeared to operate, and this system required much higher concentrations to achieve "saturation." This biphasic pattern was termed the "dual isotherm of uptake" by Epstein and was thought to reflect the operation of two separate classes of carriers, each following distinctly different Michaelis-Menten kinetics. In the low concentration range (<0.5 mM), Mechanism I was hypothesized to be a high affinity, low velocity transporter. In the high concentration range (>1.0 mM), Mechanism II was proposed to operate as a low affinity, high velocity transport mechanism with lower substrate specificity.

1.2 Site of K⁺ Uptake

Although numerous alternative hypotheses have been presented to explain the kinetic data (see 5, 20, 21, 40, for the various viewpoints), we consider that Epstein's basic concept is correct, except that System II is probably a first-order kinetic system. Hence, the kinetic studies established that K⁺ is transported into the symplast of the root via at least two separate systems.

A further point of dissension in the root ion transport literature concerns the pathway of ion movement from the soil solution into the symplast. Although most workers feel that both epidermal and cortical cells may be able to absorb K⁺ from the apoplastic solution, this has been questioned by several workers. Vakhimistrov (54) suggested that the cortical cells play an insignificant role in ion uptake, with most transport occurring at the epidermal cells. Using a theoretical model for Rb⁺ uptake into barley roots, Bange (2) found that reasonable kinetics were generated only when transport was restricted to the epidermis. The recent work of van Iren and Boers-van der Sluijs (55) also implicates the root periphery as the site of K⁺ uptake into the symplast. Their work was based on the assumption that plasmolyzing the root disrupts the plasmodesmata and thereby isolates individual cells of the epidermis and cortex. When barley roots were treated in this way, autoradiographic localization of ⁸⁶Rb⁺ showed that uptake was generally limited to the root periphery.

Recent work in our laboratory has also addressed this issue. We were concerned that plasmolysis may alter transport processes in a number of different ways, thereby weakening the conclusions drawn by van Iren and Boers-van der Sluijs. By employing radio-actively-labeled permeant and impermeant sulfhydryl reagents (N-ethyl maleimide and p-chloromercuribenzene sulfonic acid) we demonstrated that a very close correlation existed between the dramatic inhibition of K⁺ (⁸⁶Rb⁺) uptake and the restricted presence of the sulfhydryl reagent within the root periphery (22).

The results of our study indicated that although cortical cells when studied as isolated protoplasts, possessed the capacity to absorb ions, in the intact root situation K⁺ influx at low concentrations appears to be limited to the epidermis and hypodermis. Such a situation would add yet a further level of complexity to understanding the cellular mechanisms that are involved in ion transport and its regulation.

1.3 Mechanisms of K⁺Transport

Numerous studies have demonstrated that K⁺ uptake into root tissue is extremely sensitive to metabolic inhibitors. However, the problems of elucidating underlying mechanisms on the basis of such studies is all too obvious. Nevertheless, by combining electrophysiological, kinetic and inhibitor studies, it has been possible to determine that in the presence of low concentrations K⁺ uptake is an active process (44).

Most of the early and present-day studies have attempted to correlate K^+ uptake with H^+ movement(s) across the plasmalemma. Much of the pioneering work on plant ATPases linked their function to K^+ uptake. The first evidence for transport ATPases came from work with correlated K^+ influx in corn, wheat, oat, and barley roots with K^+-stimulated ATPase activity located in a microsomal membrane preparation isolated from these roots (14,15). Potassium-stimulated ATPase activity believed to be associated with plant plasma membranes was further characterized in subsequent studies (18, 24-27). The enzyme showed a pH optimum of 6.5, a requirement for Mg^{+2}, and was further stimulated by monovalent cations, particularly K^+ and Rb^+. Leonard and Hodges (24) showed that the complex kinetics for K^+ absorption in oat roots was quite similar to the kinetics observed for K^+ stimulation of ATPase activity.

Because of the correlation between plasma membrane-associated, K^+-stimulated ATPase activity and K^+ influx, and the similarity between the sequence for monovalent cation stimulation of ATPase activity ($K^+ > NH_4^+ > Rb^+ > Cs > Li^+$) and the specificity of mono-valent cation uptake into roots (52), it has been suggested by a number of workers that this ATPase is involved in K^+ uptake, possibly as a H^+/K^+ exchange system. More recently, it has been demonstrated that microsomal membrane vesicles isolated from tobacco callus, and believed to be plasmalemma in origin, exhibited a stimulation of K^+-ATPase activity in the presence of either nigericin, or valinomycin plus a protonophore (50). These results again suggest that a plasmalemma ATPase is involved in a K^+/H^+ exchange process.

As mentioned above, studies on K^+ and H^+ fluxes in excised roots have indicated a fairly close coupling between K^+ influx and H^+ efflux. It has been observed that substances which inhibit or stimulate net H^+ efflux, presumably by acting on a H^+-translocating ATPase (DCCD, fusicoccin, etc.), generally have a correspondingly similar effect on K^+ influx (6,16,28). This coupling of H^+ and K^+ fluxes is again consistent with the view that a plant ATPase may be involved both in H^+ transport and K^+ influx.

However, because of the mild stimulation of ATPase activity by K^+, the relatively high K^+ concentrations (50 mM) needed for stimu-lation, and the lack of evidence for K^+ uptake in isolated membrane vesicles, the direct involvement of the plasma membrane ATPase in K^+ transport has been questioned. Furthermore, the actual cellular origin of the membranes involved in ATPase activity and ion trans-port has been the subject of debate. It is only quite recently that ion transport has been demonstrated with microsomal membranes iso-lated from plant roots. Using microsomal membranes isolated from corn and presumably oriented "inside-out," several groups have dem-onstrated the existence of an ATP-dependent proton pump (4,8,37,49, 51). However, it is unclear whether the membrane vesicles exhibiting H^+ transport originate from the plasma membrane. In fact, several groups have shown that the membrane vesicles involved in proton pumping are separable from plasma membranes by density gradient centrifugation, and have suggested that the membranes are tonoplast in origin (8,35). Bennett and Spanswick (4) demonstrated that a membrane fraction, isolated from corn roots, contained a Cl$^-$-sensitive, H$^+$-translocating ATPase, distinctly different from the K$^+$-stimulated ATPase of the plasmalemma and believed to be located in the tonoplast. Subsequently, it was shown that this tonoplast pump was insensitive to vanadate and inhibited by NO$_3^-$, while the plasma membrane ATPase was insensitive to NO$_3^-$. O'Neill and Spanswick(43) have recently characterized, solubilized, and reconstituted a H$^-$-ATPase from red beet membranes. The ATPase exhibited H$^+$ translocating ability and was NO$_3^-$-insensitive, K$^+$-stimulated, and vanadate sensitive, indicating a plasma membrane origin. These studies emphasize the problems that have been encountered in the attempt to identify the origin of membrane vesicles isolated from plant tissues. Because of the demonstrated H$^+$ pumping capacity of these membranes, and a lack of K$^+$ transport ability, the current focus appears to be that the plant plasma membrane ATPase(s) acts as a H$^+$ rather than K$^+$/H$^+$ pump.

1.4 The Intact Root Revisited

Based on the above information, it should be obvious that we need to re-examine the anatomical and metabolic aspects of the root from the perspective of spatial and temporal regulation/modulation of ion transport processes. The ultimate focus of these studies should be the elucidation of the transport processes, and the related regulatory phenomena that function to provide the mineral nutrient requirements of the intact, growing plant. Within our laboratory, we have conducted most of our kinetic (K$^+$ uptake) studies on excised corn roots. We are presently developing microanalytical techniques that will allow us to enter the second phase of our program; namely, a spatial and temporal characterization of K$^+$, Cl$^-$ and H$^+$ transport function within the intact corn root. In the following section we outline these techniques, and during the workshop we will detail the construction and application of microelectrodes to root physiology.

2. MICRO-ANALYTICAL TECHNIQUES

In previous studies on HCO$_3^-$ transport in the giant alga, Chara corallina, we employed micro^3-pH electrodes to detect the pH gradient at the cell surface (32). The tips of these electrodes were approximately 0.5 μm in diameter and were of the Thomas-type construction (53). The micro-pH electrodes that we are using in our corn root experiments have the same tip dimensions, but are based on neutral carrier liquid membranes (1,36,45). Use of such electrodes, with 0.5 μm diameter, enables us to map the H$^+$/OH$^-$ activities along the surface of the root with an extremely high level of precision. In addition, we can insert the microelectrode into the vacuole of epidermal, hypodermal or cortical cells.

2.1 Theory of Operation for Ion-Specific Microelectrodes

Micro-pH electrodes, and ion-specific electrodes in general, are devices that determine the activity of a given ion in solution via potentiometric measurements. Neutral carrier-based liquid membrane electrodes utilize an organic solution of neutral, ion-complexing agents (ionophores) as the ion-selective liquid "membrane." This membrane separates an internal filling solution, which contains a fixed concentration of the ion of interest, from the external solution. For any given ion activity in the external solution, an activity gradient will initially exist across the liquid membrane. In the ideal situation, the membrne will selectively allow only the ion in question to cross the membrane, as it moves in response to its activity gradient. When equilibrium is reached, an electrical potential will be set up across the ion-selective membrane which will exactly balance the activity gradient. Ideally, this potential difference (EMF) is a linear function of the logarithm of the activity of the ion in question and can be described by a modified form of the Nernst equation

$$EMF = E_0 + \frac{2.303\ RT}{z_i F}\ \log a_i \tag{1}$$

where E_0 is a temperature-dependent constant,
    R the gas constant,
    T the absolute temperature,
    F the Faraday constant,
    z_i the valence of the ion (ith species),
    a_i the activity of the ith ion in the external medium.

At 25°C for a univalent ion, the term $2.303\ RT/z_i F$ equals 59.2 mV. Thus, for an ideal ion-selective electrode, a 59 mV change in potential difference would be expected for a 10-fold change in the external activity of the ith ion.

At low ion activities, particularly in the presence of interfering ions, deviations from this equation can be significant. Therefore, Nicolsky and Eisenman (9,39) developed a semiempirical equation to describe the response of ion-specific electrodes under a wide range of conditions:

$$EMF = E_0 + \frac{2.303\ RT}{z_i F}\ \log [a_i + \sum_j K_{ij}(a_j)^{z_i/z_j}] \tag{2}$$

where a_i and z_i are the activity and valence of the primary ion i,
    a_j and z_j are the activity and valence of the interfering ion j,
    K_{ij} is the selectivity coefficient and is a measure of the preference of the ion-specific electrode for the ion in question over any interfering ion. The summation sign indicates that the

contributions to the electrode EMF by all interfering ions must be taken into account. The greater the specificity of the ion-specific electrode for the primary ion, the smaller the selectivity coefficient will be. An ideally selective electrode would yield a value of $K_{ij}=0$. Under these conditions, equation (2) would reduce to equation (1). It should be noted that the selectivity coefficient is empirically derived for each ion-specific electrode, and it may vary as ionic strength and ionic composition are changed (36,38,53). It must be emphasized that K_{ij} is not a selectivity constant, but should be considered a coefficient of selectivity for the ionic conditions which were used for its determination.

There are two basic techniques for determining K_{ij}. The fixed interference method involves the measurement of potentials in solutions containing a fixed amount of interfering ion, and varying activities of the primary ion. If potentials are plotted versus log a_i, a linear response is observed at higher activities of the primary ion. As the activity of the primary ion is reduced, there is a gradual onset of interference. When this occurs, the plot of EMF versus log a_i exhibits an increasing deviation from linearity. As a_i is reduced further, the response of the electrode to the primary ion is "swamped out" by the response of the electrode to the fixed level of interfering ion. At this point, the plot of EMF versus log a_i becomes horizontal. The intercept of the extrapolated linear response with that of the horizontal response to total interference will yield a particular activity for the primary ion (a_i^*). At this activity, the electrode EMF due to the primary ion, $E_0 + (2.303 RT/z_iF)$ log a_i^*, will be equal to the EMF by the fixed amount of interferant. $E_0 + (2.303 RT/z_iF)$ log $K_{ij}(a_j)^{z_i/z_j}$.
Therefore:

$$2.303 \frac{RT}{z_iF} \cdot \log a_i^* = \frac{2.303 RT}{z_iF} \log K_{ij}(a_j)^{z_i/z_j} \qquad (3)$$

This simplifies to:

$$a_i^* = K_{ij}(a_j)^{z_i/z_j} \qquad (4)$$

Therefore, since the fixed activity of the interfering ion (a_j) is known, and the activity of the primary ion at this particular intersection (a_i^*) has been determined, the selectivity coefficient can be calculated.

The second method, the separate solution technique, involves the measurement of the electrode EMF in two separate, pure solutions. One solution contains a particular activity of primary ion, while the other contains the interfering ion at the same activity. If the measured values are E_1 and E_2, respectively, the value of K_{ij} can be calculated from the Nicolsky-Eisenman equation:

$$\log \bar{K}_{ij} = \frac{E_2 - E_1}{2.303 \ RT/z_i F} + (1 - \frac{z_i}{z_j}) \ \log a_i \qquad (5)$$

which, for $z_i = z_j$, simplifies to:

$$\log K_{ij} = (E_2 - E_1) \ \frac{z_i F}{2.303 \ RT} \qquad (6)$$

For a more complete treatment of the Nicolsky-Eisenman equation and selectivity coefficients, the following references are recommended (9,23,38,39).

2.2 Ion-Specific Microelectrode Construction and Characterization

Single-barrelled borosilicate glass capillary tubing (Catalog # 1B100F-4; WP Instruments, Inc., New Haven, CT, USA) was cleaned by sonicating in chromic acid for 30 minutes; the glass was then rinsed thoroughly and dried overnight at 200°C. Glass micropipettes were then drawn to a tip diameter of 0.5-1.0 μm using a vertical pipette puller (Model 700C, David Kopf Instruments, Tujunga, CA, USA). The micropipettes were placed, tip up, in a drilled aluminum block and dried overnight at 200°C. After cooling, a drop of tri-N-butylchlorosilane was placed on the aluminum block, the entire setup was covered with a glass beaker, and the silane vapor was allowed to react with the glass micropipettes at 200°C for 30 minutes.

The tip and shank of the micropipette was then filled with neutral carrier-based liquid proton sensor (Proton Cocktail 82500, Fluka Chemical Corp, Hauppauge, NY, USA) by placing a laboratory-made glass filling pipette into the back of the micropipette and inserting it as far as it would travel into the shank (usually 50 μm from the tip). (The filling pipette was pulled from the laboratory-made capillary glass [thick-walled; 1.5 mm O.D.] to a fine and ex- tremely long [5 cm] tip.) As the filling pipette was inserted into the microelectrode shank, liquid H^+ exchanger would usually flow into the microelectrode. Otherwise, the liquid could be forced in- to the micropipette by carefully pressurizing the back portion of the filling pipette. A second filling pipette was then used to backfill the microelectrode with filling buffer (40 mM KH_2PO_4 and 15 mM NaCl, adjusted to pH 7.0 with NaOH).

The micro-pH electrode and a 3 M KCl-filled reference electrode were connected to a high input impedance electrometer (Model FD 223, WP Instruments, Inc.; input resistance = 10^{15} Ohms) and all measure- ments were made inside a Faraday cage. The micro-pH electrodes were quite stable, with tip resistances of approximately 10^{10}-10^{11} Ohms, and as illustrated in Figure 1, they exhibited fast response times (90% response time < 5 sec).

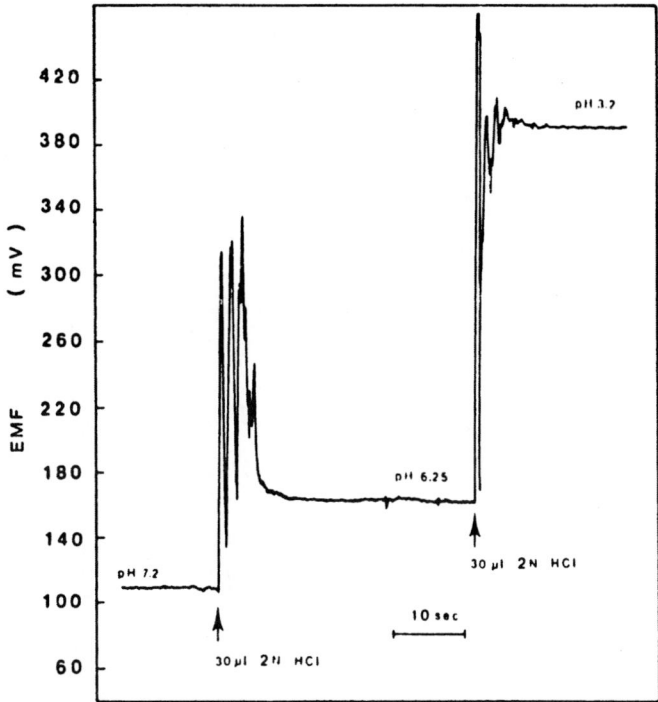

Figure 1. Response time of the micro-pH electrode. The spikes on
the electrode response following injection of the HCl
relate to the mixing profiles of the system as it is
stirred by the magnetic flea.

In order to use the micro-pH electrode for experimental purposes,
each individual electrode must be calibrated against standard
buffers of known pH values. Figure 2 indicates the linearity
observed over a wide range of pH values for two different buffer
concentrations (2 and 80 mM). The slope of the calibration curve
was approximately 57 mV per 10-fold change in H^+ activity, which is
quite close to that predicted by the Nernst equation (57.2 mV at
23°C). In addition, it can be seen that the electrode response is
only slightly sensitive to changes in the ionic strength of the
medium. This is significant for measurements of vacuolar pH, where
uncertainty concerning vacuolar ionic composition makes it difficult
to match the ionic strength and makeup of the calibration buffers
with that of the vacuolar solution. Selectivity coefficients were
calculated for H^+ versus Na^+, K^+, and Ca^{+2} by the fixed interference
method and are presented in Table 1. No interference from these
ions in typical intra- and extracellular concentrations was observed
for physiological pH values. We were quite pleased to observe that
micro-pH electrode resistances, response times, selectivities, and
calibration responses were similar or superior to those previously
presented for similarly constructed electrodes (1,46).

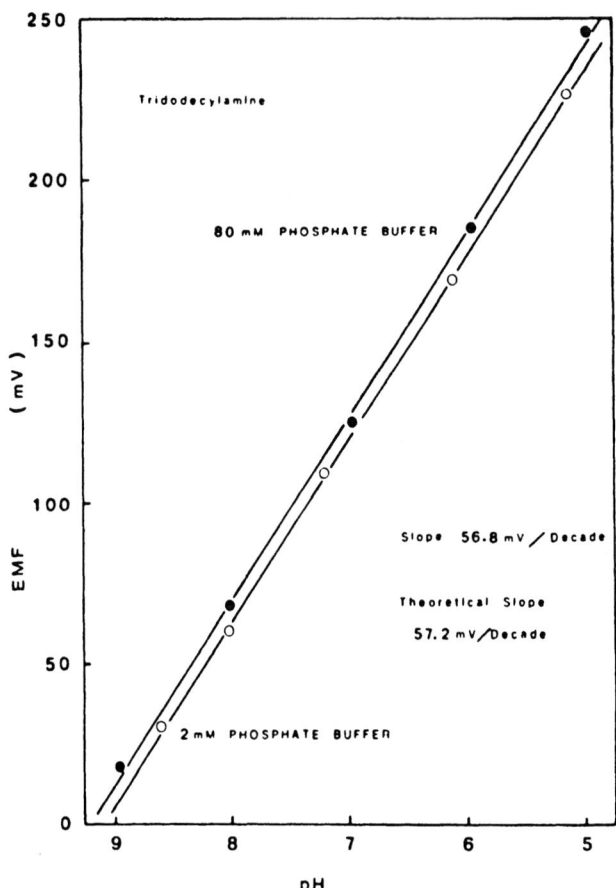

Figure 2. Calibration curves for a micro-pH electrode obtained under two buffer strengths.

Table 1

Selectivity coefficients for micro-pH electrodes determined by the fixed interference method

Interfering ion[a]	Selectivity coefficient
Na^+	1×10^{-12}
K^+	8×10^{-12}
Ca^{+2}	6×10^{-12}

[a] Concentration of interfering ion was 0.5 M.

Micro-K⁺ electrodes were also constructed using the methods outlined above. The only differences were that valinomycin-based liquid K⁺ sensor (Potassium cocktail 60031, Fluka Chemcial Co., Hauppauge, NY, USA) was used as the K⁺-specific membrane, and the microelectrodes were backfilled with 0.5 M KCl. Again, the micro-K⁺ electrodes were quite stable, with tip resistances of 10^{10} to 10^{11} Ohms. As seen in Figure 3, they responded quite quickly to changes in K⁺ activity. In fact, the response time (< 5 sec) appeared to be limited by the process of mixing in the solution.

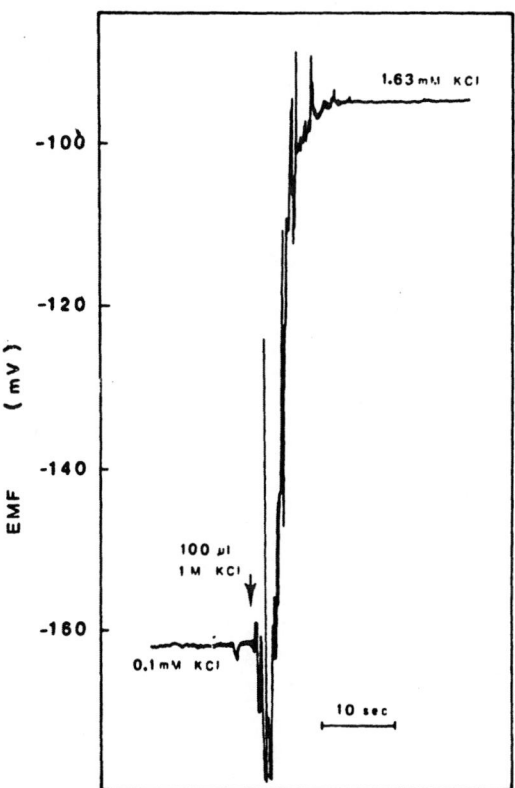

Figure 3. Response time of a micro-K⁺ electrode. The initial spikes observed during addition of 100 µl of 1 M KCl are due to electrical interference which is a consequence of the high electrical resistance of these microelectrodes.

The electrode exhibited a linear response over a wide range of K⁺ concentrations (10^{-4} M to 1 M), with a slope of 55 mV per 10-fold change in K⁺ concentration (Figure 4). It is interesting to note that although ideal ion-specific electrodes respond linearly to changes in the logarithm of ion activity, these micro-K⁺ electrodes responded linearly to changes in the logarithm of K⁺ concentration over a wide range. Indeed, they responded in a manner predicted by the Nernst equation to K⁺ concentrations as high as 1 M. At this concentration, the activity coefficient for K⁺ (as KCl) is significantly less than 1.0 in the bulk solution, it must be approximately 1.0 at the membrane-solution interface.

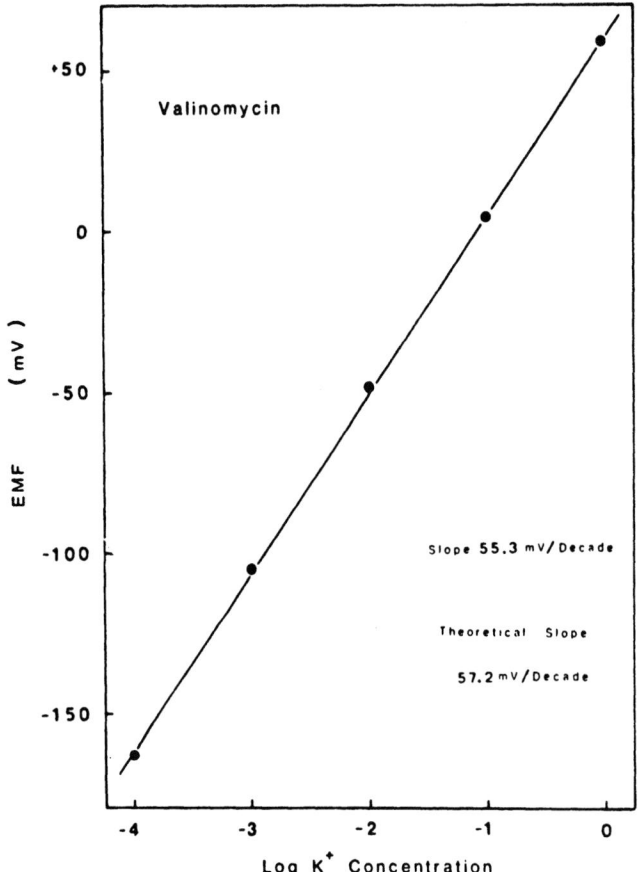

Figure 4. Calibration curve for a micro-K⁺ electrode.

Nonetheless, the observation that these microelectrodes are relatively insensitive to changes in ionic strength increases their value, particularly for vacuolar measurements.

The selectivity of the micro-K⁺ electrodes for K⁺ over Na⁺ was determined by the separate solution method. It should be useful to illustrate how this was determined. As presented earlier, the calculation is based on the Nicolsky-Eisenman equation. From equation (5) it follows that:

$$E = E_0 + \frac{2.303\ RT}{z_K F}\ \log\ (a_{K^+} + K_{KNa}\ a_{Na^+}) \qquad (7)$$

We first measured the potential difference in 0.1 M KCl; i.e.,

$$E_1 = E_0 + \frac{2.303\ RT}{F} \log a_{K^+} = -10\ mV \tag{8}$$

Then, the potential difference was measured in 0.1 M NaCl; i.e.,

$$E_2 = E_0 + \frac{2.303\ RT}{F} \log (K_{KNa}\ a_{Na^+}) = -204\ mV \tag{9}$$

Subtracting equation (8) from (9) and substituting the values for a_{K^+} and a_{Na^+} yields:

$$E_2 - E_1 = \frac{2.303\ RT}{F} \log K_{KNa} \qquad or:$$

$$\log K_{KNa} = (E_2 - E_1)\ \frac{F}{2.303\ RT} \tag{10}$$

Solving for K_{KNa} gives:

$$\log K_{KNa} = [-204-(-10)mV]/55\ mV = -3.527$$

$$K_{KNa} = 10^{-3.527} = 0.0003$$

Therefore, the micro-K^+ electrodes that we constructed from the Fluka K^+ cocktail were about 3300 times more selective for K^+ over Na^+, and for measurements of both intracellular and extracellular K^+, Na^+ will not influence the activity determination. An investigation of the influence of pH on the micro-K^+ electrodes was also conducted (Table 2). Varying the pH value between 4 and 8 had no influence on the potential difference obtained from a solution of 10 mM KCl. Again, as with the micro-pH electrodes, it was quite pleasing to learn that the various electrode characteristics tested yielded responses that were equal and usually superior to those presented in the literature for similar microelectrodes (42,57).

Table 2
Influence of pH on the selectivity of the micro-K^+ electrode.

Measurements of the electrode response was made in 10 mM KCl and 2 mM buffer at the indicated pH value.

pH	Electrode response (mV)
4	-58
5	-57
6	-58
7	-59
8	-57

Microelectrodes specific for Cl⁻ can be made in a manner similar
to that outlined above for protons and potassium ions. All microelectrode measurements were made using a high precision Narashige
manipulator mounted on a Leitz Orthoplan microscope. This recently
developed manipulator allows for very precise cellular impalements
with essentially no backlash. Thus we can insert our microelectrodes
into the cells of the epidermis, hypodermis, and cortex of corn
roots with a minimum of tissue damage, and therefore, increase the
probability that we are measuring vacuolar pH values that reflect
normal physiological conditions.

2.3 Membrane Potential Measurements

In conjunction with the ion-specific microelectrode studies,
we make routine measurements of the membrane potential (i.e., the
electric potential measured between the bathing medium and the
vacuole of the particular cell of interest). Again, using the high
precision Narishige manipulator which we have mounted on the stage
of a Leitz Orthoplan light microscope, potentials associated with
cells of the epidermis, hypodermis or cortex can be obtained. In
these studies, standard 3M KCl-filled microelectrodes (tip diameter
0.5 µm) are used and the potential is measured using a WPI model
750 amplifier (WPI Instruments, New Haven, CT, USA).

For these studies it is necessary to have the experimental system housed on a vibration-damped table. The cells of the root are
quite small (10-20 µm) and so any physical movement of the microelectrode that would be caused by vibrations would prevent an effective
seal of the plasmalemma around the microelectrode. In our system,
we use a Vibraplane inertia isolation table (Kinetic Systems Model
1201, 70 Lincoln Street, Brighton, MA, 02135 USA) to reduce these
unwanted vibrations.

2.4 Extracellular Electric Field Measurements

Bioelectric phenomena have been documented in roots by several
workers. The early studies of Bruce Scott are particularly note-
worthy for their thorough and accurate execution (47,48). Scott,
like later workers, believed that the small electric fields were
generated by currents passing through the root and the medium. In
a later section we will demonstrate that you do not need to have
actual current loops to generate an extracellular electric field.
However, at present, membrane transport processes that are electrogenic in nature are thought to establish the rather small, external
electric fields as a result of their activity.

It was not until Jaffe and Nuccitelli (19) developed an innovative technique, called the vibrating probe[1], that more sophisticated

---

[1]This system is now available commercially from the Vibrating Probe
Company, 238 Faro Avenue, Davis, CA, 95616, USA (916-753-2938).

research became possible on the nature of these fields. Their new technique allowed the detection of very small voltage gradients; the system's sensitivity is 10^{-8} volts! The operating principle of this technique has been fully detailed by Jaffe and Nuccitelli (19), and so we will give only a brief description to familarize the reader with the basic concepts[2]. A scale drawing of the vibrating probe and its housing is presented in Figure 5. The small platinum-black electrode (tip diameter 10 to 30 μm) is caused to vibrate over a known distance (usually 20 to 25 μm) by varying the power supplied to a piezoelectric element via a voltage oscillator. The frequency at which the probe is vibrated is a function of the resonance frequencies of the system; for our studies, we routinely used 290 Hz. The voltage difference is measured by the platinum-black tip at the extremes of the vibration amplitude, relative to the reference potential (see Fig. 5 for the location of the reference position on the shank of the glass electrode). A phase-sensitive lock-in amplifier (Model 5101, EG&G Princeton Applied Research, P.O. Box 2565, Princeton, NJ 08540, USA), set to the vibration frequency of the probe, is used to process the voltage signal.

The theory that underlies the conversion of the measured voltage gradient into the putative current density that would exist within the center and direction of vibration may not be as straightforward as thought previously. Generally, an Ohm's Law situation is invoked, and in this situation, if the amplitude of vibration (Δr in cms) is small relative to the size of the cell or tissue being investigated, the current density (I, in Amps cm^{-2}) is given by:

$$I = -\frac{1}{\rho} \frac{\Delta v \, \hat{a}_r}{\Delta r} \tag{11}$$

where Δv is the voltage difference (detected between the extremes of vibration),
$\hat{a}_r$ is the unit vector in the radial direction, and
ρ (Ohm-cm) is the specific resistance of the experimental medium.

An example of typical experimental values obtained during a measurement on an intact corn root are as follows: Δv, 1.0 μvolts; ρ, 1364 Ohms-cm; Δr, 2.5 x 10^{-3} cm.

---

[2]For a full treatment of the theory of operation of the vibrating probe, the reader is referred to Lionel F. Jaffe and Richard Nuccitelli, An ultrasensitive vibrating probe for measuring steady extracellular currents, J. Cell Biol. 63:614-628, 1974.

416

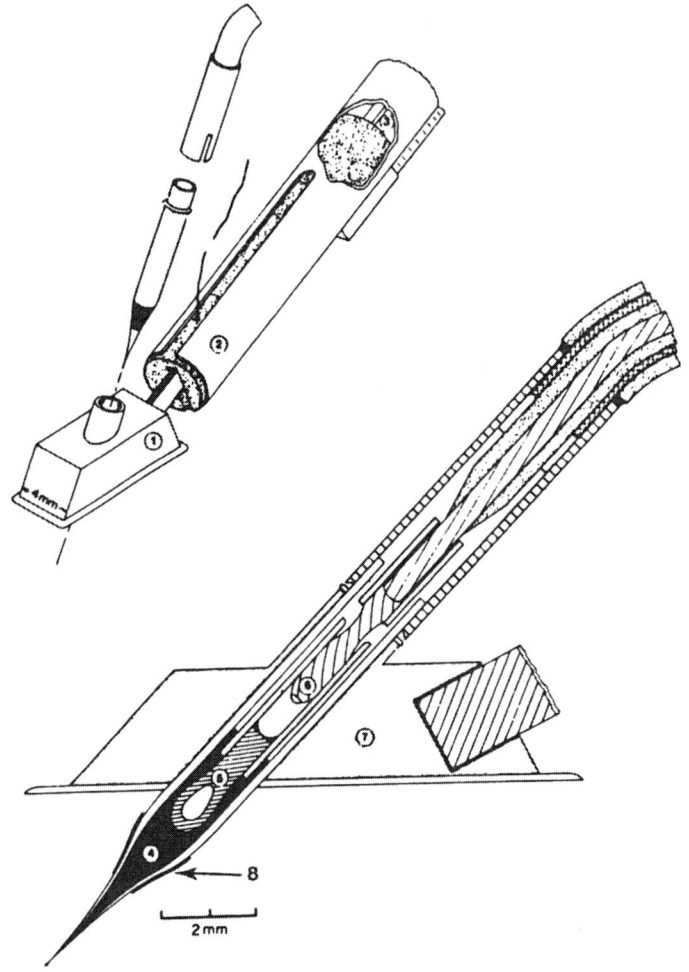

Figure 5. A scale drawing of the vibrating probe with construction
            details:
            1) lucite boat
            2) 6-gauge stainless steel tube
            3) Bender (piezoelectric) power cable
            4) Cerrotru solder filling inside of pipette
            5) No.9 sewing needle eye
            6) gold-plated, braided pin (Malco No. 096-0524-0000,
               Chicago, IL, USA
            7) lucite boat with coverglass menicus setter attached
               to bottom
            8) reference on shank of glass electrode housing.
Courtesy of Nuccitelli (41), with permission, copyright © 1983,
Alan R. Liss, Inc.

From equation (11), the current density can be calculated as follows:

$$I = - \frac{1}{1364 \ (\text{Ohms-cm})} \times \frac{1 \times 10^{-6} \ (\text{RMS volts}) \times 2.83}{2.5 \times 10^{-3} \ (\text{cm})} \ \text{Amps cm}^{-2}$$

$$= 0.83 \ \text{Amps cm}^{-2}$$

Current densities in this range have been obtained by several investigators (3,56). In one of the pioneering vibrating probe studies conducted on roots, Weiseneel, et al. (56), obtained preliminary evidence that for growing barley root hairs, H^+ fluxes appeared to be the main source of the extracellular "current."

At this point we should indicate the we have a problem in extrapolating from the voltage gradient, as measured by the vibrating probe, to the putative physiological processes occurring at the root surface. The basis for this problem comes from the experimental and theoretical work that we have conducted on the giant algal cell, Chara corallina (13). Work done in collaboration with Jack Ferrier (University of Toronto) clearly showed that extracellular electric fields provide only a small contribution to the thermodynamic gradient acting on the OH^- transport system of C. corallina. In this situation, the main driving force is an activity gradient, and the process of diffusion dominates the flux equation.

The thermodynamics of the situation can be illustrated by examining the simplest situation[3] which can be expressed by the Nernst-Planck equation; i.e.,

$$J_i = - \frac{u_i RT}{F} \frac{da_i}{d\chi} + \frac{z_i F \ a_i}{RT} \frac{dE}{d\chi} \tag{12}$$

where J_i is the flux of species "i" (moles/cm^2-sec)
u_i is the mobility of i (cm^2/volt-sec)
R is the gas constant (joules/mole-°K)
T is the absolute temperature (°K)
F is the Faraday (joules/mole-volt)
z_i is the valence of the ion
$da_i/d\chi$ is the activity gradient (moles/cm^4), and
$dE/d\chi$ is the voltage gradient (volt/cm).

---

[3]For an explanation of the more complex situation, the reader is referred to J.M. Ferrier and W.J. Lucas, Plasmalemma transport of OH$^-$ in Chara corallina, J. Exp. Botany 30:705-718, 1979; and J.M. Ferrier, Extracellular ion transport, J. Theoret. Biol. 85:739-743, 1980.

Equation (12) reduces to the Ohm's Law situation only when da_i/dx is very small compared with dE/dx. When this is the case, equation (12) becomes:

$$J_i = -u_i \, a_i \, z_i \, \frac{dE}{dx} \tag{13}$$

Curent density is given by:

$$I = -z_i \, F \, (J_i) \tag{14}$$

Substitution of equation (13) into (14) yields:

$$I = -z_i \, F \, (u_i \, a_i \, z_i \, \frac{dE}{dx})$$

$$= -u_i \, a_i \, z_i^2 \, F \, (\frac{dE}{dx}) \tag{15}$$

and it can be shown by unit analysis that $u_i \, a_i \, z_i^2 \, F = 1/\rho$. Thus equation (15) represents the Ohm's Law case, i.e.,

$$I = -\frac{1}{\rho} \, (\frac{dE}{dx}) \tag{16}$$

In unstirred medium, the Chara OH^- transport system develops a very large da_{OH^-}/dx and, in a relative sense, a small dE/dx, so it should not be possible to simplify equation (12) to give the Ohm's Law situation. Yet, when the vibrating probe is applied to the Chara OH^- transport system, it detects electric fields from which one can calculate apparent current densities and thus putative OH^- fluxes that are comparable (or much larger) to those obtained by numerical or diffusion analysis of the unstirred system (13, 29-31,33). Our present evaluation of this discrepancy is that at the frequency at which the vibrating probe operates, it creates a small volume-element in which thorough mixing occurs. As a consequence, the local da_{OH^-}/dx is reduced to a near zero value and, to maintain the steady state flux condition that exists on either side of the mixing volume-element, a dE/dx is created. If this is a valid assessment of the extracellular system, it would mean that the vibrating probe could be used to measure localized ionic fluxes, but could not be used to yield in vivo voltage gradients or associated electrical currents.

Unfortunately, as far as we can tell, the theoretical equations which have been shown to correctly analyze these types of experimental systems (12, 13) have had no impact on those presently using the vibrating probe technique. This, as Jack Ferrier (12) has pointed out, is serious because it must give rise to an incorrect theoretical picture; namely, that the ionic currents give rise to concentration profiles, etc., rather than membrane-transported species generating activity gradients which, by virtue of the nature of the transport system (i.e., rate of transport and the relative diffusion coefficients of the transported and non-transported species) establish small electric fields that are a consequence of the electroneutrality principle.

A final example will now be used to illustrate the problems of extrapolating from measurements of extracellular electric fields to putative transport events. Consider a cortical or epidermal cell of the root that is located in the "metabolic" zone in which imported sugars are being converted into organic acids, and these are being used in the vacuole for osmoticum. The protons generated from this organic acid synthesis could be transported out of the cell via a H^+-ATPase, and K^+ could be taken in via either a coupled transport process (H^+/K^+ exchange) or K^+ channels. Even if this exchange process were electroneutral, the H^+ gradient established at the root surface would tend to dissipate more rapidly than K^+ would tend to move down its activity gradient. This would be due to the difference in the ionic mobilities of the H^+ and K^+ ions. As a consequence, a small negative electric field would tend to develop at the root surface which would look like a region of inward current!!

Thus, to have a better chance of correctly interpreting the root situation, we feel that we must make many interrelated measurements involving:

1. extracellular electric fields using both vibrating probe and stationary electrodes (since the spatial resolution of the small vibrating probes is now approximately 10 μm, it should be possible to detect both the mechanism by which a particular ion is transported across the plasmalemma [i.e., via an electrogenic or neutral process] and the degree to which individual cells differ in their transport function);

2. external activity gradients of specific ions (H^+, K^+, Na^+, and Cl^-);

3. biophysical parameters associated with the potential across the plasmalemma of specific cells;

4. analysis of localized K^+ fluxes;

5. biochemical analysis of the "status" of the root cells in the spatial position of interest;

6. cytological changes within the tissue of interest.

2.5 X-Ray Microprobe Calibration

As stated earlier, it is our intent to eventually apply our knowledge of ion transport in roots to field situations. This will be done in two phases. Firstly, we will see if we can extrapolate from our experiments conducted in experimental bathing media to soil-grown (sand or vermiculite) roots. Secondly, we will use micro-electrode techniques (specific for K^+, Na^+, and Cl^-) to obtain the activities for these ions within the root tissue (i.e., epidermis, hypodermis, cortex, etc.). These values will be for vacuolar locations, and the exact cells in which the measurements were made can be identified in both the light and scanning electron microscopes.

Root tissue will be frozen in slushy nitrogen and examined using our Hitachi S800 that is fitted with a -170°C cold stage and Kevex 8000 X-Ray micro-analysis system. We may also be able to obtain frozen-hydrated thin sections, which would greatly improve the spatial resolution during elemental analysis. However, these techniques are only now being developed and are extremely preliminary in nature. In any event, our aim is to calibrate the X-Ray microprobe so that we can obtain more valuable information concerning the activities of K^+, Na^+, and Cl^- throughout the cells of the root and provide this information at the subcellular level of vacuole, cytoplasm and cell wall.

If the X-Ray microprobe work is successful, we should be able to study field-grown plants and compare them with hydroponically and growth chamber grown material.

3. CONCLUSIONS

Initially, research in the field of plant (root) ion transport yielded considerable insight into the physiology of transport. However, in recent years, although much attention has been paid to mathematical and graphical manipulations of uptake data, further progress has been slow.

We feel that if breakthroughs are to be made concerning the mechanisms by which ions are transported into and across roots, they will come by integrating experimental approaches derived from a wide range of research areas. These will include transport kinetics, membrane biochemistry, biophysics, genetics, etc. Recent advances in microelectrode technology now provide a powerful additional tool for studies in this field. The combined use of ion-specific (K^+, Na^+, Cl^-, Ca^{2+}) microelectrodes, micro-pH electrodes, electrophysiological probes, and extracellular ion current mapping (vibrating probe system), along with radioisotope flux studies, should greatly extend our knowlege of both the spatial aspect of transport along the root, and the nature of the basic transport mechanisms involved.

ACKNOWLEDGEMENTS
This work was supported by research grant PCM 81-17721 from the National Science Foundation. We would also like to acknowledge the

considerable support provided by the College of Agriculture and Environmental Sciences and the Graduate Division, who, in collaboration with the National Science Foundation (PCM 81-13554 and PCM 83-19724) funded the development of our analytical SEM facility. We would also like to thank the following individuals: Carl Scheffey and Lionel Jaffe of the Vibrating Probe Facility, Marine Biological Laboratory, Woods Hole, MA, for providing equipment to be used during the laboratory demonstrations; Richard Nuccitelli and Bill Busa for their invaluable assistance during the initial stages of our microelectrode studies; Leslie Sunell for her help on the SEM aspects of this work.

REFERENCES

1. Ammann, E., Lanter, F., Steiner, R.A., Schulthess, P., Shijo, Y. and Simon, W. Neutral carrier based hydrogen ion selective microelectrode for extra- and intracellular studies. Anal. Chem. 53:2267-2269, 1981.

2. Bange, G.G.J. Diffusion and absorption of ions in plant tissue. III: The role of the root cortex cells in ion absorption. Acta Bot. Neerl. 22:529-542, 1973.

3. Behrens, H.M., Weisenseel, M.H. and Sievers, A. Rapid changes in the pattern of electric current around the root tip of Lepidium sativum L. following gravistimulation. Plant Physiol. 70:1079-1083, 1983.

4. Bennett, A.B. and Spanswick, R.M. Optical measurements of ΔpH and $\Delta\psi$ in corn root membrane vesicles: kinetic analysis of Cl⁻ effects on a proton-translocating ATPase. J. Membrane Biol. 71:95-107, 1983.

5. Borstlap, A.C. Invalidity of the multiphasic concept of ion absorption in plants. Plant Cell Environ. 4:189-195, 1981.

6. Cheeseman, J.M., Lafayette, P.R., Gronewald, J.W. and Hanson, J.B. Effect of ATPase inhibitors on cell potential and K^+ influx in corn roots. Plant Physiol. 65:1139-1145, 1980.

7. Clarkson, D.T. Ion Transport and Cell Structure in Plants, McGraw Hill, Maidenhead, UK, 1974.

8. DuPont, F.M., Giorgi, D.L and Spanswick, R.M. Characterization of a proton-translocating ATPase in microsomal vesicles from corn roots. Plant Physiol. 70:1694-1699, 1982.

9. Eisenman, G. Glass Electrodes for Hydrogen and Other Cations: Principles and Practice, M. Dekker, Inc., NY, 1967.

10. Epstein, E. Mineral Nutrition of Plants: Principles and Perspectives, John Wiley & Sons, NY, 1972.

11. Epstein, E. and Hagen, C.E. A kinetic study of the absorption of alkali cations by barley roots. Plant Physiol. 27:457-474, 1952.

12. Ferrier, J.M. Extracellular ion transport. J. Theoret. Biol. 85:739-743, 1980.

13. Ferrier, J.M. and Lucas, W.J. Plasmalemma transport of OH⁻ in Chara corallina. II: Further analysis of the diffusion system associated with OH⁻ efflux. J. Exp. Botany 30:705-718, 1979.

14. Fisher, J. and Hodges, T.K. Monovalent ion stimulated adenosine triphosphatase from oat roots. Plant Physiol. 44:385-395, 1969.

15. Fisher, J.D., Hansen, D. and Hodges, T.K. Correlation between ion fluxes and ion-stimulated adenosine triphosphatase activity of plant roots. Plant Physiol. 46:812-814, 1970.

16. Gronewald, J.W., Cheeseman, J.M. and Hanson, J.B. Comparison of the responses of corn root tissue to fusicoccin and washing. Plant Physiol. 63:255-259, 1979.

17. Higinbotham, N. The mineral absorption process in plants. Botanical Review 39:15-69, 1973.

18. Hodges, T.K., Leonard, R.T., Bracker, C.E. and Keenan, T.W. Purification of an ion-stimulated adenosine triphosphatase from plant roots: association with the plasma membranes. Proc. Natl. Aca. Sci USA 69:3307-3311, 1972.

19. Jaffe, L.J. and Nuccitelli, R. An ultra-sensitive vibrating probe for measuring steady extracellular currents. J. Cell Biol. 56:614-628, 1974.

20. Kochian, L.V. and Lucas, W.J. Potassium transport in corn roots. I: Resolution of kinetics into a saturable and linear component. Plant Physiol. 70:1723-1731, 1982.

21. Kochian, L.V. and Lucas, W.J. A re-evaluation of the carrier-kinetic approach to ion transport in roots of higher plants. What's New in Plant Physiol. 13:45-48, 1982.

22. Kochian, L.V. and Lucas, W.J. Potassium transport in corn roots. II: The significance of the root periphery. Plant Physiol. 73:208-215, 1983.

23. Koryta, J. Ion-Selective Electrodes, Cambridge University Press Cambridge, 1975.

24. Leonard, R.T. and Hodges, T.K. Characterization of plasma membrane-associated adenosine triphosphatase activity of oat roots. Plant Physiol. 52:6-12, 1973.

25. Leonard, R.T. and Hotchkiss, C.W. Cation-stimulated adenosine triphosphatase activity and cation transport in corn roots. Plant Physiol. 58:331-335, 1976.

26. Leonard, R.T. and Hotchkiss, C.W. Plasma membrane-associated adenosine triphosphatase activity of isolated cortex and stele from corn roots. Plant Physiol. 61:175-179, 1978.

27. Leonard, R.T., Hansen, D. and Hodges, T.K. Membrane-bound adenosine triphosphatase activities of oat roots. Plant Physiol. 51:749-754, 1973.

28. Lin, W. Potassium and phosphate uptake in corn roots. Further evidence for an electrogenic H^+/K^+ exchanger and an OH^-/P_i antiporter. Plant Physiol. 63:952-955, 1979.

29. Lucas, W.J. Analysis of the diffusion symmetry developed by the alkaline and acid bands which form at the surface of Chara corallina cells. J. Exp. Botany 26:271-286, 1975.

30. Lucas, W.J. Mechanism of acquisition of exogenous bicarbonate by internodal cells of Chara corallina. Planta 156:181-192, 1982.

31. Lucas, W.J., Ferrier, J.M., and Dainty, J. Plasmalemma transport of OH^- in Chara corallina: dynamics of activation and deactivation. J. Membrane Biol. 32:49-73, 1977.

32. Lucas, W.J., Keifer, D.W. and Sanders, D. Bicarbonate transport in Chara corallina: evidence for cotransport of HCO_3^- with H^+. J. Membrane Biol. 73:263-274, 1983.

33. Lucas, W.J. and Nuccitelli, R. HCO_3^- and OH^- transport across the plasmalemma of Chara: spatial resolution obtained using extracellular vibrating probe. Planta 150:120-131, 1980.

34. Luttge, U. and Pitman, M.G. Transport in Plants. II. Part B. Tissues and Organs Encyclopedia of Plant Physiology, N.S. Vol. 2, Part B, Springer-Verlag, Berlin, Heidelberg, NY, 1976.

35. Mandala, S., Mettler, I.J. and Taiz, L. Localization of the proton pump of corn coleoptile microsomal membranes by density gradient centrifugation. Plant Physiol. 70:1743-1747, 1982.

36. Meier, P.C., Ammann, D., Morf, W.E. and Simon, W. Liquid-membrane ion-selective electrodes and their biomedical applications. In: Medical and Biological Applications of Electrochemical Devices (J. Koryta, ed.) John Wiley & Sons, NY, pp. 13-91, 1980.

37. Mettler, I.J., Mandala, S. and Taiz, L. Characterization of in vitro proton pumping by microsomal vesicles isolated from corn coleoptiles. Plant Physiol. 70:1738-1742, 1982.

38. Moody, G.J. and Thomas, J.D.R. Selective Ion Sensitive Electrodes, Merrow Publishing Co. Watford, 1971.

39. Nicolsky, B.P. Theory of the glass electrode. I. Theoret. J. Phys. Chem. (USSR) 10:495-503, 1937.

40. Nissen, P. and Nissen O. Validity of the multiphasic concept of ion absorption in plants. Physiologia Plantarum 57:47-56, 1983.

41. Nuccitelli, R. Transcellular ion currents: signals and effectors of cell polarity. Modern Cell Biol. 2:451-481, 1983.

42. Oehme, M. and Simon, W. Microelectrode for potassium ions based on a neutral carrier and comparison of its characteristics with a cation exchanger sensor. Analy. Chimica Acta 86:21-25, 1976.

43a. O'Neill, S.D. and Spanswick, R.M. Solubolization and reconstruction of a vanadate-sensitive H^+-ATPase from the plasma membrane of Beta vulgaris. J. Membrane Biol. 79:231-243, 1984.

43b. O'Neill, S.D. and Spanswick, R.M. Characterization of native and reconstituted H^+-ATPase from the plasma membreane of Beta vulgaris. J. Membrane Biol. 79:245-256, 1984.

44. Pitman, M.G. Ion uptake by plant roots. In: Transport in Plants II. Part B. Tissues and Organs (U. Luttge and M.G. Pitman, eds.) Encyclopedia of Plant Physiology, N.S. Vol. 2, Part B., Springer-Verlag, Berlin, pp. 95-128, 1976.

45. Purves, R.D. Microelectrode Methods for Intracellular Recording and Ionophoresis, Biological Techniques Series, Academic Press, NY, 1981.

46. Schulthess, P., Shijo, Y., Pham, H.V., Pretsch, E., Ammann, D. and Simon, W. A hydrogen ion-selective liquid-membrane electrode based on tri-n-dodecylamine as neutral carrier. Analy. Chemica Acta 131:111-116, 1981.

47. Scott, B.I.H. Feedback-induced oscillations of five-minute period in the electric field of the bean root. Ann. NY Acad. Sci. 98:890-900, 1962.

48. Scott, B.I.H. and Martin, D.W. Bioelectric fields of bean roots and their relation to salt accumulation. Australian J. Biol. Sci. 15:83-100, 1962.

49. Stout, R.G. and Cleland, R.E. Evidence for a Cl⁻-stimulated MgATPase proton pump in oat root membranes. Plant Physiol. 69:798-803, 1982.

50. Sze, H. Nigericin-stimulated ATPase activity in microsomal vesicles of tobacco callus. Proc. Natl. Acad. Sci. USA 77: 5904-5908, 1980.

51. Sze, H. and Churchill, K.A. Mg^{2+}/KCl-ATPase of plant plasma membranes is an electrogenic pump. Proc. Natl. Acad. Sci. USA 78:5578-5582, 1981.

52. Sze, H. and Hodges, T.K. Selectivity of alkali cation influx across the plasma membrane of oat roots. Cation specificity of the plasma membrane ATPase. Plant Physiol. 59:641-646, 1977.

53. Thomas, R.C. Ion-sensitive Intracellular Microelectrodes. How to Make and Use Them, Biological Techniques Series, Academic Press, NY, 1978.

54. Vakhmistrov, D.B. On the function of the apparent free space in plant roots. A study of the absorbing power of the epidermis and cortical cells in barley roots. Soviet Plant Physiol. 14:103-107, 1967.

55. Van Iren, F. and Boers-van der Sluijs, P. Symplasmic and apoplasmic radial ion transport in plant roots. Planta 148: 130-137, 1980.

56. Weisenseel, M.H., Dorn, A. and Jaffe, L.F. Natural H^+ currents traverse growing roots and root hairs of barley (Hordeum vulgare L.). Plant Physiol. 64:512-518, 1979.

57. Wuhrmann, P., Ineichen, H., Riesen-Willi, U. and Lezzi, M. Change in nuclear potassium electrochemical activity and puffing of potassium-sensitive salivary chromosome regions during Chironomus development. Proc. Natl. Acad. Sci. USA 76:806-808, 1979.

WATER STATUS INSTRUMENTATION

WATER AND PLANTS

K. Schurer

TFDL - Wageningen
The Netherlands

1.   THE MANY ASPECTS OF WATER

1.1 Introduction

Water is vital to life on earth. No one will dispute that statement. We can therefore consider the question "do plants need water?" to be answered in the affirmative and digress to a discussion of the next two questions: "what do plants need water for?" and "how can we get an insight into the presence or availability of water?"

To start with the first question, plants need water for different purposes like:
- maintaining cell turgor and, thus, structural integrity
- transport of nutrients from the soil to the leaves
- transport of assimilates from the leaves for use elsewhere or for storage
- photosynthesis where water is bound to carbon dioxide from the atmosphere
- temperature regulation through the use of excess heat for evaporation.

Most of these uses resort to the same source, but in different ways. The main flow is from the soil through the roots, the stem, and the leaves to the atmosphere. However, a flow can still exist without evaporation when cell turgor has to be restored. On the other hand, cells may for some time supply water for evaporation when the soil is exhausted. The cells clearly act as a buffer.

As to the question about gaining some knowledge of presence and availability of water, we can now see that we shall first decide on the exact nature of our interest: do we want to know water concentration in the soil, in the tissues or in the atmosphere, or do we want to study dynamics of water transport through the soil, through the plants or into the atmosphere? Of course a plant grower is chiefly interested in the overall result of growing good plants, but "good" is not easily defined as a post-harvest quality, and it is even much less accessible to measurement during the growing process. Instrumental data necessarily refer to details of the growing process, and it is the responsibility of the researcher to decide what details he should study.

Once we have chosen the quantity to be measured, we have to further decide on range of values, rate of change, possible interference, accuracy and stability required.

1.2 Soil Water

We shall refrain arbitrarily from a discussion of precipitation and drainage and start with soil moisture. Water can be present in the soil in widely varying amounts and with various degrees of binding to the solid matter. Usual quantities for soil moisture are:

water content: the ratio of weight loss and dry weight of a sample when it is oven-dried until a constant weight is reached.

water activity a_w: the ratio of the equilibrium vapor pressure p_w over the sample to the saturation pressure p_{ws} over pure water at the same temperature t

$$a_w = p_w(t)/p_{ws}(t)$$

soil water potential Ψ_w: effort to bring water from a free water reservoir to the soil:

$$\Psi_w = \frac{RT}{V_w} \ln (p_w/p_{ws})$$
.

with Ψ_w usually in bar (1 bar = 10^5 Pa), R the molar gas constant (≈ 8.31441 J mol^{-1}K^{-1})*, T the temperature in K and V_w the molar volume of water. Water potential takes values from $-\infty$ to 0, wilting of plants occurs below -15 bar.

head of water h_w: the positive value of water potential, but expressed in cmH$_2$O (1 cmH$_2$O = 98.0665 Pa).
pF value: pF is the logarithm of h_w: pF = log h_w

Consistent with SI units are only a_w and Ψ_w (when expressed in Pa). This discussion of quantities is both simplified and incomplete, but it provides us with useful definitions of the most commonly encountered concepts. More comprehensive treatments can be found in the literature (1,2).

---

* Units of measurement are SI units unless explicitly stated otherwise.

In the following, some important methods of measurement are summarized.

For water content, the weigh-and-dry method is an obvious choice, though it is time-consuming and requires careful procedures during sampling and handling. Other methods are neutron scattering that measures hydrogen concentration, and thermal conductivity. The neutron scattering method is hardly suited for routine use, because of the stringent safety requirements imposed by most governments. In the thermal conductivity method, water content is derived from the cooling rate of a heated rod. The method is simple and rapid. Calibration is necessary for each soil. Water content is averaged over a depth of a few centimeters.

For water potential or water activity, the Peltier psychrometer is an often used laboratory method. A soil sample is placed in a small container. The equilibrium humidity in the container is measured with a psychrometric method or with a thermocouple dew point meter. The method requires very careful procedures. Equillibration takes 24 to 48 hours. Complete instruments are commercially available from Wescor, Inc. Suitable for field use is the gypsum block method where the resistance between two electrodes depends on soil water potential. The method can cover the whole range of interest, but response is slow and accuracy rather low. The tensiometer is a method for high water potentials (0 to -1 bar). The suction pressure in a water filled porous cup is measured with a pressure gauge or a pressure transducer. Several companies offer this kind of equipment.

Permittivity of the soil depends on water potential. Capacitance measurements in the soil have suffered from interference by varying amounts of solutes and by temperature effects. Recent improvements have resulted in an attractive alternative to other methods. Most of the methods for water potential measurement can also be calibrated for the determination of water content. Water flow in the soil is very hard to measure, since velocities are extremely low and the flow pattern is three dimensional. Only forced flow through porous soils may reach a measurable level.

1.3 Tissue Water

Water content of plant tissues can be determined with the weigh-and-dry method in much the same way as water content of soil samples.

Water potential can be determined with the psychrometer or the dew point method on intact plant material. The volume of the container should be kept small enough to avoid drying of the sample. Equilibration can be much faster than with soil samples.

Water flow in plant stems has been measured in several different different ways. Chemical and radioactive labeling of the water has been attempted with limited success. Experimental difficulties with injection and detection are not easily solved. The magnetohydrodynamic method (3) looks promising. However, no further experiences are available in the literature. Nuclear magnetic resonance (NMR) has been investigated as a tool for the measurement of water content and sap flow (4). The method is elegant, but weight, dimensions and cost may be prohibitive for any but the most fundamental applications. NMR is the only method that measures mass flow. The heat pluse method has found application for trees and to a lesser extent for green plants. Recent developments will be discussed in a later section.

1.4 Water Vapor

Usually a distinction is made between humidity for water vapor as a component of an atmosphere, and moisture for water in another liquid or a solid. The measurement of humidity is called hygrometry. Similar to the difference in water potential as the driving force for liquid water flow is the difference in vapor concentration, the driving force for vapor transport. The water vapor content of a gas can be expressed by any of some twelve different kinds of quantity. The choice depends on the phenomenon studied, the methods used, and often, on convention. Often used quantities are:

Vapor pressure. The partial pressure of the water vapor component of the gas mixture is called vapor pressure, p_w. At any temperature t below the critical temperature (374°C) there is a maximum value to p_w, the saturation vapor pressure p_{ws}. At temperatures below freezing, saturation may be with respect to ice (p_{is}) or to water.

Dewpoint. Temperature at which $p_w = p_{ws}$ (t_d) is called dewpoint temperature t_d. Frostpoint temperature t_f is similarly defined by $p_w = p_{is}$ (t_f).

Relative humidity. The ratio of actual vapor pressure p_w and saturation vapor pressure p_{ws} (t) at the same temperature is called relative humidity ϕ:

$$\phi = p_w(t)/p_{ws}(t).$$

Relative humidity (r.h.) is often expressed as a percentage.

Vapor concentration. Mass of water vapor divided by volume of the system is vapor concentration c_w. From the ideal gas law the relation of c_w and p_w is seen to be:

$$c_w = 2.167 \times 10^{-3} \times p_w/T$$

with c_w in kg/m^3, p_w in Pa, absolute temperature T in K. The constant is the quotient of molar mass of water M_w and molar gas constant R.

Mixing ratio. The ratio of vapor concentration to mass concentration of dry gas is called mixing ratio x. The relation of x to vapor pressure p_w and atmospheric pressure P is:

$$\underline{x} = 0.622 \times \frac{\underline{p}_w}{\underline{P} - \underline{p}_w}$$

where the constant is the ratio of $\underline{M}_w$ and $\underline{M}_{air}$.

Saturation ratio. Corresponding to saturation vapor pressure, there is a saturation value $\underline{x}_s$, of the mixing ratio. The ratio of $\underline{x}$ to $\underline{x}_s$, is called the saturation ratio $\underline{\Psi}$:

$$\underline{\Psi} = \underline{x}(\underline{t})/\underline{x}_s(\underline{t}).$$

The relation to relative humidity is:

$$\underline{\Psi} = \underline{\phi}\ (\underline{P}-\underline{p}_{ws})/(\underline{P}-\underline{p}_w).$$

The difference between $\underline{\Psi}$ and $\underline{\phi}$ becomes significant at higher temperatures and low humidities.

Wet bulb temperature. The equilibrium temperature of a wet surface under adiabatic conditions is called the wet bulb temperature $\underline{t}_w$. The relation between $\underline{p}_w$ and $\underline{t}_w$ is given by the psychrometer equation:

$$\underline{p}_w = \underline{p}_{ws}\ (\underline{t}_w) - \underline{A.P.}\ (\underline{t}-\underline{t}_w)$$

with $\underline{A}$ the psychrometer coefficient in K^{-1}. The difference $\underline{t}-\underline{t}_w$ is called wet bulb depression.

Several other quantities have been occasionally used to express humidity. An account is given in several textbooks. The most complete survey may well be found in the proceedings of the Humidity and Moisture Conference in 1963 (5). A discussion of theoretical advances and of methods of measurement will be given in the section on humidity.

1.5 Evaporation

The major part of the water flowing through plants is used for evaporation as a means of temperature control. The transition from liquid to vapor takes place inside the leaf. The magnitude of the vapor flux is controlled by the degree of opening of the stomata. Opening and closing of the stomata is affected by water stress, carbon dioxide concentration in the leaf and local leaf temperature.

Evaporation from a crop is measured gravimetrically(lysimeter) or from the water budget of an evaporation - or a photosynthesis - chamber (2). For single plants and leaves several different designs have been described. Potometers are used for detached leaves and stems.

Stomatal resistance is measured with a porometer. A small chamber is placed over the leaf. Then, the rate of change of humidity in an originally dry chamber is observed, or the flow rate of dry air is measured, that maintains a constant humidity in the chamber (2,6,7,8). Instruments with an on-line processor are commercially available. The porometer chamber disturbs the local conditions. Continuous measurements are not possible. A measurement should be completed in a few minutes.

2. CAPACITIVE MEASUREMENT OF SOIL MOISTURE

2.1 Method

A very specific property of water among the components of soils is its high relative permittivity ε' (dielectric constant). For most of the water, ε' varies from 60 (loosely bound water) to 80 (free water). The values of ε' of inorganic and organic matter in soils range from 4 to 10. Measurement of ε' can thus be used for the determination of soil water content. The exact relationship is established in a calibration for the given soil type.

Measurement of a small capacitance in a highly conducting medium (pollution of irrigation water, abundant use of fertilizers) is best achieved with a high frequency method. The simplest procedure is to use the capacitance as the frequency determining element in a resonant circuit. Common errors are:
- shift of the resonant frequency by the conductive component of the impedance
- self-inductance between capacitance and oscillator circuit
- phase errors due to the use of real (non-ideal) components.

Solutions for these shortcomings have been realized (9,10). The sensor capacitance is determined from the ratio of the resonant frequencies of an LC-circuit with and without the sensor capacitance connected in parallel to the C in the circuit. The ratio is rather insensitive to oscillator drift and yet retains all information on soil water.

The circuit operates at about 16 MHz; the frequency shift is about 500 kHz for sensor capacitances from 50 to 150 pF (dry to saturated soil). Parasites and non-ideal behavior of components are compensated in two series resonant circuits. Automatic amplitude control takes care of the remaining conductivity effects. The control signal can be used as an indication of soil conductivity.

2.2 Use in Practice

Though the method is simple, fast and straightforward, some precautions should be observed. When the probe is inserted in the soil, a close contact between soil and probe should be achieved, but soil structure should remain unchanged. It may take some time for the disturbed soil to settle again before stable readings are obtained. In field experiments, the method has been used successfully in soils with varying contents of electrolytes up to brackish and heavily fertilized soils.

The connecting cable should not affect the frequency of the oscillator circuit, that is it should have a transfer function of 1.

Thus, cable length can only be chosen in multiples of half the wave-length. The wavelength at 16 MHz is approximately 12.5 m. The quality of the cable should be such that its properties are not affected by bending, temperature variations, pressure, or a prolonged stay in the soil. Thus, cable costs soon become an appreciable part of total sensor costs. Besides, the cable is the most vulnerable part of the sensor. Temperature has a considerable effect on permittivity, especially on that of water. A temperature sensor is therefore incorporated in the probe, and a compensation is provided for automatic systems.

Further work will be done to improve the reliability of the system. To that end the resonant circuits, the temperature compensation and the A/D conversion will be miniaturized and built into the probe. Thus, present problems with cables and with temperature correction can be completely eliminated.

3.   WATER VAPOR

3.1 Fundamentals

Saturation vapor pressure formulations have been published, derived from experimental data, thermodynamics and a combination of the two (11,12,13,14). The differences in the range -40 to +50°C never exceed 0.2%. This is of no consequence except for the most demanding measurements. The formulations of Wexler take account of the most recent experimental data, and give p_{ws} and p_{is} on the current version of the International Practical Temperature Scale IPTS-68.

These formulations refer to the simple liquid-vapor system. Saturation pressure in a more-component system like atmospheric air is slightly different. Usually this difference is expressed by writing the saturation pressure in air as $f \times p_{ws}(t)$, with $p_{ws}(t)$ the tabulated saturation pressure and f a correction factor, also called enhancement factor, that can be calculated from thermodynamical data. For a total pressure of 0.1 MPas and temperatures between -40 and +40°C, f varies from 1.0038 to 1.0050. Functional equations for f in the range 0.1 to 2 MPa and -50 to +100°C have been derived by Greenspan (15). Thus, relative humidity (to give an example) should be calculated as:

$$\phi = p_w(t)/(f \times p_{ws}(t)).$$

When accuracy justifies or requires the inclusion of f, a value of 1.004 will suffice for normal envirnomental conditions.

The psychrometer equation

$$p_w(t) = p_{ws}(t_w) - A \times P \times (t-t_w),$$

where A is the psychrometer coefficient and P the atmospheric pressure, has been the object of many studies to assess the correct values of A. A comprehensive survey of older work is given by Harrison (16). A rigorous thermodynamic treatment by Wylie (17) and experimental work by Wylie and Lalas (18) and Schurer (19) have revealed that A has a value of 5.9 to 6.4 x 10^{-4} K^{-1}. This change is significant for relative humidities of 0.5 and less.

Saturated salt solutions are often used to produce atmospheres with a well-defined relative humidity. This is based on Raoult's law that states that the reduction in saturation vapor pressure over a solution is proportional to the molecular or ionic concentration of the solute. The method is simple to use; some salt and water in a closed container suffice, but accuracy is often disappointing. Uncertainties stem from several causes:
- no thermal equilibrium in the container
- uniform vapor pressure not yet reached
- hygroscopic salt has absorbed water, solution is no longer saturated
- spread in literature data on relative humidity over salt solutions (20).
Application of Raoult's law leads to approximate values only, because the degree of dissociation in the saturated solution is uncertain. Since many of the values in common use are old and of uncertain origin, there is a demand for new determinations. Procedures for the correct use of saturated salt solutions should also be further developed.

3.2 Methods

The dew point meter is often thought of as a laboratory instrument, intended primarily for the most exacting measurements. Though the optical dew point meter is probably the best wide range instrument available, its use is by no means limited to laboratory environments. Modern automatic dew point meters are designed to be reliable routine instruments, that do not require special care or skill.

In an evaluation of dew point meters, it was found that all automatic instruments with optical detection of dew or frost gave results reproducible to within ± 0.2 K (-10 to +40° C dew point). The Swiss make MBW accepts for some reason a readout accuracy of ± 0.8 K. Other systems of dew detection do not yet reach this performance. Dew point meters require a difference of about 5 K between air temperature and dew point. Since most dew point meters use a sampling system, this requirement is easily met by heating the piping and the sensor head.

It is advantageous to convey the sample air to the instrument under pressure so that leaks will not lead to the entrainment of air with a different dew point.

At dew points between 0 and -25° C, the deposit consists of water droplets at the outset. After some time, a sharp increase in temperature marks the transition to ice. The transition occurs sooner at lower temperatures: at -5° C water may exist for days but at -20°C ice is usually formed within one or two hours. Below -25°C the deposit is formed as ice immediately.

A separate class of dew point sensors are the heated LiCl-sensors. Here, the saturation pressure over a saturated LiCl solution is equilibrated against atmospheric vapor pressure by heating the sensor. In some ranges, limited by changes in the amount of water bound, sensor temperature is a unique function of vapor pressure or dew point. Accuracy was found to be within ca 1.5 K with good long term stability when uninterrupted heating could be provided. Maximum air speed around the sensor should not exceed 0.5 m/s, otherwise heating power may be inadequate.

The psychrometer has a long history as a simple but reliable instrument for humidity measurements. Older instruments worked on an intermittent basis (clockwork ventilator) and temperatures were read on mercury in glass thermometers. In modern psychrometers for automatic measurement or control, an electric ventilator is used, and temperatures are measured with Pt resistances, thermistors or thermocouples.

Two types of psychrometer are in use, the unventilated type and the ventilated one. Since A depends on many parameters, among which the ventilation speed at the wet surface, precise measurements are only possible with the second type. The unventilated type is used for comfort measurements, and in a special configuration for the determination of water potential (section 1.2).

Relative humidity determinations with a psychrometer are particularly sensitive to errors in the wet bulb depression. The two thermometers should, therefore, be paired to within 0.1 K. A new instrument should always be checked for compliance with this condition.

Very important for the operation of a psychrometer is the use of a clean wick. The wick is best replaced once a week. It should not be touched with bare hands, and when that inadvertently occurs, the wick should be rinsed with ample distilled water.

Less suited for electrical readout, but stil widely used is the hair hygrometer. The instrument is capable of an accuracy of ca 2% r.h. over the range 40 to 95% r.h. Lower humidities cause a change from the wet curve to the dry curve and consequently large errors.

The instruments need regular cleaning to keep the moving parts free from dust and dirt, and regular checks on correct reading, followed by readjustment when the misreadings exceed 2% r.h. Regular means once every six months, and more often when conditions are adverse (e.g. in stables).

Electrical sensors of the conductance type are losing ground. Their impedance varies more or less exponentially with relative humidity. The temperature coefficient is usually in the order of 0.3 to 0.4% r.h. per K. Problems are observed with high humidities, hysteresis and long term stability. New developments are ceramic sensors that can be cleaned and dried by a heating cycle to restore the original characteristics.

438

In a test, the instruments of Novasina and of Rotronic were found to be accurate and stable; all other instruments of this group are outperformed by those of the next group. This type of sensor should be recalibrated every three to six months.

Capacitance type sensors have come into widespread use in the past ten years. The basic idea is to use a thin layer of plastic with electrodes to both sides. Water content of the plastic layer varies with relative humidity, and the resulting capacitance changes are converted to voltage or frequency signals that vary more or less linearly with r.h. Problems are observed with high humidities, hysteresis and long term stability. At high humidities, the relatively large amount of water adsorbed on the sensor (Figure 1) may penetrate the plastic and cause swelling and prolonged water uptake, resulting in a gradual shift of the indicated humidity humidity over a period of a few days.

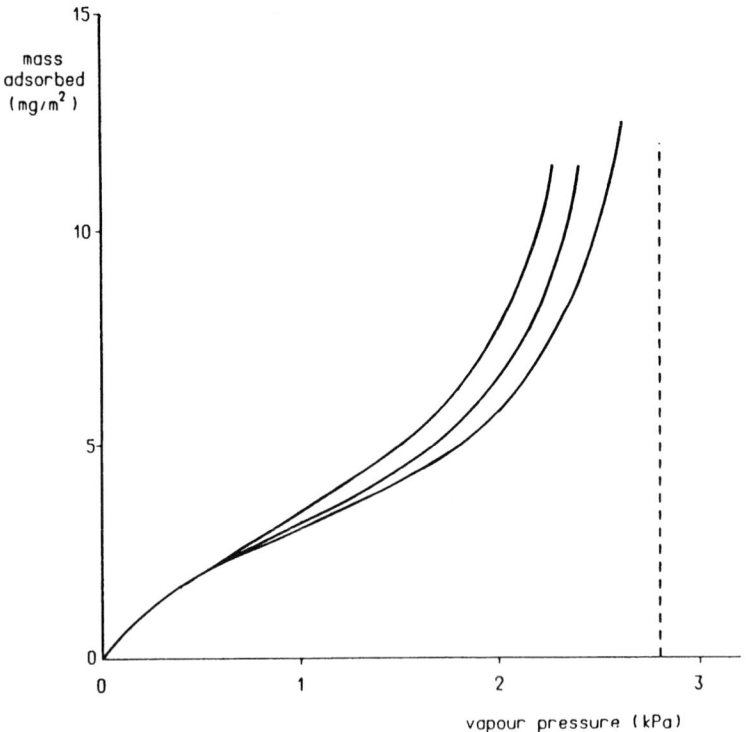

Figure 1. Adsorption isotherms of water at 23°C (three successive runs), after Bowden and Thrussell (28) -- saturation vapor pressure at 23°C.

The effect is unpredictable and, hence, uncorrectable. The manufacturer should find a plastic that does not suffer from this
effect. Indications are that some newer sensors have significantly
better high humidity performance.

The same story holds for hysteresis. Its existence is easily
recognized, but a prediction of the magnitude of the effect is uncertain. Hysteresis therefore sets a limit to the accuracy of the
sensor. Potentials of modern signal processing make it attractive
to choose sensor materials with lowest hysteresis, even at the
expense of strict linearity or zero temperature coefficient.

Long term stability is better than with conductance sensors.
Results of long term tests were fair with Vaisala, good with Valvo
and Coreci. First results with Rotronic seem good, with Endress
+ Hauser not quite satisfactory.

Infrared absorption can well be used as a hygrometric method.
Since absorption coefficients are sensitive to pressure and temperature, it is best to work under controlled conditions in the infrared analyzer. Calibration of the instrument should be done at
four or five points in the range at least, since the curve is not a
straight line, nor has a simple analytical form. Infrared absorption in a specially adapted version is the most promising instrument
for fast humidity measurements, needed for the assessment of turbulent transport. No such instruments are marketed yet, and development work may indeed take some more time.

Detailed results of the evaluation program at TFDL have been
published (21,22). Some selected test results are given in Table
3.1. More results are due for the Moisture and Humidity Symposium
of 1985 (29).

3.3 Calibration

Many sensors need regular calibration. For a start, intervals
should not be taken much longer than indicated. Experience can
then be a further guide. In case of doubt or after an untoward
event, an immediate check is expedient.

Sometimes calibration sets are supplied with the instrument.
There is, however, no general agreement on the precise values of
the equilibrium humidities over saturated salt solutions (20).
For best results sets require a constant temperature environment
and much patience from the observer.

Another approach is a direct comparison of the sensor with a
portable standard hygrometer. Since many instruments have adjustments for both zero and span, such a one-point comparison is often
not conclusive. A better way is to use a humidity generator, a
test chamber and standard hygrometer to obtain checks on different
points of the scale. In many countries such facilities are available, mostly at the national standards institute, sometimes at
institutes that have specialized on this type of measurement, like
CETIAT in France and TFDL in the Netherlands.

Table 1. Some results of tests on hygrometers

Dew Point Meters

Type	range (°C) nominal	range (°C) tested	accuracy (°C) nominal	accuracy (°C) test	variation of reading (°C) for t_{amb} from 15° C to 35° C
EG&G 440	-45 to +60	-20 to +15	± 0.4	< 0.3	< 0.1
EG&G 660	-40 to +20	-42 to +23	± 0.2	< 0.2	< 0.1
EG&G 992	-70 to +20	-50 to +20	± 0.2	< 0.2	< 0.1
Gen Eastern 1200 APS	-40 to +20	-41 to +23	± 0.2	< 0.2	< 0.1
MBW DP3	-40 to +40	-41 to +23	± 0.8	< 0.8	< 0.2
SEREG HPC2	-40 to +20	-50 to +23	± 1	< 0.2	< 0.1
Sulzer	-15 to +45	-16 to +23	± 0.1	< 0.2	< 0.1

Conductivity Sensors

Type	range (% r.h.)	accuracy(± in % r.h.)at 20%	50%	80%	90% r.h.	hysteresis[1] (% r.h.)	drift per year (% r.h.)
Novasina	20-90	<3	<2	<2	<3	<1	<2
Rotronic[2]	0-100[3]	2	2	2	2	<1	<2

Capacitance Sensors

Type	range (% r.h.)	accuracy(± in % r.h.)at 20%	50%	80%	90% r.h.	hysteresis[1] (% r.h.)	drift per year (% r.h.)
Coreci	5-98	<2	<2	<2	<2	<3	<2
Endress+ Hauser[4]	0-100	n/a	n/a	n/a	n/a	<4	n/a
Rotronic[4]	0-100	n/a	n/a	n/a	n/a	<3	n/a
Vaisala	5-95	<2	<2	<2	<2[5]	3	3-10
Valvo (Philips)[2]	10-90	<2	<2	<2	<3[6]	2	1

[1] Hysteresis as maximum difference between readings at 50% r.h. for a cycle 20-80-20% r.h.
[2] Results as obtained after adjustment of instrument.
[3] Heated sensor; conductivity sensors do not tolerate contact with liquid water.
[4] Tests on single sensors; first test period not yet finished.
[5] Heated sensor; for a normal sensor this figure is 10.
[6] Uncertainty increases rapidly for humidities over 90%.

Humidity calibrations are never an easy job. No matter what commercially inspired papers promise (23), good calibrations take time, good equipment and skilled personnel.

4. HEAT PULSE METHOD

4.1 Introduction

Water flow for transpiration is carried by xylem-vessels through the roots and stems of plants. Number and location of the xylem-vessels are plant specific. In tree stems, the whole outer portion of the wood conducts water, but in green plants, only one or a few layers of xylem-vessels are present. Sometimes these vessels are evenly distributed around the circumference of the stem (as e.g. in wheat), sometimes concentrated in a few bundles (as e.g. in potatoes).

A much smaller flow for the transport of assimilates is carried downward by phloem-vessels. In tree stems the phloem-vessels are found in the inner layer of the bark; in green plants, they alternate with the xylem-vessels.

Flow rates to be measured are of the order of 0.1 mm/s in trees, up to 1 mm/s in most green plants, and up to 10 mm/s in plants like cucumber and papyrus.

4.2 Principle

In the heat pulse method, a heat pulse is applied to the plant stem to label the flowing water. The time needed for the pulse to reach some down stream point is inversely proportional to the flow rate of the water. There are, however, some factors that complicate this simple picture:

- heat is conveyed to the measuring point by conduction through the tissues (mainly water). This heat flow may considerably exceed the convective heat flow.
- heat travels in the water flow not only by convection but also by conduction. Thus the first sign of the heat pulse will arrive too soon and the pulse will be broadened.
- there is a heat exchange between flowing water and tissues that may result in a shift of the temperature maximum in the water.
- the heat pulse may have to travel over some distance through plant tissues on its way to and from the flowing water. Since conduction is not necessarily at right angles to the flow, the exact distance the heat pulse has traveled with the flow is uncertain.

To eliminate the first complication, it is customary to use two temperature sensors instead of one, and to observe the difference of the two temperature signals. Different geometries for the two sensors are used with respect to symmetry.

In the asymmetrical arrangement, the heat pulse will first reach the nearest sensor. The difference signal goes to an extreme on one side. Then, when the heat pulse passes and the other sensor is heated, the difference signal goes back to zero and further to an extreme on the other side. The position of the zero crossing depends on the rate of sap flow. This method is most suited in situations where convective heat flow is an appreciable part of total heat flow, i.e. generally speaking in trees. The usual procedure for measurements in trees is to bore holes into the sapwood and to insert heater and temperature sensors directly in the water conducting xylem tissue.

Green plants are much more vulnerable. Boring holes would soon result in necrosis of the cells around the damaged spot and consequently in blockage of the conducting vessels. Heater and temperature sensors can thus only be placed on the outside of the stem. In most green plants, only a small fraction of the water present is actually flowing. Zero-crossings thus contain very little information on the rate of flow. It is therefore expedient to use the symmetrical arrangement when working with green plants to completely eliminate the effects of heat conduction from the measurements.

4.3 Theory

A fair theoretical model is available for the asymmetrical arrangement with bore holes in tree stems (24,25). This model assumes laminar flow and a thickness of the conducting layer that is large compared to the dimensions of the temperature sensor. A suitable choice has to be made for a parameter that describes the ratio of the area where water flows and the area occupied by non-moving wood.

An attempt to formulate a model for the symmetrical arrangement was only partly successful. For a series of measurements with the measuring head in one position an adjustment of parameters was possible, that resulted in a qualitative agreement between model and experiment. Replacing the head resulted in a new situation that required a new adjustment of parameters. Both model and experiments showed that both the magnitude of the temperature excursion and the time between application of the heat pulse and occurrence of the temperature-maximum depend on the rate of sap flow. However, the model lacked the power to predict the average flow velocity with any acceptable accuracy.

Figure 2 shows a block diagram of the model. In the first step, the input pulse s is transferred by conduction to a point 0 in the xylem-vessels. In the second step, heat travels convectively to point 1 with the sap flow. In the third step, the heat pulse travels through the plant tissues to the temperature sensor as the output signal y. If we make the reasonable assumption that the three processes can be described by linear differential equations,

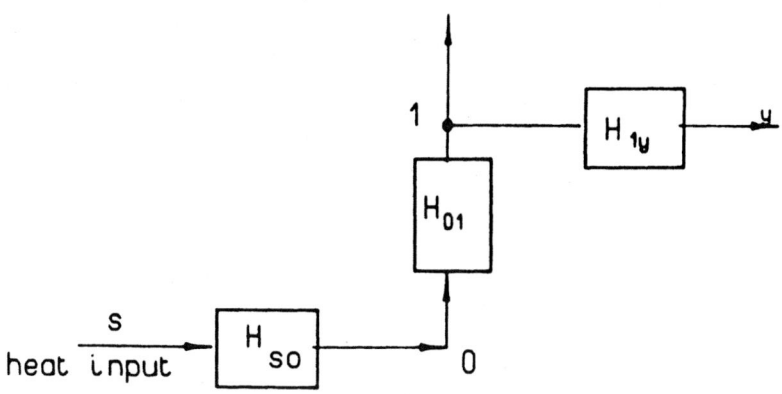

Figure 2.   Diagram of heat transfer in single sap flow sensor.
           H transfer function, $\underline{s}$ source signal, $\underline{y}$ output signal.

we can formally write

$$\underline{y} = H_{s0} \times H_{01} \times H_{1y}\underline{s}$$

where the transfer functions H_{ij} are found in the usual way as the
result of Laplace transformations. In a linear system the product
of transfer functions of the separate process steps can be represented by one overall transfer function H:

$$H = H_{s0} \times H_{01} \times H_{1y}$$

and

$$\underline{y} = H\underline{s}$$

Uncertainties are the thermal contact between heater and stem and
between stem and sensor (part of H_{s0} and H_{1y}, respectively) and the
location of points 0 and 1, and thereby, the transfer function H_{01}.
The thermal path from the cuticula to and from the xylem vessels
depends on the position of the measuring head with respect to the
vessels. This adds an extra uncertainty to H_{s0} and H_{1y}.
     The uncertainties of the symmetrical arrangement can be largely
eliminated by the use of two pairs of temperature sensors instead
of one (26). The signal at the nearest pair is taken as the input
signal $\underline{u}$ for a model, the signal at the outer pair is the output
signal $\underline{\overline{y}}$.
     Figure 3 shows the block diagram.

444

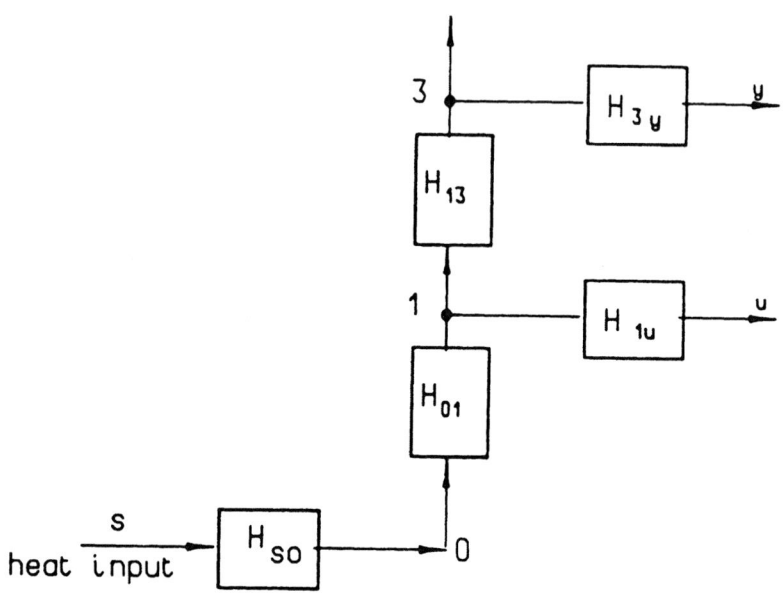

Figure 3. Diagram of heat transfer in double sap flow sensor.
0-1-3 flowing water, $\underline{s}$ source signal, $\underline{u}$ system input,
$\underline{y}$ system output.

Now,

$$\underline{u} = H_{1u}\, \underline{x}_1$$

if $\underline{x}_i$ is the signal in the flow at point i, and inversely:

$$\underline{x}_1 = H_{1u}^{-1}\, \underline{u}$$

With the assumption of a linear system and of equality of H_{1u} and
H_{3y}, we have

$$\underline{y} = H_{3y} H_{13} H_{1u}^{-1}\, \underline{u} = H_{13}\, \underline{u} \tag{4.1}$$

That means, that the relation between input and output signal is
determined by properties of the flow only, and that we have eliminated the uncertainties of the simpler model. The remaining problem
is in fact very similar to the one treated by Marshall(24). He got
good results with a convection-dispersion model. Obviously, the
model should contain other terms beside the convective term because
of heat conduction, mixing inside the flow and different flow velocities for different vessels. Since the one dimensional convection-
dispersion differential equation has been a good choice in similar

problems, we shall also try to describe our heat transport by

$$\frac{\partial T}{\partial t} + \frac{v \partial T}{\partial z} = \frac{D \partial^2 T}{\partial z^2} \qquad (4.2)$$

with T temperature (K), t time (s), v flow rate (m s^{-1}), z axial distance in the plant stem (m) and D dispersion parameter (m^{2}s^{-1}). If we could generate a heat pulse in point 1 ($z = 0$) at $t = 0$, we would obtain the impulse response solution for $T(z,t)$ at $z > 0$, $t > 0$ as the solution of (4.2). However, a heat pulse at the input s will be affected by dispersion on its way to point 1. Since the application of a heat pulse inside a green plant is not possible, we may as well consider a more general class of input signals T $(0, t)$. Stationary random signals instead of single pulses will give us the additional advantage of continuous observation. Analytical solutions of (4.2) cannot be given for these more general input signals. We shall, therefore, use a simulation model to assess an approximate solution of (4.2) (Figure 4).

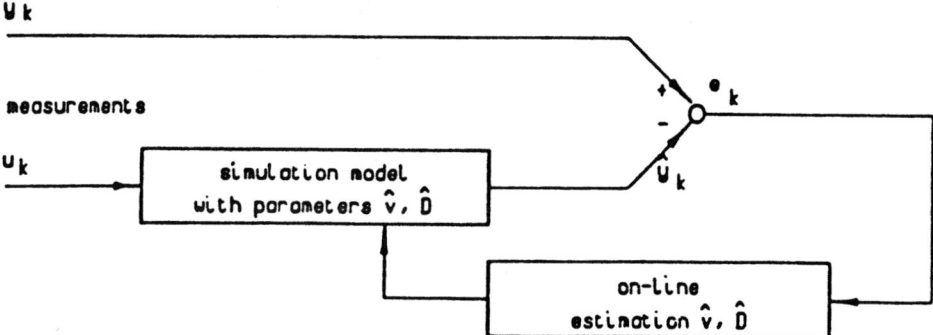

Figure 4. Combined simulation and estimation of flow parameters.

The model computes an output signal $\hat{y}_k$ (the circumflex means computed value) for the input signal u_k (index k means value at time $t = k \times \Delta t$ with Δt the sampling interval).

The technique of the measurement implies that we have discrete data, sampled with time intervals Δt, rather than a continuous flow of data. As a consequence, computations have been done with difference equations, and not with differential equations. With a large enough data set and short time intervals, this is not an essential limitation. We find good results with a set of 500 data points taken at 0.5 s intervals.

The solution of (4.2) is approximated by an iteration procedure for flow rate v and a gain parameter g, that accounts for heat losses between 1 and 3. The computation starts with arbitrarily chosen values of v, g and a parameter L, that gives the number of steps in the computation between 1 and 3. The parameter L is related to the dispersion parameter D in (4.2). For a given set of N data (usually 500, taken at 0.5 s intervals) a quadratic criterion function J is computed:

$$J = \sum_{N_1}^{N_1 + N} \underline{e}_k^2$$

with $\underline{e}_k = \underline{y}_k - \underline{\hat{y}}_k$. With a method based on steepest descents, $\underline{v}$ and g are optimized to give the lowest value of J. In each iteration an extra step is included to compute J for three values of L: the chosen value, L-1 and L+1. The value that gives the lowest criterion value is selected for the next iteration step.

The iteration stops when the gradient of J in $(J,\underline{v},g)$ space drops below a predetermined level. Then, after a computation time of N_c sample intervals, N_1 is updated to N_1+N_c and the estimation of parameters is resumed. Thus, with a short delay, an on-line estimation of parameters is performed.

4.4 Instrumentation

Design criteria for a sap flow sensor intended for multipoint measurements on green plants in the field were:
- good thermal contact with the stem
- no need for boring holes
- local temperature rise less than 5°K
- no sharp points on the sensor
- low weight
- shielding for wind and radiation
- low cost
- low power consumption.

Though it was clear from the start that the heat pulse method would best answer these requirements, the practical consequences were not at once unambiguously determined.

After some experiments with thermistors where sometimes difficulties were observed with thermal contact and sometimes with local pressure causing damage to the stem, it was decided to use classical vacuum deposition techniques to construct an essentially flat sensor. Temperature measurement is effected with Au-Te thermocouples (27). Tellurium was chosen because it gives the highest thermal e.m.f. of all materials that can be vacuum deposited without special techniques like high frequency or electron-beam heating. A disadvantage of thin-film thermocouples is their vulnerability, and a special disadvantage of tellurium is its high electrical resistance and its pronounced aging.

At first an evaporated gold-strip, 0.5 mm wide, was used as the heater. After some experiments, the vulnerable gold-strip has been replaced by a nickel-chromium wire with a diameter of 0.12 mm mounted with both ends through the epoxy substrate (Figure 5).

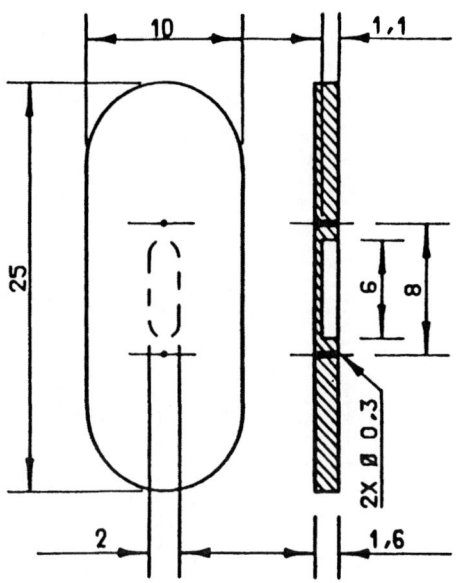

Figure 5.   Epoxy holder for heater wire

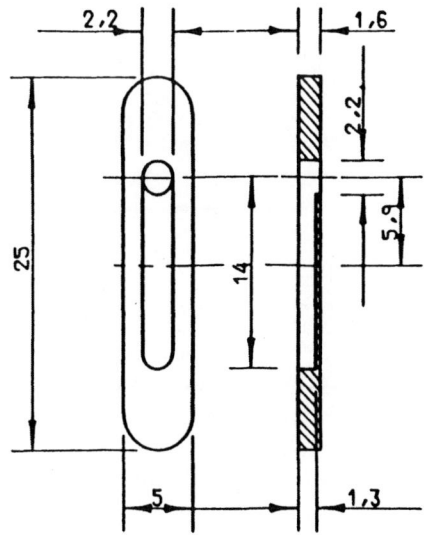

Figure 6.   Epoxy holder for temperature sensor. The AD 590 is
mounted with its substrate on a 0.3 mm epoxy layer.

With the advent of integrated circuits that produce a current proportional to absolute temperature (PTAT), a new design was made that retains heating by a single nickel-chromium wire, but uses commercial PTAT's (AD 590) in a flat pack configuration for temperature measurement. For best results, the PTAT's are paired on the criterion of equal slope of the temperature curve. The PTAT's are mounted on a heat sink in such a way that they have a fast response without losing too much of their sensitivity (Fig.6). Figure 7 shows the aluminum base of the sensor head with one pair of temperature sensors.

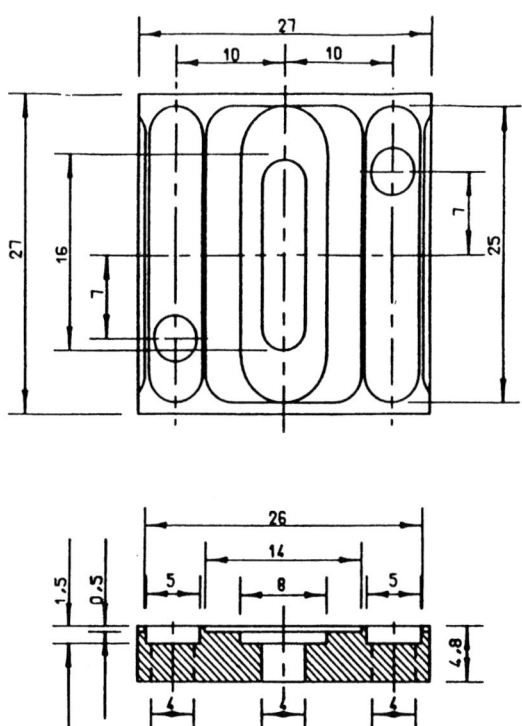

Figure 7. Aluminum base plate for sensor head with one pair of temperature sensors

The holders with the heater and the temperature sensors are mounted flush with the top of the aluminum plate. The surface of the sensor, where it comes in contact with the plant stem, is given a protective layer with an enamel lacquer.

The sensor head is mounted in a spring-loaded clip-on device (Fig. 8). For good thermal contact, the stem should have a straight internodal section of at least 30 mm. The force exerted on the stem is between 1 and 4 N for stem diameters between 3 and 15 mm. To help in positioning the sensor on the stem, a 2 mm deep V-groove is made in the PVC block facing the sensor head. The weight of the device is about 160 g. A clamp is provided to mount the device on a laboratory stand, if required. Figure 8a shows the assembled sensor.

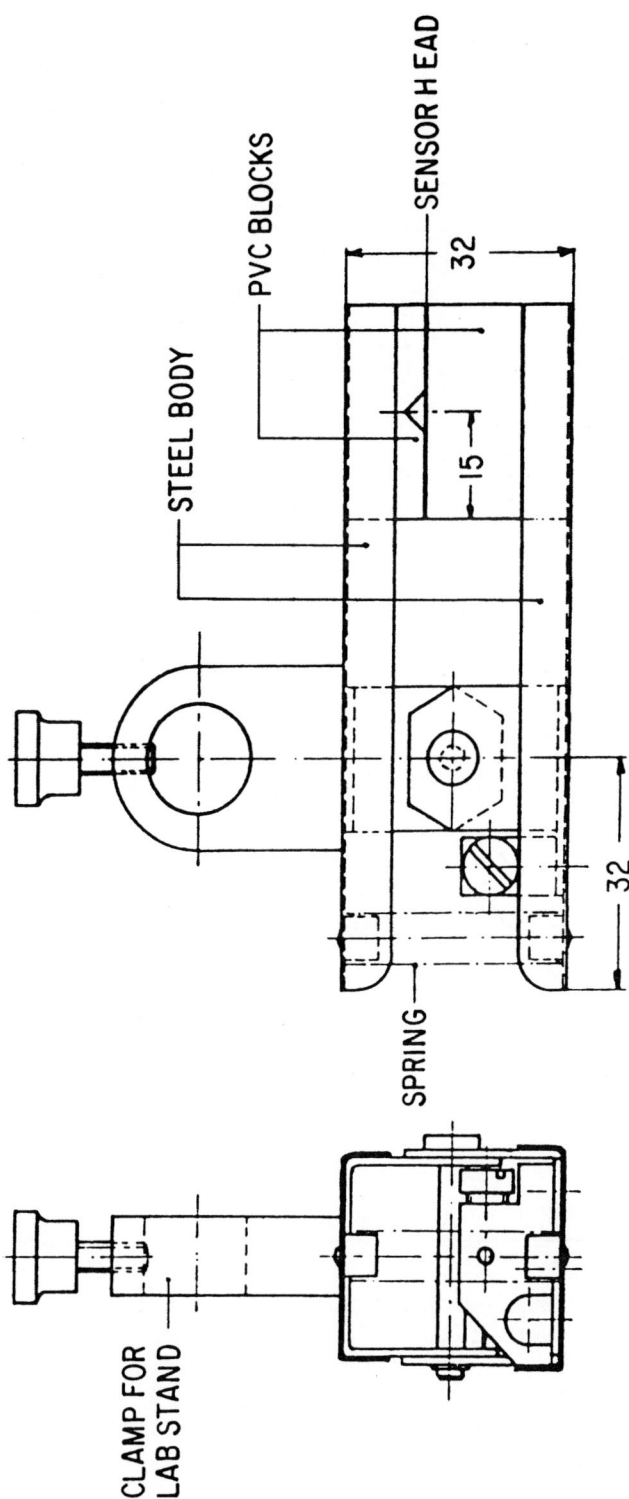

Figure 8.   Construction of the sap flow sensor

450

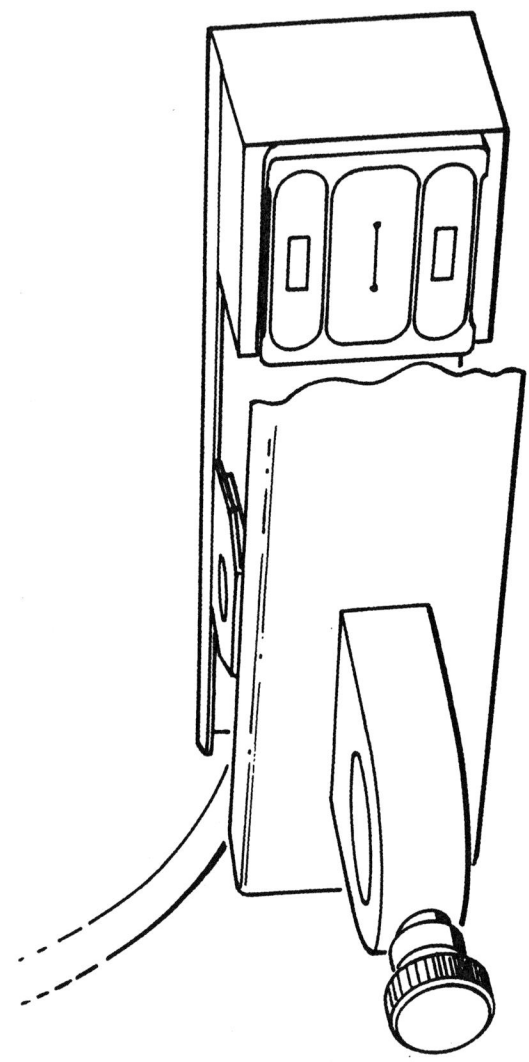

Figure 8a. Assembled sap-flow sensor.

    To make the sensor head computer-compatible, an analog pre-processing unit is incorporated as a thick-film circuit in the sensor head. This preprocessing unit comprises two buffer amplifiers, a difference amplifier and a low-pass filter with a 3 dB point at 3 Hz. The overall amplification is 20 times; the output signal is at a level of volts. Figure 9 gives a schematic circuit diagram of the thick-film circuit for one pair of temperature sensors.
    Separation between temperature sensors and heater was 10 mm in most sensor heads. This distance has been arbitrarily chosen, but

it proved rather satisfactory with a variety of plants like wheat,
cucumber, rose, potato, tomato, red pepper, pelargonium and apple
shoots. Halving the distance resulted in loss of much of the
information.

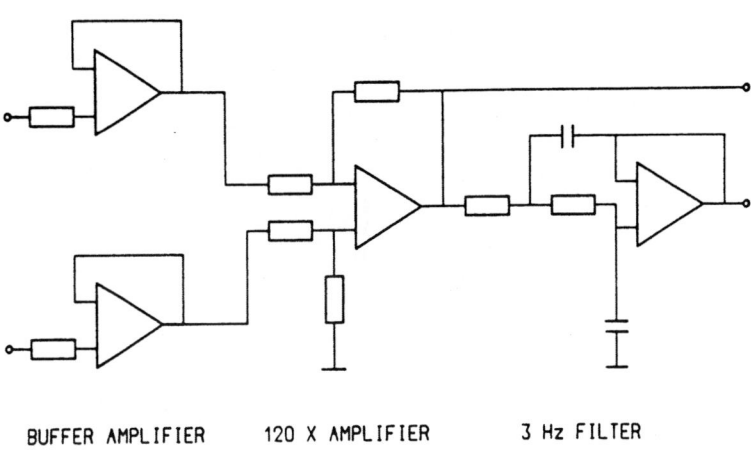

BUFFER AMPLIFIER   120 X AMPLIFIER   3 Hz FILTER

Figure 9.   Diagram of thick-film circuit for one pair of tempera-
             ture sensors

     In the new design with a double set of temperature sensors,
distances of 8 mm for the inner pair and 13 mm for the outer pair
have been used. The new sensor head is hardly larger than the old
one, and the smallest distance is still large enough to expect
sensible results (Figure 10).
     At present all digital processing is performed on a minicom-
puter (DEC PDP 11-44), but work has started for the development of
a dedicated microprocessor system.

4.5 Results

     We found no damage to the stems when the sensors were care-
fully positioned. Heating had no visible effect on the stem for
currents up to 0.6 A, administered for 0.5 to 20 s. The temperature
difference to be measured is then of the order of 10 to 100 mK.
     Ambient effects are effectively shielded when a tissue and
some aluminium foil are wrapped around the sensor. The older design
has been successfully used in plant breeding experiments in test
chambers and in the field to test resistance against water stress
of potato seedlings. Use as part of a climate control system in a
greenhouse is envisaged. In a comparison with NMR measurements at

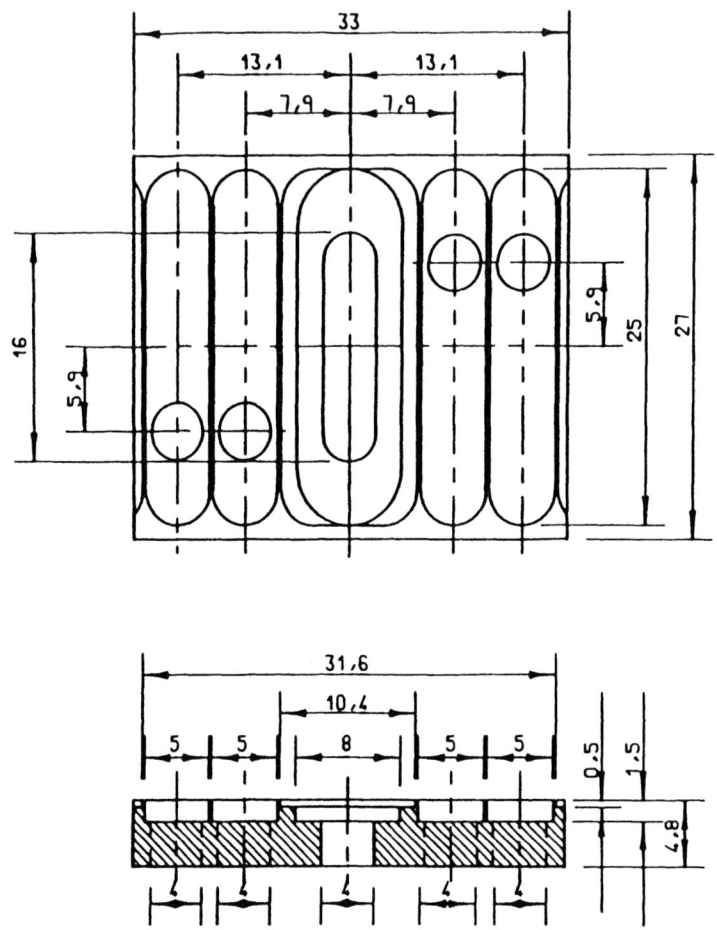

Figure 10. Aluminium base plate for sensor head with two pairs of
temperature sensors

high flow rates (up to 10 mm/s), good agreement was found (4), when
the quotient of maximum signal and time between heat pulse and
occurrence of the maximum was plotted against NMR results (Figure
11).

    The new design has been tested on flows through a plastic
capillary tube. Thermal data and weight loss data agree to about
5% for flow velocities of 0.3 to 0.9 mm/s. The system has been
shown to work on a tomato plant, but quantitative data are not yet
available.

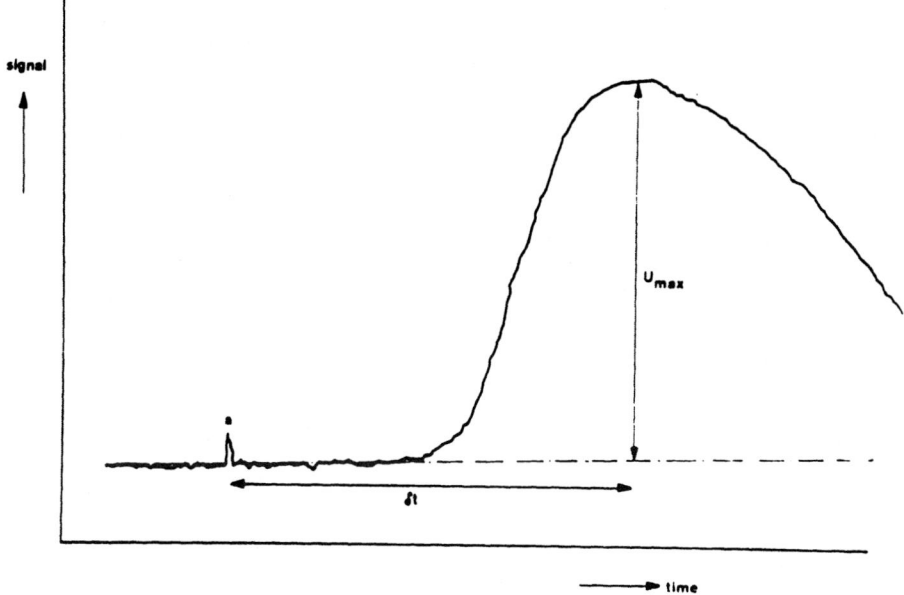

Figure 11. Typical output signal for a heat pulse given at point 'a'

A stepwise change of the flow rate was indicated in a few minutes. Since our data set comprises 500 measurements, made at 0.5 s intervals, a response time of about 4 minutes to a step change is to be expected anyway. Further experimentation will comprise tests with higher sampling rates and smaller data sets, but, since sap flow in plants is a rather slow responding phenomenon, a response in four minutes may prove quite fast enough.

A systematic revision of the procedures will result in a much more efficient program, that can be loaded into a microprocessor system to give a real field instrument.

ACKNOWLEDGEMENTS

The section on soil moisture describes work done at TFDL by the late Mr. P.G.F. Ploegaert and by Mr. M.A. Hilhorst. Mr. Hilhorst kindly provided a preprint of his symposium paper. Evaluation and calibration of hygrometers is largely the work of Mr. G.J.W. Visscher, who also read the manuscript.

LIST OF SYMBOLS

a activity
$\overline{A}$ psychrometer coefficient (K^{-1})
$\underline{c}$ mass concentration (kg/m^3)

D   dispersion parameter (m^2/s)
$\overline{E}$ difference signal
$\overline{f}$ enchancement factor
$\overline{g}$ gain parameter
h_w head of water (cmH_2O)
$\overline{H}$ transfer function
J   quadratic criterion function
k   running interval number
L   number of steps in a computation
$\underline{M}$ molar mass (kg/mol)
$\overline{p}$ vapour pressure (Pa)
pF   logarithm of h_w
P   atmospheric pressure (Pa)
$\overline{R}$ molar gas constant ($\approx$ 8.31441 J mol $^{-1}K^{-1}$)
$\overline{s}$ source signal
$\overline{t}$ temperature (°C)
$\overline{t}$ time (s)
$\underline{T}$ temperature (K)
$\overline{u}$ input signal
v   flow rate (m/s)
$\overline{V}$ molar volume of gas ($\approx$ 22.41383 x 10^{-3} m^3/mol, T = 273.15 K, P $\ast$ 101 325 Pa)
x   mixing ratio
$\overline{y}$ output signal
z   axial distance (m)
$\overline{\Delta}$ small step
ε' real part of relative permittivity
$\overline{\phi}$ relative humidity
$\overline{\psi}_w$ water potential (Pa)
$\overline{\psi}$ saturation ratio

SUBSCRIPTS

d   dew point
i   ice
k   value at time k x $\Delta \underline{t}$
s   saturation
w   water, wet bulb

REFERENCES

1. Slatyer, R.O. Plant-water relationships. Academic Press, London and New York, 1967.

2. Slavik, B. Methods of Studying Plant Water Relations. Springer-Verlag, Berlin, Heidelberg, New York, 1974.

3. Sheriff, D.W. A new apparatus for the measurement of sap flux in small shoots with the magnetohydrodynamic method. J. of Exp. Botany 23:1086-1095, 1972.

4. Van As, H. Nuclear magnetic resonance, water and plants. Ph.D. Dissertation, Wageningen University of Agriculture, 1982.

5. Wexler, A. (ed.) Humidity and Moisture, 4 Vols., Reinhold Publ. Corp., New York, 1965.

6. Stigter, C.J. Leaf diffusion resistance to water vapour and its direct measurement. 1. Introduction and review concerning relevant factors and methods. Meded LandbHogesch. Wageningen 72-3:1-47, 1972.

7. Stigter, C.J. and Lammers, B. Leaf diffusion resistance to water vapour and its direct measurement. III. Results regarding the improved diffusion porometer in growth rooms and fields of Indian corn (zea mays). Meded. LandbHogesch. Wageningen 74-21:1-76, 1974.

8. Visscher, G.J.W., Griffioen, H. and van Leeuwen, C.H. Investigations on a diffusion porometer with a fast humidity sensor. Neth. J. Agric. Sci. 26:366-372, 1978.

9. TFDL. Dutch Patent No. 173.099, 1983.

10. Hilhorst, M.A. A sensor for the determination of the complex permitivity of materials as a measure for the moisture content. In: Sensors and Actuators, P. Bergveld (ed.), Proceedings S&A Symposium of the Twente University of Technology, Kluwer Technical Books, Deventer, Antwerp, 1984.

11. Goff, J.A. and Gratch, S. Low pressure properties of water from -160 to 212°F. ASHVE Trans. 52:95, 1946.

12. Deutscher Wetterdienst. Aspirations Psychrometer Tafeln, Friedr. Vieweg & Sohn, Braunschweig, 1976.

13. Wexler, A. Vapor pressure formulation for water in range 0 to 100°C. A revision. J. Res. Nat. Bur. St. 80A:775-785, 1976.

14. Wexler, A. Vapor pressure formulation for ice. J. Res. Nat. Bur. St. 81A:5-20, 1977.

15. Greenspan, L. Functional equations for the enhancement factors for CO_2-free moist air. J. Res. Nat. Bur. St. (US) 80A:41-44, 1976.

16. Harrison, L.P. Fundamental concepts and definitions relating to humidity; Some fundamental considerations regarding psychrometry; Imperfect gas relationships. In: Fundamentals and Standards, Vol. 3, pp. 3-69; 71-103; 105-256, 1965.

17. Wylie, R.G. Psychrometric wet elements as a basis for precise physico-chemical measurements. J. Res. Nat. Bur. St. 84:161-177, 1979.

18. Wylie, R.G. and Lalas, T. The WMO Psychrometer. CSIRO, Australia, 1981.

19. Schurer, K. Confirmation of a lower psychrometer constant. J. Phys. E. Sci. Instrum. 14:1153, 1981.

20. Greenspan, L. Humidity fixed points of binary saturated aqueous solutions. J. Res. Nat. Bur. St. 81A:89-96, 1977.

21. Schurer, K. Het meten van luchtvochtigheid (humidity measurement). Koeltechniek 74:187-194, 1981 (in Dutch).

22. Schurer, K. Comparison of sensors for measurement of air humidity. In: Proceedings of ISOPOW III, D. Simatos and J.L. Multon (eds.), Martinus Nyhoff Publ., Dordrecht, Boston, Lancaster, pp. 647-660, 1985.

23. Parker, D.T. Hygrometers - meeting the needs of an expanding market. Journal A 24:29-31, 1983.

24. Marshall, D.C. Measurement of sap flow in conifers by heat transport. Plant Physiology 33:385-396, 1958.

25. Swanson, R.H. A thermal flowmeter for estimating the rate of xylem sap ascent in trees. In: Flow, Its Measurement and Control in Science and Industry, Vol. 1, R.B. Dowdell (ed.) Instrument Society of America, Pittsburgh, pp. 647-652, 1974.

26. Van Zee, G.A. and Schurer, K. On-line estimation of the rate of sap flow in plant stems using stationary thermal response data. J. of Exp. Botany 34:1636-1651, 1983.

27. Schurer, K., Griffioen, H., Kornet, J.G. and Visscher, G.J.W. Measurement of the rate of water flow in plants. Netherlands J. of Agri. Sci. 27:136-141, 1979.

28. Bowden, F.P. and Throssell, W.R. Adsorption of water vapour on solid surfaces. Proc. Roy. Soc (London) 209A:297, 1951.

29. Visscher, G.J.W. and Schurer, K. Some research on the stability of several capacitive thin film (polymer) humidity sensors in practice. In: Moisture and Humidity 1985, Instrument Society of America, Research Triangle Park, N.C., pp. 515-523, 1985.

STEM DIAMETER AND ELECTROCHEMICAL MEASUREMENTS

William Gensler

University of Arizona
Tucson, Arizona 85721

1. INTRODUCTION

This discussion will focus on the measurement of plant stem diameter and apoplast electrochemical status. Emphasis will be directed primarily towards the instrument techniques themselves under field conditions as opposed to the physiological interpretation of the data. For the latter see references 1,3,5,6,7,9,10,11, 13,15.

Both of the above methods access plant status directly as opposed to indirect methods which focus on the soil and/or the environment. Both methods are amenable to remote long term automatic acquisition using the most recent developments in low power electronics.

2. STEM DIAMETER INSTRUMENTATION

Basic Components. Instrumentation to measure the stem diameter can be considered in terms of the transducer itself, the housing to attach the transducer to the plant, the cable connecting the transducer to the electronics and the electronics itself. Each of these aspects will now be considered in detail.

2.1 Transduction

The basic transducer selected for the stem diameter measurement is the linear variable differential transformer. The advantages of this device are its size, relative immunity to environmental changes, weight, power requirements and resolution. The major disadvantage is cost. The Schaevitz MHR series is well suited to plant measurements, for example, the 100 MHR has a rated total range of 5000 micrometers (18). Its weight including the core is 6 grams.

The principle of operation is quite straightforward. A transformer primary is wound contiguous to a pair of secondary windings in the manner shown in Figure 1.

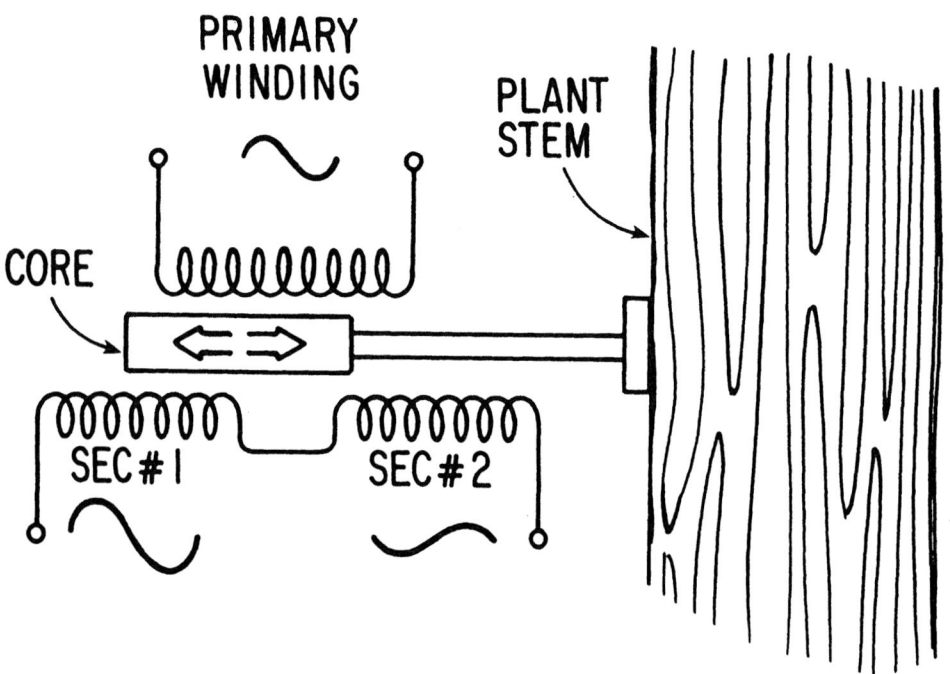

Figure 1. Basic Stem Diameter Measurement Technique using a Linear Variable Differential Transformer (LVDT).

The primary is driven by a sine wave of electrical potential. The magnetic circuit linking the primary and secondary coils passes through a magnetic core. If the core is placed midway between the two secondaries, the coupling between the primary and each secondary is equal and opposite. The voltage out of the combined secondaries is zero. As the core moves to one side, coupling between one secondary increases and coupling between the primary and the other secondary decreases. The output voltage takes on a some non-zero value. A similar situation ensues when the core moves in the opposite direction. The output sine wave can be read directly as an AC voltage using a digital multimeter. A more common technique is to feed the output into a synchronous demodulator and produce a DC voltage proportional to the displacement of the transducer core. Sine wave frequencies of 2500 and 10,000 hertz are common. Amplitude of the primary sine wave varies depending on the electronics available; one to three volts peak to peak is normal. Power level is relatively modest. The input impedance of the primary at 2500 hertz

is in the order of 150 ohms. Position resolution is determined
largely by the electronics, since there is no physical connection
between the core and primary winding.

2.2 Housing

A major problem in the application of LVDT's in plant sensing
is the method of attachment to the plant. Since the measured variable is long term displacement, great care must be exerted to insure
a valid, continuous measurement under field conditions for periods
in the order of months and even years. To achieve this continuity,
the housing employed is mounted directly on the stem.

The primary design characteristics are weight, insensitivity
to wind loading, and stability in the presence of variable moisture
and temperature levels. The most recent housing design is shown in
Figure 2.

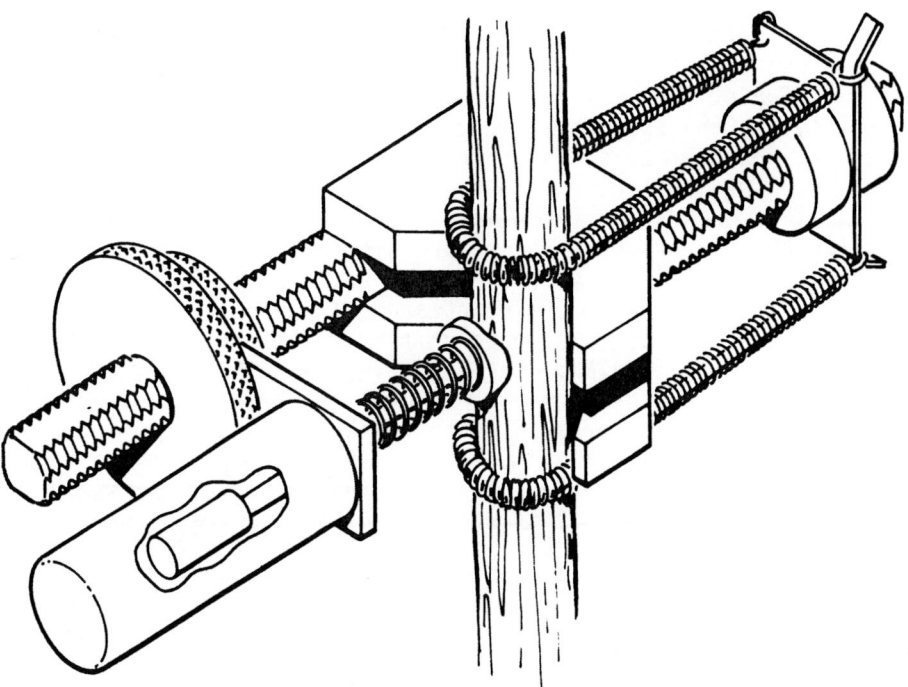

Figure 2. Diameter Transducer and housing attached to the stem of
a plant. Note the cutaway view of the core of the LVDT.

Two "V" brackets form a slot in which the stem is located. The
brackets are separated by a threaded rod which supports the LVDT
itself. A second threaded rod protrudes from the V brackets in the
opposite direction. This latter bracket supports the spring mounting plate. The stem is pressed into the slot by means of two extension springs. An arm protrudes from the core and presses against the

stem. The arm is pressed against the stem by means of a compression spring. This permits measurement of positive and negative displacement. The V brackets, threaded rods and arm are made of G10, a fiberglass impregnated epoxy. This material has a thermal coefficient of expansion of 9 micrometers/meter/degree C. In addition, it is virtually insensitive to moisture, sunlight and fungus. The use of G10 in both the arm and the threaded rod supporting the LVDT results in partial compensation of changes in displacement due to temperature. In the most recent designs, a maximum uncompensated distance of 2 centimeters yields a displacement of 2.7 micrometers for a daily temperature variation of 30 degrees centigrade.

The LVDT support plate is made of aluminum and attached to the body of the LVDT by means of epoxy. Knurled aluminum nuts are used to fix the location of the LVDT relative to the plant. The hole in the support plate for the threaded rod is broached to set the location of the LVDT relative to the plant in the vertical orientation. The spring support plate is made of brass, and the nuts holding the plate are made of nylon. These latter two materials were selected mainly for their insensitivity to the environment. Thermal expansion of this portion of the housing is not significant.

Stainless steel springs are used in both cases. The expansion springs were designed to accommodate a stem diameter from 3 to 20 millimeters. The lower dimension is largely a function of the plant's ability to sustain the weight and torque associated with the housing and transducer. The compression spring employed has a spring constant of 0.335 N/mm and a free length of 15.75 mm. This produces a maximum force on the plant stem of approximately 2.7 N. If one assumes the force is exerted uniformly, this results in a maximum pressure of 0.092 N/mm squared. The extension spring has a spring constant of approximately 0.025 N/mm and a free length of 50 mm. The maximum spring force is approximately 1.6 N. Assuming a 1 mm surface area and a 20 mm diameter, this yields a pressure of 0.04 N/mm squared on the stem surface. Visual inspection of the stem surface indicates no apparent damage from either of these spring forces.

The extension springs were designed so that no adjustment is necessary as the plant passes from a diameter of 5 millimeters to a diameter of 20 millimeters. The location of the LVDT must be adjusted when the stem diameter change is approximately 6 millimeters. In cotton, this requires two adjustments over the entire growing season. If the adjustment is made carefully, there is an unbroken series of measurements which have relative integrity from a 5 micrometer diameter to a 20 micrometer diameter.

Flex epoxy (3M-#2216) is used throughout. This has a very slow curing time but results in an adhesion that can take the rigors of field transport and installation.

The housing plus the LVDT weighed 17 grams in the most recent design. This could be reduced one more gram by chamfering the edges of the V brackets and drilling a hole in the V bracket at a location that does not compromise the mechanical strength.

2.3 Cable

Conventional four conductor #22 solid telephone "D" station wire can be used to connect the LVDT's to the centralized electronic processing equipment. This wire is molded in a star quad configuration with the primary wires set opposite to one another to minimize cable cross coupling. Coupling capacitance between adjacent wires is approximately 60 picofarads/meter. Four unshielded conductors cannot be used because of the excessive crosstalk on the normal runs required for field work. Twisted pairs cannot be used if they are in the same jacket for the same reason.

2.4 Electronic Circuitry

The electronic circuitry centers around the use of the second generation Signetics chip SE5521E (19). This chip contains the sine wave generator and the synchronous demodulator. It also contains a built-in operational amplifier (unused in this design). A complete circuit capable of multiplexing over 12 transducers is shown in Figure 3. It is driven from a single +5V source. The circuit consists of three parts: the Signetics SE5521E, a relay multiplexer, a synchronization circuit and an output amplifier circuit. A 2500 hertz sine wave is generated within the SE5521E chip and feeds through a magnetic relay contact to the LVDT primary. The LVDT secondary feeds back through a second contact on the relay and into the chip. The output of the chip is a ± DC voltage increment superimposed on a DC level of 2.5 volts. This output is then filtered in the op amp to produce a steady DC voltage. A synchronization circuit is required to set the phase of the switching voltage feeding the synchronous demodulator. This switching voltage is set at the same phase as the secondary winding output voltage of the LVDT's. The external operational amplifier acts as a low pass filter, amplifer and level adjustor. The operational amplifier circuit was designed to produce a ± 2047 millivolt linear output corresponding to ± 3250 micrometers. Resolution of the overall mechanical and electrical circuit is 2 micrometers/millivolt. The circuit is amenable to a sleep, wake up mode of operation for low power consumption.

3. ELECTROCHEMICAL MEASUREMENTS

The basic circuit associated with in vivo electrochemical measurements is shown in Figure 4. A measuring electrode is placed invasively in the plant tissue, and a reference electrode is placed in the root zone. An electrical potential appears between the two wires leading from the electrodes. This potential is coherent, reproducible and relates to environmental changes (4). Individual aspects of the technique will now be considered in detail.

462

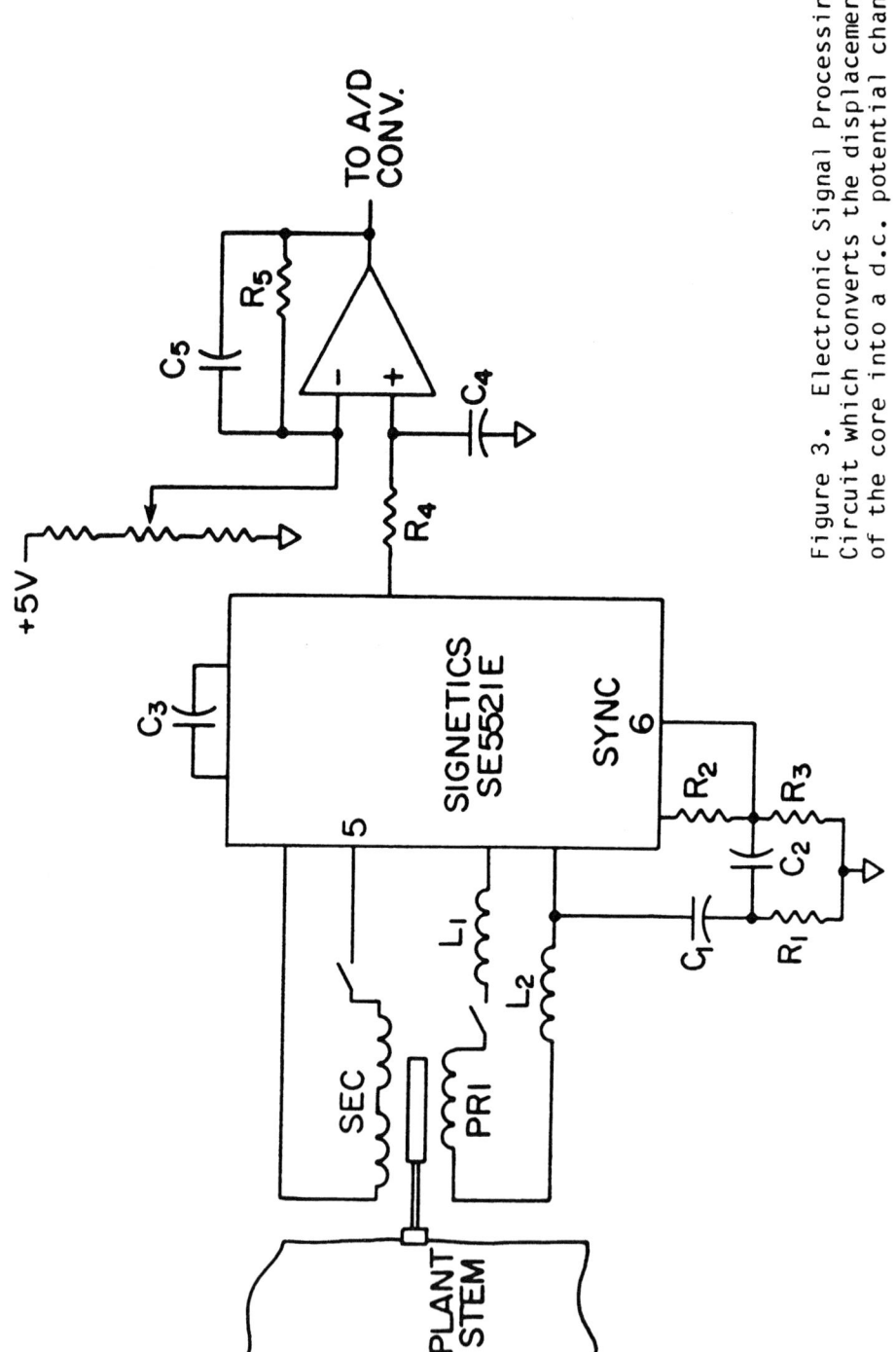

Figure 3. Electronic Signal Processing Circuit which converts the displacement of the core into a d.c. potential change.

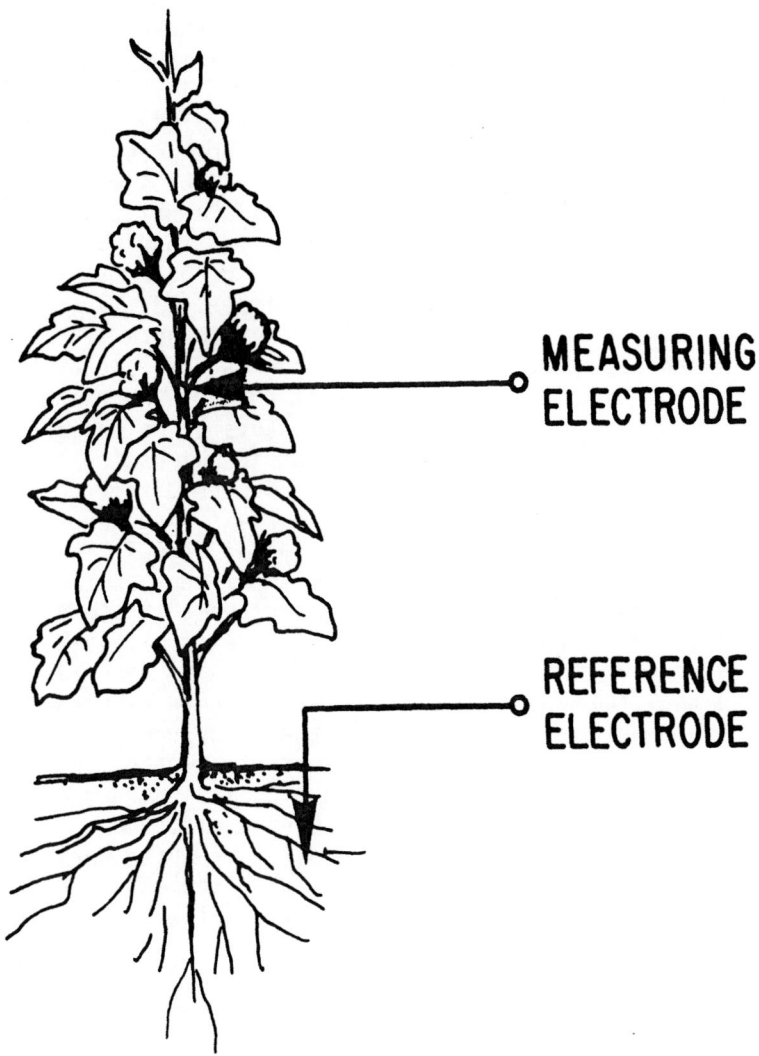

Figure 4. Basic Electrochemical Measuring Circuit. The measuring electrode is placed in the stem, petiole or peduncle. The reference electrode is placed in the root zone.

3.1 Measuring Electrode

The measuring electrode consists of a palladium rod 150 micrometers in diameter and 8 millimeters long. The tip of the electrode is pointed electrolytically to facilitate insertion in the plant and minimize the tissue reaction. Two strands of twisted #38 wire, 5 centimeters long, are soldered to the electrode. The #38 wire is, in turn, soldered to a heavier gauge wire. The latter is attached to the stem of the plant by means of velchro ties. This intermediate length of #38 wire has two functions. It thermally isolates the heavy gauge wire from the probe thereby eliminating any spurious heating of the probe. The wire also mechanically isolates the heavy gauge wire from the probe such that movement of the heavy gauge wire due to wind loading is not transferred into the probe, thereby causing rewounding of the stem. The plant tissue forms a collar around the probe at the entry point and seals the internal tissue from the outside. A normal apoplast fluid then passes along the surface of the probe. This process takes in the order of hours in a field environment. Probe penetration is approximately 3 millimeters.

The tissue response has been examined with both light and electron microscopy (14,17). If the probe is inserted at the upper shoulder of the growth-time "S" curve of cotton, the response is almost non-existent. If the insertion occurs at an earlier age, the tissue responds by forming a swelling in the region of the probe entry. The situation with pecan trees is quite different. The probes can only be placed in the buds in the spring before the tissue becomes too woody. However, the pecan tree forms a cyst around the probe. Whether this cyst is impervious to fluid transfer is unknown. The author's experience indicates that there is a basic difference in the reaction to entry of a foreign body between cotton and pecans. It appears that herbaceous plants in general do not respond to the probe insertion in the same manner as pecan trees.

The response to the insertion of the electrode is extremely consistent: a potential appears between the measuring and reference electrode that begins to rise, that is, the measuring electrode moves positive with respect to the reference electrode. The rise is approximately exponential. The duration is quite variable depending on the conditions of the plant and the environment. Tomato plants in growth chambers will take approximately three days to reach a stable potential. Cotton in the field takes about six hours.

Rewounding will occur if the probe is mechanically dislodged from its original position. The potential waveform that accrues from such a situation will repeat the original process but on a smaller scale depending on the magnitude of the disturbance.

3.2 Reference Electrode

The reference electrode is the subject of considerable research since a stable reference electrode is essential for the comparison of long term variations of the potential of the overall galvanic cell. There are several design problems associated with the reference electrode. The major problem is stability. One must achieve a constant potential between a copper wire and bulk soil for months at a time. A second problem is the avoidance of variable junction potentials caused by fluid flow out of the electrode to the soil. A third problem is moisture encroachment into the metallic parts of the electrode. Fungus buildup in the electrode over the long term

is also significant. Polarization is not a severe problem because
of the use of high impedance amplifiers.

Many designs have been attempted in order to solve the various
problems mentioned. The most recent design is shown in Figure 5.
It consists of a 1/2 inch I.D. plexiglass tube 12.5 centimeters long.
A porous ceramic disc is connected to one end of the tube. The tube
is filled with an acrylamide gel containing potassium chloride to a
height of three to five centimeters. Above this there is a solution
of potassium chloride for a distance of four more centimeters. A
chloridized silver wire electrode is immersed in the potassium
chloride solution. The silver wire is connected to a copper wire
which transfers the signal to the surface. The plug holding the
silver wire has an air well to remove any bubbles from the region of
the electrode. It also contains a hole used to fill the electrode
with potassium chloride solution during fabrication. The top of the
plug is sealed with epoxy. Above the epoxy there is a layer of
asphalt cement which functions as a moisture seal. Above the asphalt
cement there is a layer of epoxy or blocking compound (3M-4407) for
mechanical strain relief.

To make 125 milliliters of the gel, the following procedure is
required:

Make up 3 separate solutions named A, B, and C. In addition,
make up 24 ml of saturated KCl.

To make 100 ml of Solution A, mix 9.5 g of acrylamide powder;
0.5 g of N,N Methylene Bis Acrylamide; 100 ml of saturated KCl.
Decant top of solution and use. Leave the precipitate on the
bottom of the mixing vessel.

To make Solution B, mix 10 grams of ammonium persulfate with
40 ml of water.

To make Solution C, mix 1 ml of N,N,N',N'-tetra methyl
ethylene diamine (TEMED) with 1 ml of water.

Then mix together: 100 ml of Solution A, 24 ml of saturated
KCl, 1.25 ml of Solution B, and 1.25 ml of Solution C. Pour
into tubes immediately. Set up occurs in 25-30 minutes.

To make auxiliary reference electrodes containing 2.5 molar KCl,
repeat the above procedure, except use 2.5 molar KCL throughout.
Distilled and dionized water is used throughout.

The purpose of the acrylamide gel is to prevent fluid from
leaking from the electrode chamber. It yields a stable junction
potential at the interfaces.

The reference electrodes are always used in pairs. A reference
electrode is fabricated using saturated KCl and an auxiliary electrode is fabricated using 2.5 molar KCl.

466

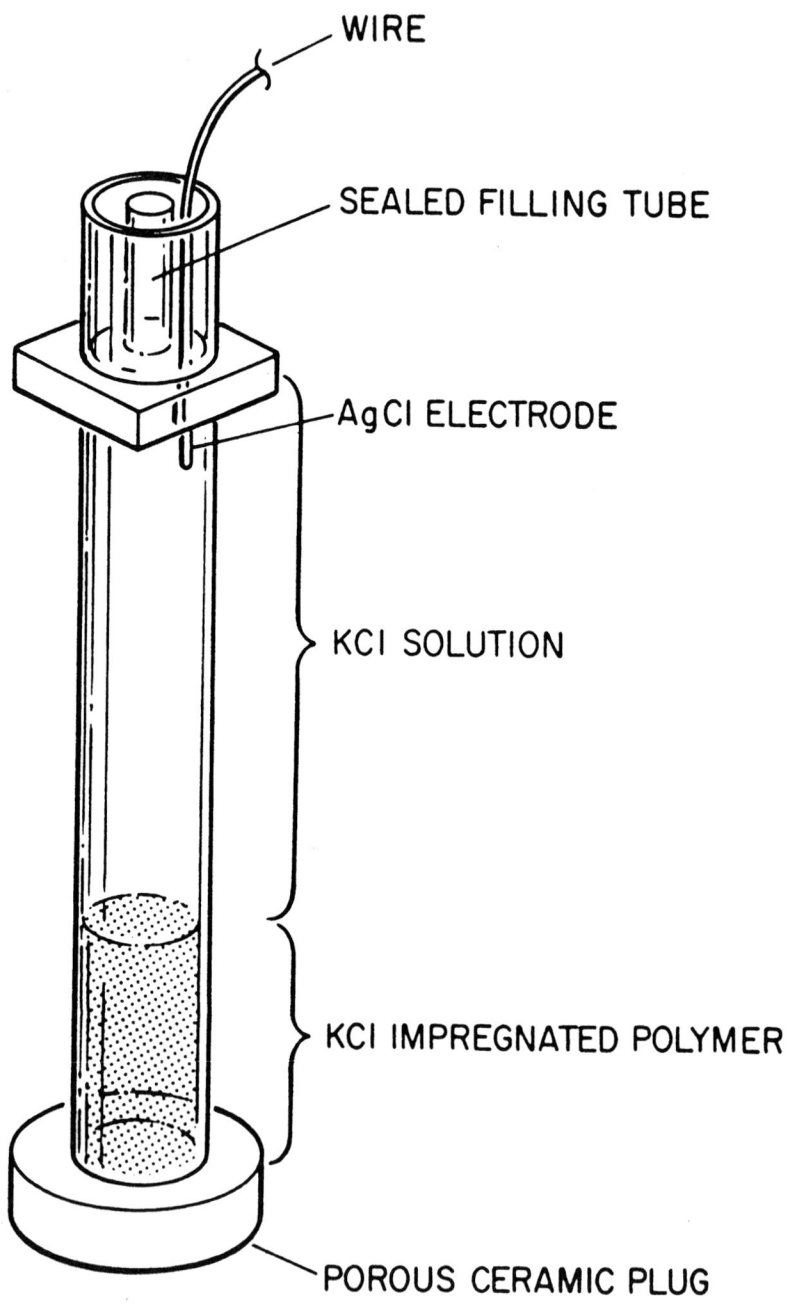

WIRE

SEALED FILLING TUBE

AgCl ELECTRODE

KCl SOLUTION

KCl IMPREGNATED POLYMER

POROUS CERAMIC PLUG

Figure 5. Reference Electrode.

The auxiliary electrode is measured against the saturated KCl electrode to yield approximately a positive 22 millivolt signal. If
this potential is constant, then there is an extremely high probability that the main reference electrode has a constant potential
between the copper wire and bulk soil. The only other situation
which would yield a constant potential between these two electrodes
is if the potential of each electrode with respect to bulk soil
moved in equal and opposite directions simultaneously. This is
highly unlikely. One is essentially building a concentration cell
with the soil as the medium between the two electrodes and measuring
the potential of the concentration cell.

The manner of chloridizing the silver wire is the subject of
much controversy in the literature (2). One suggested procedure
would be to place the silver wire and a platinum electrode in a
solution of saturated KCl plus a drop of AgCl. Apply a potential
to the cell such that the silver begins to dissolve into the solution. Then reverse the potential and develop a chloride layer on
the silver wire. Reverse the potential once again and obtain an
electrode surface free of AgCl. Then repeat the procedure until
the surface has a "thick" layer of chloride. The rationale here is
to "clean" the electrode surface by the reversing procedure before
the final layer of chloride is put on.

The saturated KCl electrode will be about 220 millivolts
positive with respect to the standard hydrogen electrode. The two
electrodes are placed in bulk soil at a depth of about one meter.
The depth is set by the expected temperature variation. The deeper
the placement, the lower the variation of potential due to temperature. A chloridized electrode of the type described above has a
charge transfer resistance of about 15-30 kilo ohms.

3.3 Total Galvanic Circuit

The measuring electrode, plant structure, soil and reference
electrode form a galvanic cell. The question that must be addressed
is the origin of the potential that accrues from the cell. More
specifically, what parts of the path give rise to the potential that
is measured across the two copper wires?

Each part of the path gives rise to an interface or junction
potential. The measured potential is the algebraic sum of these
individual potentials. Furthermore, it is not possible to measure
the value of the individual components of the overall potential.
Changes in potential of the various regions can be measured. To
determine the origin of these changes, one uses multiple probes.
Several probes are placed in a single plant such that the probes
have a common path over most of the circuit. For example, two
electrodes are placed in the stem at various heights. The common
path ensues from the stem below the lower electrode back to the
reference electrode. The potential of each measuring electrode is
measured and compared. If a change occurred in the reference electrode, the change would be simultaneously manifest in the potential

of the individual electrodes. If the potential of the soil changed, it would show up in a similar change in the potential of the two measuring electrodes. Using an analysis of this type, it is easy to observe changes in potential that accrue from parts of the circuit path common to several measuring electrodes. This analysis also indicates that the changes in the potential of the measuring electrode arise from the interface between the measuring electrode and the plant apoplast fluid, in other words, the fluid that wets the surface of the measuring electrode.

3.4 Signal Processing Circuit

The potential that arises from the circuit described above is difficult to characterize in terms of an electrical equivalent. A Thevenin equivalent of the electrode configuration indicates the output impedance is in the order of 160 megohms (16). In addition, the energy that can be drawn from the source is very limited. For this reason, the potential must be measured with an amplifier whose input impedance is in the order of 10 exp 9 ohms. Fortunately, in the present electronic environment, this is not difficult to obtain. Typical amplifiers such as the RCA 3130 or the National LM11H are suitable when connected as voltage followers. The use of a single polarity power supply is adequate for many applications. Manual data acquisition is easily implemented. Automatic data acquisition requires substantial hardware and must take into account the fragility of the source. Multiplexing of probes from many plants over the same field is accomplished in the same manner but care must be taken to insure that the plant does not become the sink for electrostatic discharge of line driven electronic equipment. The measuring electrode-electrolyte interface will integrate such discharges and will not recover to the predischarge potential for several hours if the disturbance is severe. This requires that all equipment is battery operated and any connection to line driven equipment must be optical or wireless. Solar driven equipment has been used extensively over the past 8 years to drive the data acquisition equipment. This approach has been very successful.

Electronic signal conditioning circuits are conventional. The amplifier output is fed into a dual slope A/D Converter, e.g. the ICL7109CPL. The CMOS microprocessor CDP1806A can be used for supervision of the multiplexing and signal transfers.

3.5 Telemetry

A 120 km telemetry system was designed and constructed in 1978 to implement the electrochemical data transfer from fields northwest and south of Tucson, Arizona, to the campus of the University of Arizona. This equipment was operational for four cotton seasons. Thirty-two channels of data would be transmitted every 15 minutes, 24 hours each day for 4 months. The communications hardware was based on a pulse interval modulation principle selected for minimum energy consumption. A present fourth generation telemetry system is under construction using frequency shift keying, modulation and packet data transmission. This system is a prototype for a large scale 400 field systerm, 40 hectares per field. A block diagram of the system is shown in Figure 6. The entire system is solar driven.

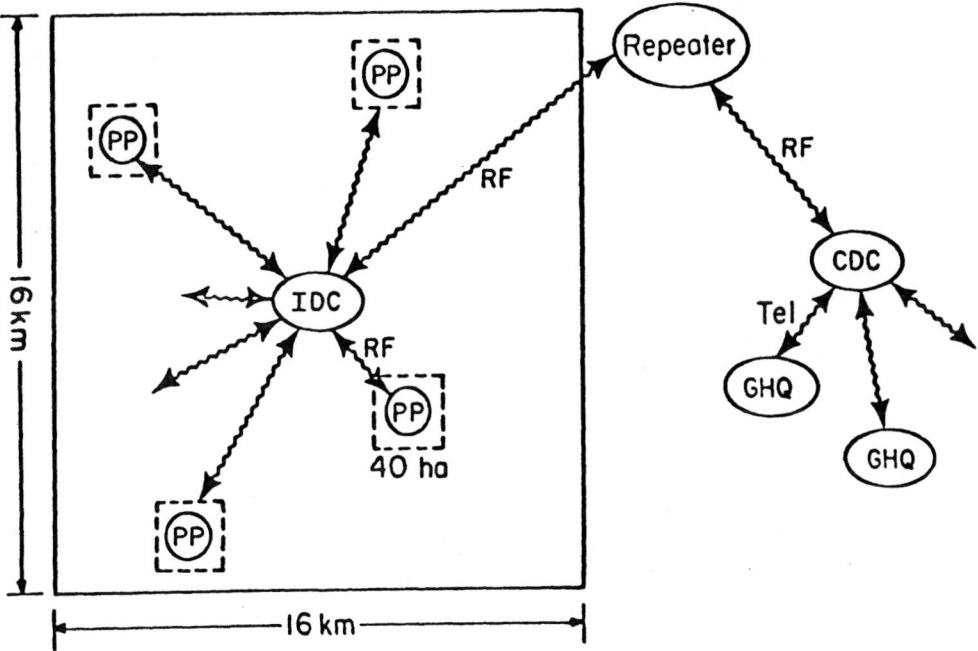

Figure 6. Water Status System. Portable Pods (PP) are placed in the fields and connected to the plants. The water status signals are transmitted by radio frequency to an off field site called an intermediate data center (IDC). The data are further transferred to a central data center (CDC) via a repeater. At the repeater the data are processed into a daily water status index. This index is set to the grower's headquarters by telephone lines.

Communication from the field to the Intermediate Data Center and then on to the Central Data Center would be by radio. Further transfer from the Central Data Center to the Grower's Headquarters or pickup truck would be by telephone or radio linked telephones.

An alternate approach under consideraion in the case of drip irrigation is to transmit the data from the field to the nearby pump computer system and actuate the drip controls directly, based on the data from the field. Such a system would be completely automatic, the grower would only monitor the field to insure proper operation.

3.6 Placement of the Measuring Electrode (14,17)

Electrodes have been placed in stems, petioles and peduncles.

Experience has indicated that the petiole of the leaf is a suitable location for a short term measurement, in the order of three weeks. In cotton, after this period of time the leaf itself changes its orientation and appearance. Potential variations decrease at this time. The potential variations of the peduncle are not as strong as in other parts of the plant. There is a decisive drop of the homeostatic potential level in the peduncle upon boll burst and subsequent drying out of the peduncle in cotton. Both the peduncle and petiole are also present during a limited part of the life of the plant. While interesting from a research viewpoint, the more suitable probe location for production agriculture is in the main stem. The main stem is present during the entire life of the plant and remains functional throughout. Probes placed in the stem do not dislodge as the stem diameter increases. In fact, the stem fixes the radial location of the probe and then engulfs the probe as the diameter increases. This action is so strong that the probe "disappears" right up to and including the solder joint. This action on the part of the plant necessitates the unusually long length of the probe.

There are several types of potential variations. The first type is the healing potential that occurs upon insertion or rewounding of the probe location (7). This is characterized by a rise in potential from some more negative value. The magnitude of the rise is variable, anywhere from 150 to 400 millivolts. The potential levels off to a steady state or homeostatic value. Recent research indicates the plant tenaciously maintains this level even in the presence of pulsed electron injection and extraction from the probe-tissue interface (16).

The homeostatic potential undergoes decisive changes due to water status shifts (4). An interesting aspect of these water related potential shifts is the timing (11). Irrigation in the morning does not evoke a potential drop until the late afternoon. The potential drop due to irrigation is very decisive, systemic and large magnitude. The plant acts very precipitously in changing the homeostatic level. The change following furrow irrigation occurs within a four minute interval. The change occurs at widely separated parts of the plant "simultaneously," that is, within the four minute resolution of the data acquisition equipment. Magnitudes of the potential shift are as high as 400 millivolts. The potential moves more positive shortly after sunrise on the day following irrigation. The drop and rise during the night time period repeats for several days following irrigation but at steadily decreasing magnitudes. This indicates some form of analog process is involved (11).

Drip irrigated cotton has been measured during the morning hours. The potential has been found to rise in the same manner as furrow irrigated cotton. In a sense, it appears to yield potential shifts similar to furrow irrigated cotton immediately following irrigation. In general, a water stressed plant will elicit a square wave potential variation wherein the potential falls in the morning and rises in the late afternoon. An unstressed plant will elicit a potential that rises in the morning and falls in the later afternoon. Rainfall in the desert has a very strong influence on the electropotential of cotton. There are dramatic shifts of potential disproportionate to the actual amount of water reaching the plant. The potential will drop sharply at the time of the rainfall and then rise very strongly the day following the rainfall.

Nocturnal potentials are the name given to very sharp drops and rises in the potential during the night period. There are indications that this type of potential shift is temperature related, but this is not verified.

A third type of potential shift occurs in midmorning. The potential will drop as the sun rises until midmorning, at which time there is a very sharp shift from a decreasing potential to an increasing potential. The rise continues until approximately noon at which time a quasi-steady state level ensues. There is also evidence that the potential variations are systemic. Figure 7 indicates the potential of two petioles along a side branch in a cotton plant. There is a decisive and contrasting pattern in these two potentials.

3.7 Origin of the Potential Shift

The origin of the potential has been investigated both theoretically and experimentally. Goldstein premised three possible sources for the potential: a redox couple or couples in the interfacial electrolyte, an image charge on the probe and unbalanced charge (13). Silva-Diaz applied the method of cyclic voltammetry to determine the presence of redox couples in the apoplast electrolyte (20). He observed that there were no distinct redox couples present but that palladium oxides were formed. Furthermore, oxygen may be a factor in the potential changes. The interpretation of the results was hampered, however, by the large value of solution resistance. Ledezma-Rascon applied passive and active loads to the probe-tissue interface of carbon and palladium electrodes in a comparative attempt to define the chemical nature of the interface (16).

3.8 Prince Road Hypothesis

The experimental and theoretical results described above together with the observations of the potential variations in the field have led the author to a hypothesis concerning the origin of the potential variations. First a review of the experimental support for the hypothesis: 1) the potential of the electrode combination of an in vitro acqueous solution undergoes a decisive and rapid drop in potential with the removal of dissolved oxygen from the solution; 2) cyclic voltametry analysis indicates Palladium Oxides may be present; 3) the probe is contiguous with the apoplast electrolyte; 4) there is a delayed, but concomitant expansion of the stem diameter and precipitous drop in potential following irrigation; 5) rewounding causes a drop and then a rise in potential; 6) the plant maintains a homeostatic potential but will rapidly and decisively change the potential on a systemic basis; 6) the drop in potential is very precipitous, but the rise in potential is much slower.

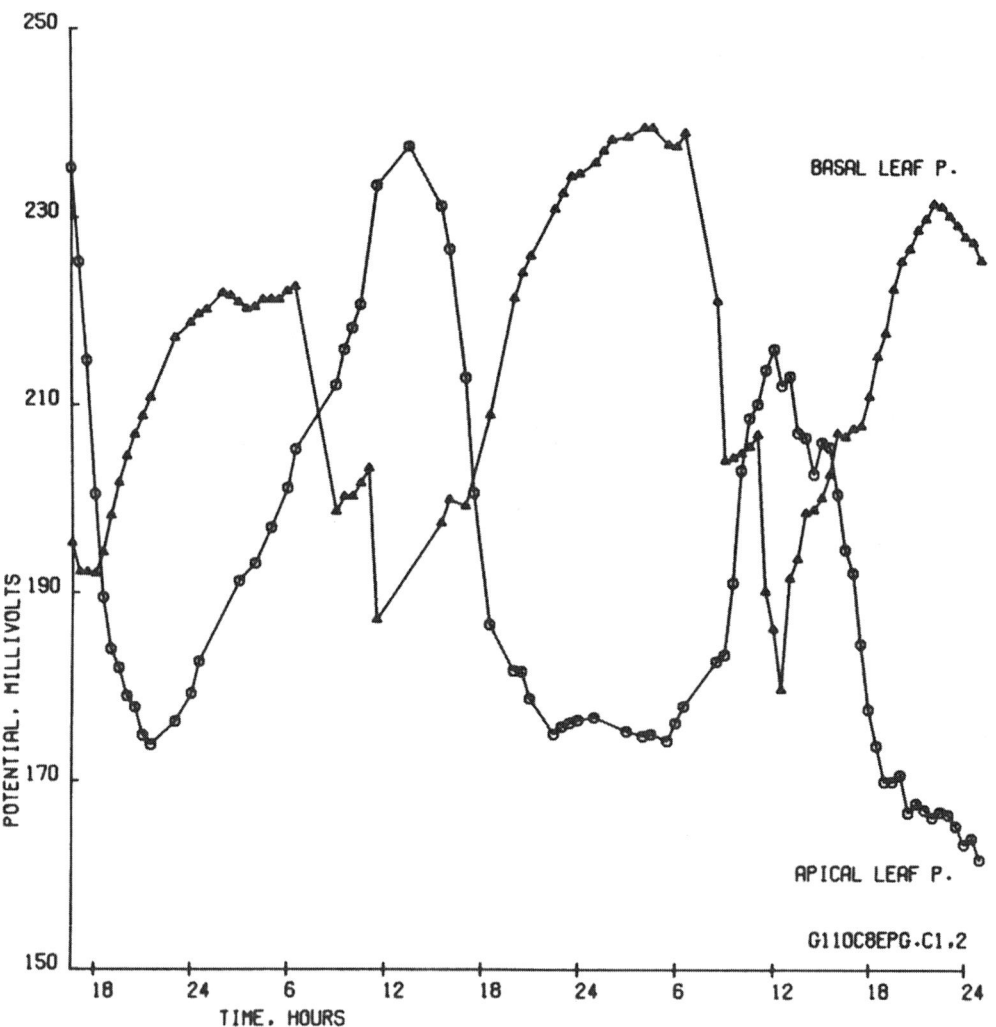

Figure 7. Electropotential of the apical and basal petiole of a side branch of a cotton plant.

Based on these observations, the hypothesis is that the potential is a measure of the oxygen concentration of the apoplast.

The observations and hypothesis are highly consistent. The plant uses the apoplast as an "oxygen storage area" outside the cell. When the need arises, the cells transfer the oxygen inside

the cell, thereby causing a precipitous drop in potential. The drop
in potential in the late afternoon on the day of irrigation is a
result of the plant requiring oxygen for the energy consuming process of stem expansion and transport. The process occurs in the
late afternoon when there is a decreasing radiation load. The plant
simply waits through the day until a more propitious set of environmental conditions ensues.

Rewounding yields a condition wherein the plant needs oxygen
for metabolic processes associated with healing. The physical
process of wounding mixes compounds together that were previously
held separately. The subsequent reactions are oxygen consumptive.
Hence, the decrease in potential.

The potential shifts associated with water status are a direct
manifestation of the oxygen status of the apoplast. When the plant
is in an unstressed condition, the potential rises in the morning as
the level of oxygen in the tissue builds up. The converse situation
ensues when the plant is in a water stressed condition. The oxygen
level of the plant drops in the morning because of the plant's
inability to maintain normal photosynthetic activity.

The drop and rise of potential through the midmorning period
is a further manifestation of this same phenomenon. The potential
drops as the oxygen level drops from dawn to midmorning in a moderately water stressed plant. Finally, a switching point is reached
wherein the plant changes stomatal status and builds up the oxygen
level once again, thereby causing a rise in potential.

The hypothesis is based on the fact that the apoplast is the
"front yard" of the cell. The raw materials and finished products
of cellular activity must pass through this region.

The hypothesis must be scrutinized and verified or disproved
by further experimentation. If valid, the use of penetrating probes
furnishes a very powerful tool for analysis of plant cellular activity on a continuous, non-destructive basis under field conditions.

3.9 Microelectrodes

A natural confluence of several unrelated fields has resulted
in the development of in vivo microelectronic probes for use in
field agriculture. Electronic components have become increasingly
smaller and capable of processing information at lower and lower
energy levels. Semiconductor fabrication techniques permit precise
manufacture of multifunctional chips of very small size. Electrochemical methods have reached the point of development wherein
assays of minute volumes are possible. These individual subjects
can be brought to bear on the problem of invasive in vivo plant
sensing by the use of invasive planar microelectrodes in agriculture.
A small glass or titanium wafer, approximately 2 by 10 millimeters
by 250 micrometers, is the substrate for an electronic circuit
fabricated by conventional semiconductor microelectronic methods.
The wafer is inserted into the plant in the manner shown. On the
surface of the wafer are "windows" which look out to the apoplast

electrolyte. These windows can be made of palladium for acquisition of the above described passive homeostatic potentials. In addition, multiple windows can be fabricated to perform active, in vivo, electroanalytical assays such as coulostatics and cyclic voltammetry. The possibilities are myriad.

The construction of probes of this type follows the highly sophisticated procedures of planar semiconductor fabrication. Rubylith masks are placed on top of photoresist to define a pattern of metallic windows and channels to transfer the signal from the window to the solder pad. A second process then covers portions of the pattern to insulate the entire structure except the windows. The front of the probe is beveled to facilitate penetration. Individual probes are cut out of the wafer using diamond dicing machines. Figure 8 illustrates a probe with three windows on a glass wafer substrate.

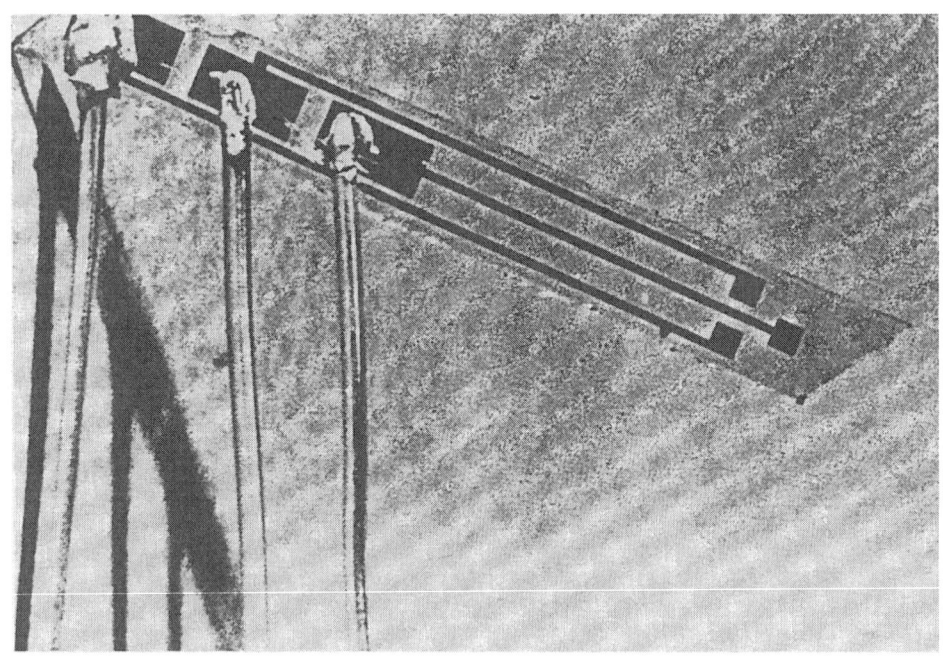

Figure 8. Experimental three window probe on a glass substrate. Window size is 400 by 400 micrometers.

The major design constraints are the necessity of a very narrow probe to minimize the perimeter of the entry. The second constraint is the requirement of a metallic band located around the entry point such that the plant tissue will form a continuous bond with the probe material. This will prevent any water loss and yield a normal electrolyte in the vicinity of the probe surface. A metallic band is necessary because of the apparent lack of adhesion between plant tissue and silicon, silicon dioxide, glass and plastic. The mechanical strength of the substrate is also of importance. If the perimeter of the probe is minimized, the probe material must be capable of sustaining the mechanical stress associated with probe insertion. A number of approaches are under consideration at the present time. A sandwich approach uses a rigid metal base on top of which is placed the planar layers. Another approach is the use of a probe placed within a hypodermic needle. The former acts as a rigid sheath to carry the probe into the plant. It automatically supplies the proper material for the entry point seal. There is an electrochemical constraint placed on the probe insofar as the window size must be large enough to insure an exchange current density that results in a stable potential.

" . . . and the seed sprouts
and grows, and he knows not
how it happens."

Mark 4/27

REFERENCES

1. Diaz-Munoz, F. Simultaneous stem diameter and water potential measurements. Mexicon, Cuenevaca, 1983.

2. Geddes, L.A. Electrodes and the measurement of bioelectric events. Wiley-Interscience, New York, 1972.

3. Gensler, W. Bioelectric potentials in higher plants and their relation to growth in higher plants. Ann. NY Acad. Sci. 238:2380-2399, 1974.

4. Gensler, W. Method and apparatus for electrically determining plant water status. U.S. Patent No. 3, 967, 198, 1976.

5. Gensler, W. Tissue electropotentials in Kalanchoe blossfeldiana during wound healing. Am. J. Bot. 65:152-157, 1978.

6. Gensler, W. Transition electropotentials and growth in Lycopersicon esculentum. Bioelectrochemistry and Bioenergetics 5:152-167, 1978.

476

7. Gensler, W. Electrochemical healing similarities between animals and plants. Biophys. J. 27:461-466, 1979.

8. Gensler, W. An electrochemical instrumentation system for agriculture and the plant sciences. J. Electrochemical Society, 127:2367-2370, 1980.

9. Gensler, W. Apoplastic electropotentials in cotton under variable water stress. Ann. Meeting of Am. Soc. Plant Physiol., Urbana, IL, June 1982.

10. Gensler, W. and Diaz-Munoz, F. Stem diameter variations in cotton under field conditions. Crop. Sci., 1983.

11. Gensler, W. and Diaz-munoz, F. Simultaneous stem diameter expansions and apoplastic electropotential variations following irrigation or rainfall in cotton. Crop. Sci., 1983.

12. Gensler, W. An electrochemical water status sensing system. First Conference on Electronics in Agriculture, Am. Soc. of Agr. Eng., Chicago, December 1983.

13. Goldstein, A. and Gensler, W. Physiological basis for electro-phytograms, Part 1: Theoretical considerations. Bioelectro-chemistry and Bioenergetics, 8:645-659, 1981.

14. Goldstein, A. Interface between cotton tissue and a penetrating noble metal probe. Am. J. Bot. 69:513-518, 1982.

15. Goldstein, A. and Gensler, W. In situ measurement of cell wall donnon potentials. Ann. Meeting of Am. Soc. Plant Physiol., Plant Phy. Sup. 72:61, 1983.

16. Ledezma-Rascon, E. Modeling of the bioelectric system formed by palladium and carbon electrodes inserted in cotton (Gassypium hirsutum) plants. Masters Thesis, University of Arizona, 1985.

17. Rugenstein, S. Tissue response to palladium microprobe as observed in Gassypium hirsutum, L. (Malvaeceae). Am. J. Bot. 69:519-528, 1982.

18. Schaevitz Engineering, Pennsauken, NJ, Technical Bulletin 1002A.

19. Signetics Inc., Santa Clara, CA.

20. Silva-Diaz, F., Gensler, W. and Sechaud, P. In vivo cyclic voltammetry in cotton under field conditions. J. Electrochem. Soc. 30:1464-1468, 1983.

INDEX